THE DEVELOPMENT OF THE TUNA INDUSTRY IN THE PACIFIC ISLANDS REGION: AN ANALYSIS OF OPTIONS

Edited by
David J. Doulman

David J. Doulman was a fellow and the director of the Pacific Islands Development Program's tuna project at the East-West Center, Honolulu. Before joining the program in 1985, he was chief fisheries economist for the Papua New Guinea government. Doulman received his doctorate from James Cook University, Australia, and has written widely on tuna and fisheries in the Pacific islands region. Currently, he is senior economist with the Forum Fisheries Agency.

Library of Congress Cataloging-in-Publication Data

The Development of the tuna industry in the Pacific islands region : an analysis of options.

1. Tuna industry—Pacific Area. I. Doulman, David J., 1950– .
HD9469.T83P163 1987 338.3′71758 87–6818
ISBN 0–86638–096–5

Contents

Foreword

Multinational Corporations in the Pacific Tuna Industry is an East-West Center research project embodying a range of issues, analyses, and findings. The project was undertaken from the perspective of Pacific island countries on a scale not previously attempted. The Standing Committee of the Pacific Islands Conference, in considering key research issues for the Pacific islands region, views the activities of multinational corporations and foreign investment in the Pacific tuna industry as being critical to the development of Pacific island nations. The tuna research project sought to enhance the understanding and knowledge of leaders and peoples of these countries to enable them to formulate policies for the effective control, management, and utilization of the tuna resources for their development.

Dr. David J. Doulman and his team of tuna specialist researchers provide us with a view of the Pacific tuna industry that is comprehensive at the macro level in its coverage of specific issues such as international marketing networks and financial issues. The research also takes us down to the micro level of the policies and production processes and how they relate to the various production and supply factors, be they fishing vessels, processing plants, fishing methods, or supply pricing.

Tuna is currently a major geopolitical issue in the Pacific islands region. This research project is therefore timely in its real and potential impact on the region's tuna industry. The project provides an avenue for the Pacific island nations to express the ideals of their sovereignty through their control and use of the large and valuable tuna resources found within their respective exclusive economic zones.

Charles W. Lepani
Director
Pacific Islands Development Program
East-West Center

Acknowledgments

The success of the Pacific Islands Development Program's tuna project has been due to the efforts and support of many individuals, regional and international organizations, and Pacific island governments. Without their contributions the project would not have achieved its high level of success and international prominence.

Research funding for the project came from several sources including the East-West Center, the United Nations Development Programme, the State of Hawaii, the Australian Development Assistance Bureau, and the United States Agency for International Development. The financial support of these organizations made the project possible.

Over the duration of the project I received encouragement from the former director of the Pacific Islands Development Program, Filipe N. Bole, and the current director, Charles W. Lepani. The former research coordinator, Michael P. Hamnett, provided excellent professional guidance while Jeanne Hamasaki and LeNora Wee gave administrative support. Jesse M. Floyd and Linda Lucas Hudgins assisted various stages of the project, contributing professional and administrative support.

In addition to the consultants who prepared reports for the project, I am indebted to Osamu Narasaki and Robert Young for their assistance. The flexibility and the highly competent approach of all the consultants made my task easier.

Secretarial assistance to the project was provided by Forrest Hooper and Joanne Yamane. In addition, Janie Aucoin, Titilia Barbour, Lynette Tong, and Mary Yamashiro contributed typing support. Barbara Yount edited the book and it was proofread by Edith Kleinjans. Production of the book was undertaken by Jacqueline D'Orazio and her staff. Special thanks go to Lois Bender for typesetting the manuscripts. Russell Fujita designed the book's layout and cover.

David J. Doulman
June 1987

Abbreviations and Acronyms

ACP	Asian, Caribbean, and Pacific
ADB	Asian Development Bank
AFZ	Australian fishing zone
ATA	American Tunaboat Association
ATSA	American Tuna Sales Association
BANPESCA	Banco Nacional Pesquero y Portuario, S.N.C., Institución de Banca de Desarrollo (National Fishery and Ports Development Bank, Mexico)
BFAR	Bureau of Fisheries and Aquatic Resources (Philippines)
BOI	Board of Investment (Thailand)
CDC	Commonwealth Development Corporation
CIF	Cost, insurance, and freight
CPUE	Catch-per-unit-of-effort
CSIRO	Commonwealth Scientific and Industrial Research Organization (Australia)
DO	Dissolved oxygen
DWFN	Distant-water fishing nation
EDP	European Development Programme
EEC	European Economic Community
EEZ	Exclusive economic zone
EIA	Environmental impact assessment
EIB	European Investment Bank
ESCAP	Economic and Social Commission for Asia and the Pacific
ETP	Eastern tropical Pacific
FAD	Fish-aggregating device
FAO	Food and Agriculture Organization (United Nations)

FDA	Food and Drug Administration (United States)
FDCA	Food, Drug, and Cosmetic Act (United States)
FFA	Forum Fisheries Agency
FFAP	Federation of Fishing Associations of the Philippines
FLSA	Fair Labor Standards Act (United States)
FOB	Free on board
FPLA	Fair Packaging and Labeling Act (United States)
GATT	General Agreement on Tariffs and Trade
GDP	Gross domestic product
GNP	Gross national product
GRT	Gross registered tonnes
GT	Gross tonnes
IAC	Industry Assistance Commission (Australia)
IATTC	Inter-American Tropical Tuna Commission
ICLARM	International Center for Living Aquatic Resources Management
IDA	International Development Association
IFC	International Finance Corporation
JAS	Japan Agricultural Standards
KGKK	Kaigai Gyogyo Kabushiki Kaisha (Overseas Fishing Company Ltd, Japan)
KMIDC	Korea Marine Industry Development Corporation
LIBOR	London Interbank Offered Rate
LPG	Liquified petroleum gas
MAF	Ministry of Agriculture and Food (Philippines)
MAFF	Ministry of Agriculture, Forestry and Fisheries (Japan)
MFN	Most favored nation
MIGA	Multilateral Investment Guarantee Agency
MITI	Ministry of International Trade and Industry (Japan)
MNC	Multinational corporation
MSY	Maximum sustainable yield

NFC	National Fishing Corporation (Tuvalu)
NFDL	National Fisheries Development Ltd (Solomon Islands)
Nikkatsuren	Federation of Japan Tuna Fisheries Co-operative Associations
NMFS	National Marine Fisheries Service (United States)
NOAA	National Oceanic and Atmospheric Administration (United States)
NRT	Net registered tonnage
OECD	Organization for Economic Cooperation and Development
OJT	On-the-job training
OLDEPESCA	Organización Lationoamericana de Desarrollo Pesquero (Latin American Organization for Fishery Development, Mexico)
PAFCO	Pacific Fishing Company Ltd (Fiji)
PEMEX	Petroleos Mexicanos (Mexican National Petroleum Company)
PFDF	Pacific Fisheries Development Foundation (United States)
PICOSA	Pesca Industrial Corporación, S.A. (Industrial Fish Corporation, Mexico)
PNAC	Philippine-North American Conference
PNGTF	Papua New Guinea Tuna Fisheries Pty Ltd
PPM	Productos Pesqueros Mexicanos (Mexican Fishery Products)
PTPEA	Philippine Tuna Producers and Exporters Association
SAMI	Selling Areas Marketing Inc
SBT	Southern bluefin tuna
SEAFDEC	South East Asian Fisheries Development Center
SEPESCA	Subsecretario de Pesca (Subsecretariat of Fisheries, Mexico)
SPC	South Pacific Commission
SPF	South Pacific Forum
SPFC	South Pacific Fishing Company (Vanuatu)

SPREP	South Pacific Regional Environment Programme
STL	Solomon Taiyo Ltd (Solomon Islands)
TCAP	Tuna Canning Association of the Philippines
TML	Te Mautari Ltd (Kiribati)
TOR	Terms of reference
UNDP	United Nations Development Programme
UNEP	United Nations Environment Programme
USDOL	U.S. Department of Labor
USITA	U.S. International Trade Administration
USITC	U.S. International Trade Commission
USTF	U.S. Tuna Foundation
WB	World Bank

Introduction

Background

In 1981 the Pacific Islands Development Program (PIDP) was directed by the Standing Committee of the Pacific Islands Conference to evaluate the potential beneficial roles of multinational corporations in the Pacific islands region. The minutes of the Pacific Islands Conference Standing Committee meeting of 21 April 1981 noted in part:

> In discussing the role of multinationals, the Standing Committee felt there should be a better understanding of the role of multinationals in the Pacific islands in terms of their objectives, organization, and know-how, as well as their role in development. At present there is a tendency to react with some hostility toward multinationals, which has effectively kept governments and multinationals at arm's length. At the same time, many governments have tried to develop cooperatives and local companies to achieve the same purpose as the multinationals, and, in many cases, these schemes have run into difficulties. Furthermore, the new developers tend to try to attain monopolies for themselves when, in effect, competition may be more beneficial for the countries and people of the islands.
>
> The Committee felt that there should be a study of multinationals and, more specifically, on why they succeed, how they succeed, and in what areas. With a better understanding of the role of multinationals, perhaps systems could be developed for joint or cooperative efforts between governments and multinationals that will be beneficial for both parties.

The Standing Committee addressed the question of multinational corporations again in 1984 and resolved that evaluation of the role of these corporations should be undertaken on a sectoral basis, with the tuna industry being the first sector to be examined. Staff members for the tuna project were appointed in late 1984, and the project began in January 1985.

Objectives

In accordance with the directions of the Standing Committee in 1981 and 1984, the objectives of the tuna project were to

- Analyze the current and future role of multinational corporations in the tuna industry in the Pacific islands region, and
- Evaluate the potential contribution these corporations could make to industry development in the region.

In the context of the study, multinational corporations were interpreted in a broad sense to incorporate most forms of private investment.

Research areas

The project had four major areas of research:

Overview of the Pacific tuna industry. The first step in studying the tuna industry in the Pacific islands region was a profile of tuna resources and an assessment of their potential for development. This work drew on the stock assessment research conducted by the South Pacific Commission (SPC) and research by Pacific island governments and other international and regional agencies.

World tuna markets. Success in the fishing industry depends not only on catching fish but also on knowing where and how to market the fish. In the international tuna industry particularly, this requires complete, up-to-date information on marketing opportunities, buyers' product specifications, prevailing market requirements, and price fluctuations. Success in the industry also requires a thorough understanding of the structure and dynamics of world tuna markets for frozen and processed products. Lack of this kind of information puts producers and exporters (especially in developing countries such as the Pacific island countries) at a disadvantage in their dealings with established traders. The tuna project addressed these information needs by reviewing past and present conditions in the major tuna markets: Australia, Canada, Japan, the Middle East, the United States, and Europe.

International business and foreign investment in the industry. The international tuna industry is dominated by a relatively small number of companies, but information on their activities and operations is not easily available. Research in this project area began by addressing this concern and by identifying and profiling each of the major companies involved in the world's tuna industry. Particular emphasis was placed on those companies involved in the tuna industry in the Pacific islands region. Data were presented on the level of their investment, the nature of their activities, the extent of their operations, and their cooperation and involvement with other companies in the tuna industry. An important objective in this phase of the research was to provide a better understanding of the operations and strategies of, and linkages among, the major multinational corporations involved in the tuna industry.

Development options and issues. This research focused on evaluating options for Pacific island governments with respect to financing tuna projects, packaging tuna projects, taxing tuna projects, treatment of infrastructure, and environmental and social impacts.

Research methodology

Several research methods were employed to complete the project. Data were obtained from published literature, U.S. and Japanese fisheries associations, government documents, and regional and international organizations. Information was also sought from the tuna companies involved in the Pacific islands region. In addition, this information was augmented by field research and interviews with government officials and industry leaders. Experts were contracted to undertake studies on specific issues that required specialized knowledge and firsthand experience.

Finance

The tuna project was financially supported by the East-West Center, the Australian Development Assistance Bureau, the State of Hawaii, the United Nations Development Programme, and the United States Agency for International Development. The total cost of the project was approximately $300,000.

Intern program

The objectives of the internship were to familiarize middle-level fisheries professionals with PIDP's tuna research and to provide an opportunity for them to actively participate in and contribute to the output of the project. The three countries that nominated interns were Kiribati, Tonga, and the Marshall Islands.

Collaboration

PIDP's policy is to collaborate with national, regional, and international organizations to the maximum extent possible. The tuna project recognized that close liaison should be maintained with the Forum Fisheries Agency as well as with other organizations such as the South Pacific Commission, the United Nations Development Programme, the Food and Agriculture Organization, the United Nations Centre on

Transnational Corporations, and the Commonwealth Secretariat. Liaison was also established with institutes in Australia, Canada, Japan, and the United States with a view to cooperating on research. Particularly close research collaboration was maintained with the University of Sydney's Transnational Corporations Project.

Close liaison was also maintained with government representatives in the Pacific islands region to ensure that research findings were disseminated quickly. The project director also assisted governments in other ways, for example, in advertising employment vacancies, recruiting fisheries staff, providing trade information, and obtaining background information on consultancy groups.

Staff

The project director was Dr. David J. Doulman. Prior to joining the project he was chief fisheries economist in Papua New Guinea. Dr. Doulman will continue to work in the fisheries field as senior economist with the Forum Fisheries Agency in Solomon Islands. Dr. Jesse M. Floyd worked on the project as a senior analyst from November 1984 to December 1985. Mr. Iosefa A. Maiava from Western Samoa was a professional associate with the project from January to June 1985. Dr. Linda Lucas Hudgins joined the project as a senior analyst in 1986. Dr. Hudgins was on leave from the graduate school of economics at the University of Notre Dame, Indiana. Dr. Gary C. Anders also joined the project on a short-term appointment from January to March 1987. He is a professor in the school of business and public administration at the University of Alaska/Juneau.

I. OVERVIEW

1.
Domestic Tuna Industries

David J. Doulman and Robert E. Kearney

ABSTRACT—This chapter presents an overview of the tuna resources and the domestic tuna fishery and industry in the Pacific islands region. The status of tuna stocks in the region is reviewed, and artisanal production is discussed. These sections are followed by a historical account of aggregated commercial catch data for domestic fleets in the region. Next, the operation of fleets by gear type and their catches for each country are analyzed, as well as transshipment facilities and tuna processing. Finally, the multinational operations involved in the tuna industry in the Pacific islands region are discussed.

INTRODUCTION

With the exceptions of Papua New Guinea, Solomon Islands, and Fiji, most countries and territories in the Pacific islands region consist of small individual islands or groups of islands sparsely distributed throughout the tropical central and western Pacific Ocean. Geographically, the region stretches from the Republic of Palau in the west through Micronesia, Melanesia, and Polynesia to Pitcairn Island in the east (Figure 1).

The agricultural resources and land-based economic activities in the Pacific islands region are generally restricted because good arable land is limited in most countries. Thus a dependence on marine resources has been vital to the subsistence economy and culture of most people in the region. Furthermore, a comparative lack of inshore fisheries resources has meant that many Pacific island fishermen have had to rely heavily on the subsistence exploitation of tuna and related pelagic species. This dependence has continued to the present and has become even more critical in some countries and territories because of high rates of population increase.

During the second half of the twentieth century, commercial exploitation of the region's tuna resources has brought unprecedented opportunities for development to Pacific island countries. To take advantage of these new opportunities, governments and individuals in these countries have invested in fishing vessels and shore-based facilities. Governments have also facilitated investment in similar undertakings by foreigners, including multinational corporations.

In promoting tuna industries, governments are attempting to reap more benefits from the exploitation of their tuna stocks than if they permitted stocks to be harvested only by the distant-water fishing nation (DWFN) fleets in return for access fee payments. Domestic tuna industries are being promoted (1) to strengthen and broaden the economic bases of countries in the region; (2) to generate employment opportunities, foreign exchange, and government revenue; and (3) to facilitate the training of Pacific islanders and the transfer of fishing and related technology to island countries.

Concurrent with the nurturing of national tuna industries, competition from the DWFN fleets for the region's tuna resources has developed and even intensified. This competition indicates a need for the conservation and coordinated management of these previously only lightly exploited tuna resources. This is particularly important in those areas of the region where the exploitation of these resources by industrial (national and DWFN) fleets actually or potentially has an impact on artisanal, including subsistence, production in the region.

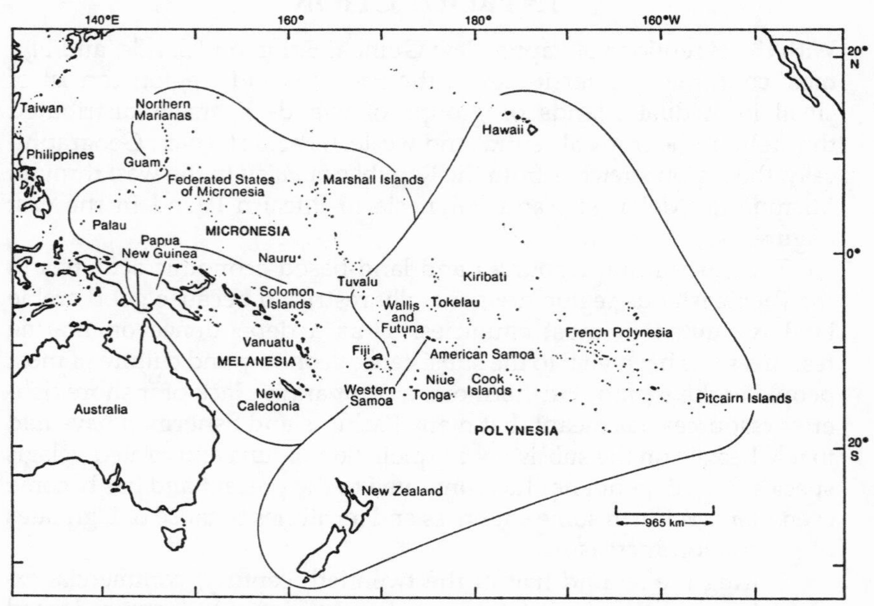

Figure 1. Pacific islands region

STATUS OF TUNA STOCKS

Although ten species of tuna are known to occur in the tropical central and western Pacific Ocean, four species account for almost the entire commercial catch. These four species of tuna are skipjack (*Katsuwonus pelamis*), yellowfin (*Thunnus albacares*), albacore (*Thunnus alalunga*), and bigeye (*Thunnus obesus*). Each of these is discussed in turn, followed by a more general discussion of less commercially important species of tuna and billfish. The mobility of tuna resources and the interaction among fisheries in the Pacific islands region are also reviewed.

Skipjack tuna. Skipjack tuna is distributed throughout the world's tropical oceans and accounts for approximately 40 percent of the world's total landings of tuna. The size of the total resource of skipjack in the tropical central or western Pacific area at any one time has been estimated to be on the order of 3 million tonnes. A total annual harvest of several million tonnes is theoretically possible. Such a harvest would be greater than the present total annual yield of between 200,000 and 300,000 tonnes.

Yellowfin tuna. Yellowfin tuna is exploited by both surface and longline fisheries throughout the world's tropical oceans. In the 1980s it has become a principal target species for purse seiners. The yellowfin tuna resources in the Pacific island region should be able to support total annual catches of at least twice the present levels. However, because of the dependence of some fisheries, particularly longlining, on larger individuals and the changes in size composition of the population, which heavy exploitation inevitably brings, the full impacts of attempts to maximize total landings are not likely to be positive.

Albacore tuna. There has been a significant decline in the longline fishing effort for albacore in the region in recent years, and the region's stocks of this species are generally considered to be presently underexploited. However, any increase in total effort would result in some decline in the catch-per-unit-of-effort (CPUE).

Bigeye tuna. Catch statistics on bigeye tuna are even more limited than those of other tuna species; however, the species is generally considered to be underexploited.

Other tuna and billfish. Other species of tuna and numerous species of billfish are taken in the tropical central and western Pacific but mostly as incidental catches in fisheries for the four major species already discussed. Some species, particularly blue marlin (*Makaira nigricans*) and black marlin (*Makaira indica*), do have special significance to the sportfishing and tourist industry. Furthermore, both of these species are considered to be heavily exploited and more likely to require conservation management than the major tuna species.

Mobility of resources and fisheries interaction

All tuna and billfish are considered highly migratory. Individuals of all the major species exploited in the tropical central and western Pacific are sufficiently mobile to necessitate international cooperation in total resource conservation and maximum total exploitation.

The magnitude of the skipjack resource in the central and western Pacific is such that it appears immune to long-term decimation by fishing pressure at levels approximating those at present. Unfortunately, the ability of the total resource to support greatly increased catches has been misinterpreted to imply that fisheries do not have any impact on each other. Even though the skipjack resource is immense, it is not infinite, and fisheries operating very close to each other in both time and space could affect changes in catch rates without an adverse influence on the total renewable resource.

It is essential to differentiate between two factors: (1) the relative immunity of the total skipjack resource to overexploitation by levels of total fishing effort that are several times greater than those currently applied and (2) the vulnerability of catch rates in small areas to changes induced by heavy localized fishing pressure.

The available scientific information is inadequate to enable precise determination of the degree of interaction among the various types of tuna fisheries and in the many different fishing areas of the tropical central and western Pacific. The results of tagging studies are indicative of interaction levels. Interaction is greatest in areas where artisanal and large-scale commercial fleets operate at the same time in the same general location. In addition, longer-term declines in catch rates as a result of interaction among fisheries would be much greater for the longer-lived tuna species such as yellowfin and bigeye than for skipjack.

Furthermore, when assessing the real impact of interaction among fisheries, governments in the Pacific islands region need to take full account of socioeconomic implications of a decline in the total catches, or catch rates, that are likely to result from increased catches by the same or other gear type. For example, the development of a large-scale commercial pole-and-line fishery in an area where an artisanal fishery has exploited the same species could be expected to cause some decline in catch rates by artisanal fishermen. The pole-and-line fishery could, in turn, be affected by the introduction of purse seining with the potential decline in pole-and-line catch rates being influenced by the intensity of the purse seining and its proximity to the pole-and-line fishing grounds, as well as by the mortality, migration characteristics, and growth of the tuna stocks being exploited. Governments need to evaluate the socioeconomic implications at the artisanal, commercial, national, and international levels of a decline in catch rates and the possible social impact that could result from increased competition for the same large but finite tuna resource, even though total catches might continue to increase.

ARTISANAL PRODUCTION

Various species of tuna are exploited by artisanal fishermen throughout the Pacific islands region, both for subsistence and, where markets exist, for sale. As a general rule, in those island countries with a robust inshore (reef) fishery, fishermen tend to target on these species in preference to tuna. This is because inshore species of fish are frequently more accessible and are considered by some Pacific islanders to have superior table qualities to tuna. The species of tuna harvested by artisanal fishermen are usually surface-swimming species, particularly skipjack and juvenile yellowfin, though larger fish are also taken with handlines used increasingly around fish-aggregating devices (FADs).

The major species of tuna found in the Pacific islands region are highly mobile, truly oceanic, and they seldom enter shallow or turbid coastal waters. Their exploitation by subsistence fishermen has traditionally required oceangoing canoes and highly developed sailing and navigational skills.

It is therefore not surprising that tuna has accounted for a significant fraction of total food consumption only in those island communities with limited food alternatives such as agriculture and coastal fisheries resources. Tuna has been relatively more important to those communities located on small islands that do not have sizable lagoons. As a result, the Micronesian and Polynesian communities have generally had a greater dependence on tuna than the Melanesians. In addition, those small island communities in the tuna-rich equatorial areas, which do not experience pronounced seasonal fluctuation in tuna abundance, tend to have become even more dependent on tuna fishing than those communities that are characterized by seasonal fluctuations in tuna abundance.

Prior to contact with the western world, the most common traditional form of fishing practiced by islanders involved the use of a barbless hook or lure in combination with a handline or a short line-and-fishing pole. The tuna hooks traditionally used by Pacific islanders were made of tortoise shell or, in some cases, wood. They were generally attached to a lure made of mother-of-pearl shells and brightly colored parrot feathers.

Most governments in the region have expressed a desire to provide income-earning opportunities for artisanal fishermen by facilitating their integration into the industrial fishery. Some countries have already taken measures toward this end; however, logistical considerations associated with artisanal fisheries development have constrained the rate of integration.

FADs are being deployed in several countries to assist in the development of the artisanal fishery in general and with the exploitation of pelagic species in particular. The capital and maintenance costs of these devices are usually borne by governments, which are sometimes provided with funds supplied under foreign aid arrangements.

Reliable and comprehensive catch data for artisanal landings of fish are universally poor, and the data for artisanal landings of tuna in the Pacific islands region are no exception. Even though the artisanal tuna fisheries in the region are of extreme social significance, the quantity of tuna harvested by artisanal fishermen is small in terms of the magnitude of the region's tuna resource. The total catches of tuna of all species by artisanal fishermen for subsistence needs and commercial purposes in the region are estimated at less than 10,000 tonnes annually.

COMMERCIAL TUNA PRODUCTION, 1977–1982

The tuna fishery is the Pacific islands region's largest fishery in terms of total catch and its most valuable commercial fishery. Depending on market conditions, species composition, and size distribution of commercial catches, the unprocessed value of tuna caught in the region is estimated to be worth $600 million to $750 million annually. The region's tuna resources are exploited by (1) artisanal fishermen for their subsistence needs and, where possible, for sale on local markets and to fishing companies; (2) fishing fleets based in the region; and (3) fleets operating from ports in foreign countries.

According to fisheries statistics published by the Food and Agriculture Organization (FAO), tuna (skipjack, yellowfin, bigeye, and albacore) caught in the Pacific islands region for the period 1977–82 accounted for an average of 20 percent of total world catches of these species. On average between 1977 and 1982, 326,662 tonnes of tuna were harvested annually in the region. FAO reports a decline in production between 1981 and 1982, but its published figures for this period are questionable, even though Papua New Guinea's pole-and-line fishery collapsed at the end of 1981.

Vessels operating in the Papua New Guinea fishery at its peak in 1978 harvested more than 48,000 tonnes of tuna, though in 1981 production dropped to 24,000 tonnes. Landings by fleets based in Solomon Islands and Fiji also recorded declines in tuna production in 1982. These cutbacks were largely attributable to the reductions in fishing effort (as opposed to lower CPUE in this same year) because of the downturn in the international tuna market. As a result of sluggish consumer demand for tuna from 1981 to 1986, fishermen and some processors have encountered difficulties in selling whole and canned tuna profitably. This situation, in turn, has forced production adjustments on tuna fishing fleets both in the region and throughout the world.

The most prolific species of tuna in the region, skipjack, accounted for an average of 64 percent of total annual catches in the region between 1977 and 1982. Skipjack tuna is commercially harvested by purse seine and pole-and-line fishing vessels and is primarily used for canning, though it is also used in the production of *katsuobushi* (smoked tuna) and related products for sale in Japan. In its fresh form, skipjack is also gaining market acceptability as a lower-priced sashimi fish, particularly in Japanese supermarkets. Average annual catches of skipjack from the region between 1977 and 1982 totaled 209,523 tonnes.

Catches of yellowfin in the Pacific islands region between 1977 and 1982 averaged 83,185 tonnes per annum, or 26 percent of the region's tuna catches. Approximately 65 percent of catches of this species from the region is taken by longliners. It is primarily destined for the sashimi market in Japan, while yellowfin taken by purse seiners (which tends to be smaller in size than longline-caught fish) is canned. Pole-and-line-caught yellowfin is usually canned.

In contrast to the production of skipjack and yellowfin, production of bigeye and albacore tuna caught in the region is relatively small by international standards. On average between 1977 and 1982, 25,227 tonnes of bigeye tuna were harvested annually from the Pacific islands region. Longline-caught bigeye is sold to the Japanese sashimi market.

On average between 1977 and 1982, catches of albacore in the region were 7,107 tonnes annually. Albacore is a whitemeat tuna that is canned.

During the period 1978–82, the region supplied 29 percent of the world's skipjack on average each year. The region's fishery is expected to remain a critical area of operation for DWFN fleets and a source of raw material for the international canning industry. However, the continuing development of a viable purse seine tuna fishery in the Indian Ocean to supply the canning industry could erode the relative international importance of the region's fishery.

As a source of supply for yellowfin, the region's tuna fishery contributed an average of 15 percent to total world production of this species during 1977–82. Catches of bigeye and albacore averaged 13 and

5 percent, respectively, as the proportions of total world catches of these species each year.

Although FAO data for 1984 are not available, total catches of tuna in the Pacific islands region are estimated to be in the range of 630,000 to 650,000 tonnes. This level of catches probably accounts for about 35 percent of the world's total production of tuna in 1984.

DOMESTIC INDUSTRIES

The development of domestic tuna industries in the Pacific islands region involves public and private investment in fleets and shore-based transshipment and processing facilities. Commercial tuna fishing fleets employing pole-and-line, longline, and purse seine techniques are based permanently in 12 countries and territories in the region. These fleets harvest tuna for sale on local fresh-fish markets, for processing at facilities within the region, and for export in the frozen round form. Shore-based facilities for handling and processing tuna are located in ten countries and territories in the region. Transshipment facilities in two countries consolidate catches made by individual vessels so that tuna can be exported to processors in economic quantities. Processing facilities for smoking and/or canning tuna exist in eight countries. Plans are under way to expand and reorganize some existing processing facilities, while several other countries are contemplating the establishment of new shore-based ventures.

Fleets

Until the late 1940s the only commercial tuna fleet based permanently in the Pacific islands region was in Hawaii. This fleet consisted of pole-and-line and longline vessels that supplied fish to the fresh-fish market and to the cannery in Honolulu, Hawaiian Tuna Packers.

In the mid-1950s, in response to the rising international demand for canning material (especially in the United States), Japanese, South Korean, and Taiwanese longline tuna fleets were based in American Samoa (1950), Fiji (1953), and Vanuatu (1957). The size and number of longline fleets in the region continued to grow in the 1960s. Two fleets of longliners were based in French Polynesia by the U.S. canning companies Star-Kist/Castle & Cooke and Van Camp. With a continuing strong demand for canning material and an increased understanding of the nature and extent of tuna stocks in the region, particularly skipjack, pole-and-line fleets were established in Fiji (1963), Papua New Guinea (1970), and Solomon Islands (1973).

Countries and territories in the region continued to foster the establishment of tuna fleets and related shore-based industries. Since the late 1970s purse seiners have been based or registered at several ports in the region, including American Samoa (1980), Guam (1980), Nauru (1982), Papua New Guinea (1981), and Vanuatu (1984). They have been based adjacent to proven fishing grounds, where facilities exist to tranship catches or, in the case of Vanuatu, to take advantage of the country's "tax-haven" status.

While the size of the purse seine fleet has increased in recent years, the size and number of longline fleets in American Samoa, Fiji, and Vanuatu have contracted, and the operations of the longline fleets in French Polynesia were terminated. At the same time, the size and number of pole-and-line fleets in the region have been relatively stable. The one notable exception is the pole-and-line fleet in Papua New Guinea, which collapsed at the end of the 1981 fishing season but was re-established in 1984 on a reduced scale.

Foreign aid has played an important role in helping several countries in the Pacific islands region to establish embryonic domestic tuna industries. Fleets in four countries (Kiribati, Solomon Islands, Tonga, and Tuvalu) have been directly established through bilateral aid programs with the government of Japan and with the British government in the case of Kiribati. The provision of vessels to these countries has been in addition to other fisheries aid supplied for the development of the artisanal sector and has been allocated to the construction of shore-based infrastructure and the training of Pacific islanders.

An approximate inventory of tuna fishing vessels based in the Pacific islands region in 1985 is shown in Table 1. An indication of the destination and utilization of catches is also given.

Although fleets are based within the region, capital ownership (vessels) remains largely in foreign hands. Approximately 90 percent of the total capital invested in the region's fleets is estimated to be owned and controlled by individuals and corporations outside the region. This fact is an important policy consideration for governments.

Longline fleets. Of the six longline fleets based in the Pacific islands region, one fleet (Hawaii) involves owner/operator vessels; two fleets (Solomon Islands and Tonga) are owned by public corporations; and three fleets (American Samoa, Fiji, and Vanuatu) involve the charter of Korean and Taiwanese vessels by canners (American Samoa and Fiji) and a fishing company (Vanuatu).

The longline fleet in Hawaii consists of 45 vessels. All vessels are less than 50 gross registered tonnes (GRT), and all operate as ice boats, which means that fishing trips do not usually exceed ten days' duration. Fish caught by these vessels is landed at Honolulu, where it is sold at auction for the local fresh-fish market. Some prime-quality

Table 1. Approximate inventory of tuna fishing vessels based or registered in the Pacific islands region, 1985

Country/territory	Vessel type	Base port	Number	Destination of landings
American Samoa	Purse seine (single)	Pago Pago	10	Van Camp and Star-Kist canneries
	Longline	Pago Pago	86	Pago Pago
Fiji	Pole-and-line	Lami	6	PAFCO cannery
	Longline	Lami	11	
	Purse seine (single)	Lami	1	
Guam	Purse seine (single)	Agana	11	Star-Kist cannery at Pago Pago and sales to Thailand
Hawaii	Pole-and-line	Honolulu	9	Sashimi market throughout Hawaii
	Longline	Honolulu	45	
Kiribati	Pole-and-line	Betio	4	Star-Kist cannery Pago Pago and PAFCO Levuka
Marshall Islands	Pole-and-line	Majuro	4	*Katsuobushi* plant at Majuro
Nauru	Purse seine (single)	n.a.	2	n.a.
New Caledonia	Longline	Noumea	2	Japanese sashimi market
Papua New Guinea	Pole-and-line	Kavieng	9	Sales on open market and sales to PAFCO cannery
	Purse seine (single)	Rabaul	1	
Solomon Islands	Pole-and-line	Tulagi	27	Solomon Taiyo cannery at Tulagi, sales on open market, and Japanese sashimi market
	Longline	Tulagi	2	
	Purse seine (group)	Tulagi[a]	1	
Tonga	Longline	Nukualofa	1	PAFCO cannery
Tuvalu	Pole-and-line	Funafuti	1	PAFCO cannery
Vanuatu	Purse seine (single)	Manta[b]	3	Star-Kist cannery at Pago Pago and albacore transshipped to Puerto Rico
	Longline	Palikulo	17	

[a]Operates in Solomon Islands about three months per year.

[b]Operating in eastern Pacific and based in Ecuador.

sashimi-grade tuna is shipped to the U.S. mainland, and in the past some has been exported to Japan.

The fleet in Solomon Islands consists of two vessels owned and operated by National Fisheries Development Ltd (NFDL). The vessels were provided to Solomon Islands government by Japan in 1981 under a bilateral fisheries development aid program. Tuna landed by the vessels is sold in Japan on the sashimi market.

Tonga's longline fleet consists of one vessel. Like the longliners operated by Solomon Islands government, this vessel was acquired by the Tongan government from Japan in 1982 under bilateral aid arrangements. The Tongan fleet concentrates on albacore, which it delivers to Fiji's cannery for processing.

Other longline fleets operating in the region involve chartered vessels from Taiwan and South Korea. These vessels primarily fish for albacore, which is landed and processed at facilities in American Samoa, Fiji, Vanuatu, and, until 1984, Hawaii. The number of vessels in these longline fleets has fluctuated depending on (1) tuna market conditions and (2) the availability of vessels for charter.

In 1985 a total of 161 longliners was based in the region: American Samoa (86 vessels), Fiji (10 vessels), Hawaii (45 vessels), Solomon Islands (2 vessels), Tonga (1 vessel), and Vanuatu (17 vessels). The longliners operating in American Samoa are chartered to Star-Kist and Van Camp, in Fiji to the publicly owned Ika Corporation, and in Vanuatu to the South Pacific Fishing Company (SPFC). However, the number of vessels in these fleets has been declining, and this downward trend is expected to continue.

Pole-and-line fleets. Pole-and-line tuna fleets are the most prolific in number and size in the Pacific islands region. Fleets are based in seven countries: Fiji, Hawaii, Kiribati, Marshall Islands, Papua New Guinea, Solomon Islands, and Tuvalu. In 1985, 60 pole-and-line vessels were estimated to be based in the region. Apart from Hawaii and Papua New Guinea, there is substantial public participation in the ownership of these fleets. The location and size of individual fleets in the region are largely determined by the availability of reliable and good-quality baitfish supplies. Live bait is required for pole-and-line fishing operations, with the result that those countries with reasonably robust baitfish stocks (Papua New Guinea, Solomon Islands, and to a lesser extent Fiji and Hawaii) have, or have had, the largest fleets.

The availability of baitfish in the region is a necessary condition for the development of pole-and-line fishing operations. In addition, this fishing method is seen by several governments as a technologically appropriate method for the region—a factor that has also played a role in its becoming so firmly entrenched. In many respects, pole-and-line fishing represents an intermediate technological step between artisanal operations and sophisticated purse seining.

Despite the technological and other advantages of pole-and-line fishing, this method of accumulating very large catches is more costly than purse seining, which is its principal competitor in supplying tuna for canning. The average production cost for pole-and-line-caught fish is estimated to be at least $100 per tonne higher than that for tuna harvested by purse seiners. However, the pole-and-line product, if handled well after capture, is of better quality than the purse seine product, and for this reason it should sell at a premium. With the downturn in the world tuna market in the 1980s and the general excess-supply situation that resulted, the price differential between pole-and-line and purse seine products has been eroded to the detriment of pole-and-line operators.

The operation of the pole-and-line fleet in Fiji is controlled by the government-owned Ika Corporation. Tuna landed by the corporation's vessels, or by those on charter, is sold under contract by the Ika Corporation to the the Pacific Fishing Company Ltd (PAFCO) for canning. The size of Fiji's pole-and-line fleet is constrained by the availability of baitfish and to some degree by the seasonality of the skipjack resources within the country's exclusive economic zone (EEZ).

The pole-and-line fleet in Hawaii consisted of nine vessels in 1985. Vessels in the pole-and-line fleet are owned by individual fishermen, and the fleet's marketing operation is organized by two cooperatives. Most vessels are between 39 and 69 GRT. Catches are off-loaded in Honolulu, and because of the seasonal nature of the fishery, wide fluctuations in daily catch rates occur between the winter and summer months.

Kiribati has four pole-and-line vessels that are owned and operated by the public corporation, Te Mautari Ltd (TML). One vessel was supplied by the British government under its aid program to Kiribati and the other three vessels by the Japanese government as fisheries aid. The vessels range in size from 20 to 120 GRT. TML proposes to eventually increase the size of its pole-and-line fleet to eight vessels. Cultivated baitfish (milkfish) is used to supplement supplies of naturally occurring baitfish to support pole-and-line operations.

Tuna landed by the fleet in Kiribati is sold to Star-Kist in American Samoa or to PAFCO in Fiji for canning. It is transported via a converted longline vessel.

Pole-and-line fishing in Papua New Guinea started in 1970. A combination of strong international demand for canning material, extensive and accessible baitfish resources, and year-round tuna fishing conditions contributed to the fishery's expansion. However, it collapsed at the end of the 1981 fishing season.

Efforts by the Papua New Guinea government to re-establish the fishery were realized when eight pole-and-line vessels commenced operations in August 1984. The vessels operating in the fishery are 59-GRT class vessels. The corporation operating in the fishery, Papua New Guinea Tuna Fisheries Pty Ltd (PNGTF), is financially and logistically backed by Japan's Okinawan prefectural government.

Catches made by the fleet in Papua New Guinea are sold overseas on the open market because PNGTF has no ties with traders or processors in Japan or elsewhere. Between August 1984 and December 1985 about 50 percent of the landings were sold to PAFCO (Fiji), while the yellowfin (which accounts for approximately 30 percent of total landings) was sold in Japan and Thailand. In 1985 nine pole-and-line vessels operated in Papua New Guinea.

The largest pole-and-line fleet currently operating in the Pacific islands region is in Solomon Islands. In 1985 the fleet consisted of 27 vessels, 13 of which were owned by either the joint-venture company, Solomon Taiyo Ltd (STL), or NFDL. The remaining 14 vessels were chartered from individual owner/operator fishermen from Okinawa. NFDL supplies all fish caught to STL for processing or export in the frozen round form. All catches made by the fleet in Solomon Islands are landed at Noro (Western Province) or Tulagi (Central Province near Honiara), where they are consolidated for processing or export. The pole-and-line vessels operating in Solomon Islands fishery are mainly 59-GRT vessels, and those vessels owned and operated by NFDL have been constructed in the country.

Tuvalu has one pole-and-line vessel (175 GRT). The vessel was provided to the government of Tuvalu in 1982 by the Japanese government as part of its fisheries aid program. The vessel is operated by the National Fishing Corporation (NFC) as a training and commercial fishing vessel.

Purse seine fleets. Experimental purse seine fishing in the Pacific islands region dates back to the 1960s. However, it was not until the mid-1970s that fishing grounds in the region were proven.

With the demonstrated viability of the DWFN purse seine operations in the Pacific islands region, island governments encouraged the basing of seiners at ports that could accommodate such vessels. In 1985, 29 purse seiners were estimated to be based and/or registered in the region. More than 70 percent of these vessels (21) are based in the U.S. territories of American Samoa (10 vessels) and Guam (11 vessels).

The seiners in American Samoa supply tuna to the Star-Kist and Van Camp canneries. In the case of Guam, the basing of vessels at

Agana is an attempt to obtain locational advantages from being close to the main purse seine fishing grounds in the western Pacific while still having the security and other financial advantages of being situated in a U.S. territory. The seiners based in American Samoa and Guam represent a variety of registrations, but all vessels are affiliated formally or informally with the American Tunaboat Assocation (ATA).

To augment longline and pole-and-line catches for processing in Fiji and thereby to provide an economic throughput, the Ika Corporation chartered one New Zealand seiner in 1985. This vessel operates in Fijian waters for the duration of the southern winter (April to October).

Nauru's purse seine fleet consists of two vessels that previously fished in Peru's anchovy fishery. Details of the operation of the vessels since their purchase by Nauru's municipal authority in 1982 are sketchy, but catches are known to have been small.

A single seiner of Honduran registration, owned by Taiwanese fishing interests, has operated in Papua New Guinea since 1981. The seiner makes exceptionally good catches, which are transshipped in Rabaul harbor for transportation to American Samoa (Star-Kist) for processing.

A group seiner of Japanese registration owned by Taiyo Fishery Ltd operates in Solomon Islands waters under contract to STL. In recent years the catches made by this vessel have been among the best recorded for any purse seine vessel in the world. Tuna landed by the seiner is either processed by STL or exported in the frozen round form.

In 1984 and 1985 three purse seine vessels obtained Vanuatu registration. The vessels are owned by Ecuadorian entrepreneurs and are based at Manta, Ecuador. The vessels fish in the eastern tropical Pacific and deliver their tuna to Ecuador for canning.

A striking feature of fishing and processing arrangements in the Pacific islands region is the extent to which several countries are cooperating in the development of their respective tuna industries. For example, the Fiji cannery is, or has been, supplied with tuna from fleets based in Kiribati, Papua New Guinea, Tonga, and Tuvalu in addition to its own fleets. This type of cooperation provides a foundation for the development of a regional tuna industry and possibly for a regionally owned cannery.

Catches

Tuna fleets of all gear types based in the region have harvested an estimated 140,000 to 160,000 tonnes of tuna and related species annually. Catches made by the region's fleets are consolidated by country in Table 2 for the period 1978–84. In 1984 the quantity of tuna harvested by the region's fleets accounted for about 25 percent of total catches made

Table 2. Estimated total catches of tuna in the Pacific islands region by fleets based permanently in the region, 1978–84

Country/territory	1978	1979	1980	1981	1982	1983	1984	Total		
				tonnes				tonnes	%	ranking
American Samoa†	30,000	30,000	30,000	40,000	50,000	50,000	50,000	280,000	29	1
Fiji[a]	10,490	7,073	8,232	9,820	7,811	7,859	8,970	60,255	6	5
Guam†	–	–	10,000	30,000	35,000	35,000	35,000	145,000	15	4
Hawaii[b]	6,288	4,638	4,238	4,487	3,156	4,867	3,734	31,408	3	8
Kiribati[c]	560	757	800	828	717	2,328	2,627	8,617	1	9
Palau†	15,000	15,000	15,000	5,000	0	0	0	50,000	5	6
Papua New Guinea[d]	48,933	26,945	36,009	26,029	3,000†	3,000†	9,000†	152,916	16	3
Solomon Islands[e]	17,773*	24,334	22,176	26,729	19,537	33,499	35,000	179,048	19	2
Tonga†	420	420	420	420	400	400	400	2,880	0.3	10
Tuvalu[f]	–	–	–	–	223	346	350	919	0.1	11
Vanuatu[g]	9,182	7,724	8,300	4,840	3,880	4,541	3,945	42,412	4	7
Western Samoa[h]	762	965	1,829	2,235	2,438	2,591	2,600	13,420	1	9
Total	139,408	117,856	137,004	150,388	126,162	144,431	151,626	966,875	100	

* Extracted from Food and Agriculture Organization. Various years.

† Estimated.

[a] Data supplied by Fiji Ministry of Fisheries.

[b] Provided by industry.

[c] Dalley 1984. Tuna sales by Te Mautari only.

[d] Papua New Guinea Department of Primary Industry (Fisheries Division). Various years.

[e] Anon 1984.

[f] Schupp 1984.

[g] Vanuatu National Planning and Statistics Office 1985. Production is assumed to equal exports.

[h] Philipp 1984.

by all tuna fleets operating in the region. At 1984 prices, the estimated unprocessed value of the combined catches made by the region's fleets was in the range of $90 million to $100 million. The retail value of catches (canned and sashimi) would have been approximately $250 million.

In contrast to catches made by the DWFN fleets operating in the Pacific islands region, tuna harvested by fleets based in the region land their catches at domestic ports for processing and/or for transshipment overseas. Landing catches at domestic ports can generate substantial indirect benefits. In most cases, these benefits are significantly more valuable to the economies of the countries in the region than the benefits associated with the primary activity of fishing. For this reason, countries want the tuna harvested in their EEZs to be at least transshipped, if not processed, at domestic ports.

Total landings by tuna fleets based at ports within the region showed an upward trend from 139,408 tonnes in 1978 to 151,626 tonnes in 1984, an increase of 9 percent (Table 2). The annual fluctuations in landings are due both to international changes in the world tuna industry and to changes in fishing effort and related factors.

Transshipment facilities

Two shore-based design-specific transshipment facilities are located in the Pacific islands region. These facilities have a total freezer capacity of 6,000 tonnes and are capable of handling 25,000 tonnes of tuna per annum (Table 3). The facilities are located in Palau (Koror) and in Vanuatu (Palikulo).

The transshipment facility in Palau was established in 1964 by the U.S. multinational corporation, Van Camp Seafood Company. The tuna landed at the facility was transshipped for processing at Van Camp's cannery in American Samoa. With the downturn in the world tuna market, Van Camp abandoned the facility in Palau in 1982, and it has not been used commercially since that time. Palauan interests are moving to purchase the facility and convert it into a sashimi transshipment facility utilizing the longline product caught in the country's waters.

The tuna transshipment facility in Vanuatu, owned by South Pacific Fishing Company (SPFC), was established in 1957 by Japanese, British, and U.S. interests. It was subsequently purchased by the Japanese multinational Mitsui (72 percent equity ownership), the Vanuatu government (19 percent), and private interests (9 percent). The facility was established to service longline fleets—initially Japanese and subsequently Taiwanese—operating from Palikulo (Espiritu Santo). Subsequent to the closure of Honolulu's cannery in 1984, albacore caught by the Palikulo fleet is being shipped to Mitsui's cannery (Neptune Packing Corporation) in Puerto Rico. The logistics and added costs associated

Table 3. Shore-based transshipment facilities in the Pacific islands region, 1985

Country/location	Established	Freezer capacity (tonnes)	Annual handling capacity (tonnes)	Product handled	Permanent employment	Ownership	Current utilization
Palau Koror	1964	2,000	15,000	Pole-and-line and purse seine product	100	Van Camp (U.S.)	Closed 1982
Vanuatu[a] Palikulo	1957	4,000	10,000	Longline product for canning (albacore)	41	Joint venture: Mitsui 72%; Vanuatu government 19%; Private interests 9%	Facility in Palikulo underutilized
Total capacity/ employment		6,000	25,000		141		

Source: Shephard and Clark 1984 and data supplied by government agencies.

[a]Mitsui withdrew from the joint venture in 1986.

with shipping tuna to Puerto Rico rather than to Hawaii have eroded the viability of operations.

Since the mid-1970s the transshipment facility in Vanuatu has been underutilized. Tuna exports declined from $13.2 million in 1978 to $7.8 million in 1983, a fall of 41 percent. This decline was reflected in the relative importance of tuna exports in Vanuatu's trade, falling from 32 percent of total exports in 1978 to 27 percent in 1983.

In late 1986 Mitsui withdrew from Vanuatu's transshipment facility. According to industry reports, the government is trying to attract a replacement foreign investor to operate the facility.

In addition to the existing transshipment facilities in the region, facilities are being either considered or constructed in several island countries. Majuro, Marshall Islands, has a $2 million wharf and cold-storage facility under construction. This facility is being financed by the Japanese government under its fisheries aid program and is expected eventually to service some 400 tuna vessels, mainly of Japanese flag, fishing the Marshall Islands waters. Later the facility is to be expanded to make Majuro a prominent regional tuna transshipment base.

The Federated States of Micronesia has been considering the establishment of a cold-storage and transshipment facility at Dublon Island, Truk State, which already has an excellent wharf for large ocean-going vessels.

In Papua New Guinea, a medium-sized transshipment facility (3,000 tonnes capacity) is being considered for either Manus or Rabaul to service purse seine fishing fleets in the country's EEZ. The Fiji government and the Fiji Chamber of Commerce are also interested in attracting American purse seiners to Suva.

Tuna transshipment from the U.S., Japanese, and Taiwanese purse seine vessels is also taking place at Guam, Tinian (Northern Marianas), and Papua New Guinea. These operations, however, are inherently temporary because they do not involve the use of shore-based facilities. In 1984 about 80,000 tonnes of tuna were estimated to be transshipped at ports in Guam, the Northern Mariana Islands, and Papua New Guinea.

The future of design-specific transshipment facilities in the Pacific islands region will depend primarily on three factors: (1) the extent to which island countries interpret and enforce the provisions of the Law of the Sea Convention (Article 62) with respect to requiring the DWFN fleets to land tuna at domestic ports; (2) the capacity of countries to provide and manage—either by government or private investors—appropriate facilities; and (3) the size and type of the DWFN fleets operating in the region.

Processing

Two types of tuna processing take place in the Pacific islands region: the processing of *katsuobushi* and the production of canned tuna and related products.

Katsuobushi. *Katsuobushi* production involves filleting, boiling, smoking, drying, and molding of skipjack. It is used as a popular food additive, soup base, etc., in Japan. The demand for *katsuobushi* outside Japan is limited, and the Japanese market since the mid-1970s has shown little growth.

Katsuobushi plants in the Pacific islands region are located in the Marshall Islands, Solomon Islands, and, until 1979 and 1981, in Papua New Guinea and Palau (Table 4). The combined daily processing capacity of all plants is 34 tonnes of wet fish input. However, the active plants—Marshall Islands and Solomon Islands—have a combined daily processing capacity of only 9 tonnes of wet fish input or an annual processing capacity of 2,300 tonnes.

The *katsuobushi* plant in the Marshall Islands was established at Majuro in 1984 as a joint venture between Marshallese and Okinawan entrepreneurs (Table 4). The venture trades as the Long Island Nankatsu Company. The initial capital investment in the facility was $200,000. The facility processes tuna landed by two pole-and-line vessels based in Majuro, and the processed product, along with chicken-feed by-product, is exported to Japan.

In Solomon Islands, *katsuobushi* processing was established by STL at Tulagi in 1974 (Table 4). The plant has a daily processing capacity of 4 tonnes or an annual processing capacity of 1,300 tonnes. Between 1980 and 1983, 831 tonnes of *katsuobushi* were exported by STL for approximately $3.1 million. Tuna processed at the *katsuobushi* plant is landed by NFDL, STL, or contracted vessels.

The inactive *katsuobushi* plants in Palau and Papua New Guinea were the largest plants in the Pacific islands region (Table 4). The plant in Palau had a daily processing capacity of 10 tonnes. The company operating the plant—Caroline Fishing Company—was a joint venture between Palauan and Japanese investors.

Papua New Guinea's *katsuobushi* plant was established at Nago Island (Kavieng) in 1971 by Gollin Kyokuyo, a Japanese/Australian corporation (Table 4). In 1979 Gollin Kyokuyo's operations were taken over by Star-Kist (Papua New Guinea) Pty Ltd, at which time the *katsuobushi* plant was mothballed.

In 1983 the Papua New Guinea government assumed control of Nago Island and the *katsuobushi* plant as part of a taxation settlement

Table 4. *Katsuobushi* processing plants in the Pacific islands region, 1985

Country/ location	Established	Daily processing (input wet tonnes)	Annual capacity (input wet tonnes)	Permanent employment (persons)	Ownership
Marshall Islands Majuro	1984	5	1,000	52	Joint venture between Marshallese and Okinawan interests (Long Island Nankatsu Co)
Palau Koror[a]	1977	10	n.a.	n.a.	Caroline Fishing Co, a joint venture between Palauan and Japanese interests (Nantaku Co affiliate of Hassui Reizo)
Papua New Guinea Kavieng[b]	1971	15	2,000	50	Government of Papua New Guinea; formerly Gollin Kyokuyo (1971–79) and Star-Kist (1980–82)
Solomon Islands Tulagi	1974	4	1,300	50	Solomon Taiyo
Total		34	n.a.	n.a.	

Source: Data supplied by countries.
[a]Not operated since 1982.
[b]Not operated since 1979.

with Star-Kist. The *katsuobushi* plant was scheduled to be reactivated in 1985 by PNGTF, but this did not eventuate.

The production of *katsuobushi* in the Pacific islands region is constrained in some countries by the lack of timber for the smoking process and by Japan's stagnant market. Nonetheless, it is a suitable form of processing for many island countries because of (1) skipjack availability, (2) its labor-intensive processing procedures, and (3) the product's storage qualities after processing. However, the financial viability of "stand-alone" *katsuobushi* plants appears doubtful.

Canning. Japanese and U.S. multinational corporations are active in tuna canning in the Pacific islands region. Until 1984 five canneries were operating in the region in American Samoa, Fiji, Hawaii, and Solomon Islands (Table 5). The Hawaii cannery closed in October 1984, but it will be reactivated in 1987 on a reduced scale.

The total processing capacity of the canneries in the region is 210,000 tonnes per year. Of this total capacity American Samoa has 155,000 tonnes (74 percent of the total capacity), Fiji has 15,000 tonnes (7 percent), Hawaii has 35,000 tonnes (17 percent), and Solomon Islands has 5,000 tonnes (2 percent).

Star-Kist operates American Samoa's largest cannery—Star-Kist Samoa Inc. The cannery has an annual processing capacity of 80,000 tonnes. Star-Kist is wholly owned by the U.S. multinational giant, the H.J. Heinz Corporation. Star-Kist established the cannery in American Samoa in 1963, and it is Heinz's largest tuna canning operation.

Van Camp also operates a cannery in American Samoa. The company trades as Samoa Packing and is a wholly owned subsidiary of the U.S. multinational Ralston Purina. The Van Camp cannery has an annual processing capacity of 75,000 tonnes.

The American Samoan canneries process the longline, pole-and-line, and purse seine product. The tuna industry in American Samoa is the territory's leading industry in terms of employment, government revenue, and export receipts. Tuna exports accounted for an average of 95 percent of the territory's total exports between 1978 and 1983 (Table 6).

The tuna cannery in Fiji is located at Levuka (Ovalau Island), about 60 kilometers from Suva. The cannery is a joint venture that trades under the name of PAFCO (Table 5). PAFCO's cannery has an annual processing capacity of 15,000 tonnes. Because of its inability to secure adequate supplies of longline- and pole-and-line-caught fish to operate at full capacity, the financial position of the cannery has been less robust than it otherwise might have been. The introduction of the New Zealand-owned and -operated purse seine vessels in 1980 was an

Table 5. Plants canning skipjack, yellowfin, and albacore in the Pacific islands region, 1985

Country/location/ company	Year estab- lished	Annual processing capacity (tonnes)	Freezer capacity (tonnes)	Storage capacity (cases)	Permanent employment (persons)	Ownership	Management
American Samoa Pago Pago Star-Kist	1963	80,000	n.a.	n.a.	n.a.	Heinz Corp (Star-Kist)	United States
American Samoa Pago Pago Van Camp	1954	75,000	n.a.	n.a.	n.a.	Ralston Purina (Van Camp)	United States
Fiji Levuka PAFCO	1964 (freezing) 1975 (canning)	15,000	2,000	150,000	350	Joint venture: 36% Fiji government; 60% C. Itoh; 4% private	Japan until December 1986
Hawaii Honolulu Wraf Corp (formerly Castle & Cooke)	1917	35,000	1,450	n.a.	400	100% Hawaii consortium (formerly 100% Castle & Cooke)	United States
Solomon Islands Tulagi (STL)	1972	5,000	1,600	n.a	270	Joint venture: 50% Solomon Islands government and 50% Taiyo	Japan
Total		210,000	n.a.	n.a.	n.a.		

Source: Data supplied by countries.

Table 6. Value of tuna and amount of tuna exports as a percentage of total exports from selected Pacific island countries, 1978–83

Country	1978	1979	1980	1981	1982	1983	Total
American Samoa[a]							
Value $m	102.1	124.6	148.0	198.4	170.7	153.1	896.9
Percent	98	99	97	99	91	86	70
Fiji[b]							
Value $m	10.6	13.8	10.9	18.6	9.6	14.1	47.6
Percent	5	5	3	6	3	6	6
Papua New Guinea[c]							
Value $m	29.8	24.9	38.4	29.4	1.9	0.3	124.7
Percent	4	2	4	4	0	0	10
Solomon Islands[d]							
Value $m	7.6	16.9	23.2	22.0	14.0	29.2	112.9
Percent	23	28	38	38	25	41	9
Vanuatu[e]							
Value $m	13.2	12.0	15.3	9.6	7.1	7.8	65.0
Percent	32	26	54	32	31	27	5
Total $m							1,277.1
Percent							100

[a]U.S. Department of Labor 1986.

[b]Fiji Bureau of Statistics 1984.

[c]Bank of Papua New Guinea. Various years.

[d]Solomon Islands government 1984b.

[e]Vanuatu National Planning and Statistics Office 1985.

attempt to overcome fish supply problems. The performance of these vessels has been below expectations.

PAFCO also has sought to purchase pole-and-line fish within the Pacific islands region. In 1984 and 1985 some of the tuna harvested by the fleet in Papua New Guinea was sold to PAFCO, but with the closure of its fishing operations in 1986 this supply source became unavailable. Attempts to purchase tuna from STL have been less successful, despite the 1984–85 U.S. tuna embargo on Solomon Islands fish and the loss of the country's principal export market for frozen tuna. STL has been reluctant to sell tuna to PAFCO because Taiyo and C. Itoh—the Japanese partners in these joint ventures—are competing Japanese corporations. This is an instance where the interests of the foreign partners in these ventures take precedence over the regional cooperation interests of Pacific island countries.

Tuna is Fiji's third most important export-earning industry after sugar/molasses (63 percent of total exports in 1983) and gold (10 percent). Between 1978 and 1983, Fiji's tuna exports contributed from 3 to 6 percent annually to the country's total export receipts and accounted for 6 percent of the total tuna exports from the Pacific islands region over this period (Table 6).

The oldest established cannery in the Pacific islands region is located in Hawaii. Tuna canning started in Honolulu in 1917, and, apart from the interruption of World War II, it operated continuously until its closure in October 1984. The cannery traded as Hawaiian Tuna Packers, but following its closure the cannery's assets were purchased by the Wraf Corporation, a Hawaii-based consortium. New lease conditions require the Wraf Corporation to operate at least a one-line tuna cannery and to support local pole-and-line fishermen. The operation of a one-line cannery and the support of a nine-vessel pole-and-line fleet are unlikely to be profitable in themselves.

Since 1972, STL has operated a tuna cannery at Tulagi in Solomon Islands. The Japanese partner in the joint venture, Taiyo Fishery Company, is Japan's second-largest fishing and fish-processing company. Taiyo has a 50 percent equity holding in STL, with Solomon Islands government holding the remaining 50 percent.

STL's cannery has an annual processing capacity of 5,000 tonnes of tuna. The tuna processed at the cannery comes from the NFDL pole-and-line fleet, the joint-venture fleet, and the vessels contracted to NFDL. The tuna industry is the most important industry in Solomon Islands. As a proportion of total exports, tuna exports ranged from 23 percent of the total in 1978 to 41 percent in 1983 (Table 6).

STL has announced plans to relocate the cannery from Tulagi to Noro. This move is consistent with the government's decentralization policy and is aimed at siting the cannery closer to prime bait and fishing grounds. Logistically, however, because of Noro's relative isolation in comparison with that of Tulagi, additional costs and supply difficulties probably will be encountered in operating the cannery.

The relocation of the cannery to Noro will triple its processing capacity. The total cost of constructing the new cannery will be approximately $7 million, which is being funded by the Asian Development Bank.

STL's canned tuna has a high market reputation and is principally sold in the United Kingdom under upmarket labels such as John West. Like Fiji, Solomon Islands is a signatory to the Lome Convention and is therefore eligible for concessional trade access to the markets of the European Economic Community's member countries. Within Solomon Islands, an attempt has been made by government and industry to replace imported canned fish with STL's lower-grade, darkmeat tuna packs. To facilitate import replacement, tariffs are imposed on canned fish imports to make the domestically packed fish more price-competitive.

MULTINATIONAL CORPORATIONS

Some of the world's leading multinational fishing, tuna processing, and trading companies have operated, or continue to operate, in the Pacific islands region. The tuna potential of the region was identified by the Japanese early in this century, but it was not until the 1950s that the full potential of the region's tuna resources was recognized. Since then, the region has become the focus of international tuna interest, with a range of multinational corporations indirectly financing or directly initiating ventures. The primary goal of these corporations, particularly in the 1960s and 1970s, was to obtain tuna from the region for trade and/or for processing.

The corporations, through their international networks, have been able to muster financial resources to provide expertise and to locate or provide markets for a venture's output. In the past this contribution was highly valued by island countries that were hesitant to proceed alone into what they perceived as a highly complex and "closed-shop" tuna business. But with the realization that big business does not have a monopoly on industry information and marketing and that island countries have received fewer benefits from ventures than were initially promised and expected, these countries are now more confident and pragmatic in assessing tuna development and fishing proposals.

Some island countries have had their tuna development plans disrupted by the activities of multinational corporations. While Pacific island countries aspired to the development of stable and long-term industries that could bring them reasonable financial and economic returns, this goal was not shared by many of their multinational investors with their short-term objectives.

Many of the tuna ventures associated with multinational corporations in the region involved minimal investment. In most cases, the ventures focused on securing raw materials for processing, though this was not the case for American Samoa, Fiji, and Solomon Islands. A number of ventures—for example, those of the U.S. corporations—involved direct investment in fishing activities to ensure throughput for their canneries in American Samoa and elsewhere. However, as a general rule, the multinationals minimized their financial exposure in the region so that they could relocate their operations at little or no financial loss should market conditions change—as they did for the canning industry in the early 1980s—or should cheaper alternative raw materials become available.

Although Korean and Taiwanese fleets operate in the region on a distant-water basis and in some cases are chartered by Japanese and U.S. companies, no Korean or Taiwanese corporations have established tuna ventures in the region.

Australia and New Zealand—countries that have traditionally played a key investment role in the Pacific islands region—have not become involved in tuna ventures in the region. The reasons are that in the past tuna has not been an important industry in either country and because Australian and New Zealand per capita consumption of tuna is small.

Three U.S. corporations and ten Japanese corporations have participated in some capacity in the tuna industry in the Pacific islands region. The U.S. corporations have had a total of nine ventures in the region, though only two of these were active in 1985 (Table 7).

The only U.S. tuna ventures remaining in the region are located in a U.S. territory. This is largely because of (1) the length and special relationship between a territory and the United States and (2) the security that U.S. investors have in a U.S. territory.

The approach of Japanese tuna corporate operations in the Pacific islands has been different from those of the U.S. corporations (except for American Samoa) in that the Japanese operations have tended to be more stable and predicated on end uses other than canning. Moreover, the involvement of diversified Japanese trading companies in the industry has meant that a broader perspective of the fishing industry has been adopted, though their tuna fishing/processing ventures were expected to be profitable units in themselves.

Table 7. Pattern of multinational corporation involvement in the tuna industry in the Pacific islands region

	Corporations												
	United States			Japan									
Country	Castle & Cooke	Star-Kist	Van Camp	Hassui Reizo	Ho-koku Marine	Itoh	Kyo-kuyo	Mitsu-bishi	Mitsui	Nan-katsu	Nip-pon Suisan	Ocean Suisan	Taiyo
American Samoa		x*	x*										
Federated States of Micronesia												x	
Fiji						x*							
French Polynesia	x	x	x										
Guam			x										
Hawaii	x												
Marshall Islands										x*			
Palau			x	x									
Papua New Guinea		x			x	x	x	x			x		
Solomon Islands													x*
Vanuatu									x				

* Active in 1985.

Ten Japanese multinational corporations have participated in tuna ventures in six countries in the region as single investors, joint-venture partners, or consortium participants. This strategy contrasts with that of the U.S. corporations, which operate as single investors in ventures. For island countries, the consortium approach, as seems popular with Japanese corporations, has advantages because it increases the expertise available to a venture, reduces the risk exposure of individual investors, and, most important, provides a natural check and balance on the activities of each partner.

Of the ten Japanese multinationals that have participated in the tuna industry in the Pacific islands region, four remained active in 1985 (Table 7). The active corporations—C. Itoh, Mitsui, Nankatsu, and Taiyo—are among Japan's top fishing/processing and trading corporations. C. Itoh is leaving Fiji, and indications are that Mitsui wants to reduce its involvement in the fishing industry worldwide, despite its significant fishing/processing investments. However, the commitment of the Japanese corporations to their ventures in the region compared with that of their U.S. counterparts, has been more long term and more closely aligned with the interests of Pacific island countries.

CONCLUSION

Tuna is important to all Pacific island countries as a source of food and as a resource capable of sustaining commercial development. Because many island countries have limited avenues for land-based industries, governments assign a high priority to tuna development.

Tuna resources in the region are large and generally underexploited. However, there are pockets within the region where fishing effort is heavy, and more information is required as to the extent and impact of competing fisheries on tuna stocks. But despite the existence of underexploited tuna resources in the region, it will be the international tuna market and its ability to absorb increased production that will govern the rate of further expansion in the fishery.

Traditionally, artisanal production has been more important in Micronesia and Polynesia than in Melanesia, and consequently the Micronesians and Polynesians have tended to be superior oceanic fishermen. Pacific island fishermen have developed specialized fishing gears and practices to harvest tuna. Recently, FADs have been widely introduced in the region, largely to assist artisanal fishermen to improve their catch rates.

The Pacific islands region plays a significant role as an international source of tuna supply with at least 20 percent of the world's total catches coming from the region each year.

To increase financial and economic returns to island countries from the region's tuna stocks, many countries have fostered the establishment of fishing, transshipment, and processing industries. In most cases, these industries have been established in cooperation with Japanese or U.S. corporations, some of which are world leaders in the tuna industry. In some cases, joint ventures involving Pacific island governments have been established, while other ventures, especially those involving U.S. corporations, have been private investment undertakings.

Longline, pole-and-line, and purse seine fleets are permanently based in 12 countries and territories in the islands region. These fleets deliver tuna for processing to facilities in the region and for export to foreign processors and other specialized markets. Shore-based facilities for transshipping and processing tuna (canning and/or smoking) exist in ten island countries. The processing capacity of at least one facility is to be expanded, and several other countries are evaluating the possibility of developing shore bases of various types.

Thirteen Japanese and U.S. multinational corporations have been involved in some capacity in the Pacific islands region since the 1950s. Apart from American Samoa and Hawaii, where tuna canneries were operated, the U.S. ventures in the region focused on acquiring fish for processing at these canneries or at canneries in other locations in the United States, including Puerto Rico. Japanese involvement in the region included two joint-venture canneries, a joint-venture tuna transshipment facility, and several fishing operations. While Japanese corporations have usually sought to form a consortium and in some cases have included government participation in tuna ventures, U.S. corporations in contrast have adopted a "go-it-alone" approach. Thus U.S. tuna operations in the Pacific islands region, apart from those in American Samoa, have tended to be inherently less stable and less enduring than the operations of the Japanese corporations.

2.
Distant-Water Fleet Operations and Regional Fisheries Cooperation

David J. Doulman

ABSTRACT—This chapter describes and analyzes distant-water fishing operations in the Pacific islands region. It traces the expansion of the fishery, fleet operations, and catches in the exclusive economic zones of South Pacific Forum member countries and dependent territories. Access arrangements between distant-water fishing nations and Pacific island countries are reviewed. Information relating to agreements, classified by fishing method and country, is presented, with particular emphasis on Japan's access agreements in the region. The issue of regional fisheries cooperation among Pacific island countries is also discussed, including the role of the Nauru Group.

INTRODUCTION

The Pacific islands region has 14 independent and self-governing countries together with 8 British, French, New Zealand, and U.S. territories. The independent and self-governing countries are members or observers of the region's principal political grouping, the South Pacific Forum (SPF), and its Solomon Islands-based fisheries development and management arm, the Forum Fisheries Agency (FFA).

All countries and territories in the Pacific islands region declared exclusive economic zones (EEZs) between 1977 and 1984. Depending on individual circumstances, countries declared economic zones, fisheries zones, or, in some cases, both. The total area of the region's EEZs exceeds 30 million square kilometers, with the combined EEZs of SPF member countries accounting for approximately two-thirds of the total area (Table 1).

Large and commercially proven tuna resources exist within the Pacific islands region, with many of the established fishing grounds falling within the EEZs of SPF member countries. Island countries focus on these resources as a means of encouraging distant-water fishing

Table 1. Selected statistics and information relating to Pacific island countries

Country	Status of 200-mile zone[a]	Year declared[b]	Zone area (000 km²)[c]	Zone area (%)	Law of Sea signatories[d]
South Pacific Forum member countries[e]					
Cook Islands	economic	1977	1,830	6.0	yes
Federated States of Micronesia[f]	fishing	1979	2,978	9.7	yes
Fiji	economic	1981	1,290	4.2	yes
Kiribati	economic	1983	3,550	11.6	no
Marshall Islands	fishing	1979	2,131	7.0	yes
Nauru	fishing	1978	320	1.0	yes
Niue	economic	1978	390	1.3	no
Palau	fishing	1979	629	2.1	yes
Papua New Guinea	fishing/economic	1978	3,120	10.2	yes
Solomon Islands	fishing/economic	1978	1,340	4.4	yes
Tonga	economic	1979	700	2.3	no
Tuvalu	economic	1984	900	2.9	yes
Vanuatu	economic	1978	680	2.2	yes
Western Samoa	economic	1977	120	0.4	no
Subtotal			19,978	65.3	
Dependent territories					
American Samoa[g]	economic	1977	390	1.3	

French Polynesia[h]	economic	1978	5,030	16.5
Guam[g]	economic	1977	218	0.7
New Caledonia	economic	1978	1,740	5.7
Northern Mariana Islands[g]	fishing/economic	1978–83	1,823	6.0
Pitcairn Island[i]	fishing	1980	800	2.6
Tokelau[j]	economic	1977	290	0.9
Wallis and Futuna[h]	economic	1978	300	1.0
Subtotal			10,591	34.7
Total			30,569	100.0

[a,b]Moore 1985.

[c]South Pacific Commission. 1984a.

[d]United Nations. 1983. Signatories as of 10 December 1982. The Trust Territory of the Pacific Islands (Federated States of Micronesia, Marshall Islands, Palau, Northern Mariana Islands) signed the convention as an observer.

[e]Federated States of Micronesia, Marshall Islands, and Palau have applied or are eligible to apply for Forum membership.

[f]Federated States of Micronesia consists of four states: Kosrae, Pohnpei, Truk, and Yap.

[g]U.S. territories.

[h]French territories.

[i]British territory.

[j]New Zealand territory.

activity within their EEZs and, where possible, the development of domestic fishing and processing industries.

Distant-water fishing nations (DWFNs) enter into contractual arrangements with island countries in order to harvest the region's tuna stocks. Distant-water fishermen benefit from these arrangements by gaining secure and stable access to rich fishing grounds, while island countries derive revenue and, in some cases, fisheries aid.

The Pacific islands region is an internationally important tuna fishing area. In 1985 the world's catches of skipjack tuna (*Katsuwonus pelamis*), the major canning species, stood at about 700,000 tonnes. Approximately 64 percent of the total was harvested by distant-water purse seine fleets operating in the western sector of the islands region.

Distant-water tuna fishing operations in the Pacific islands region were pioneered by Japanese fishermen. Initially, distant-water operations were confined to pole-and-line fishing, but in time longline and purse seine fisheries were developed. As Japanese fishermen demonstrated the viability of distant-water operations, tuna fleets from other nations joined the fisheries.

EXPANSION AND DEVELOPMENT

Around 1910 Japan's tuna fleets started to move offshore in search of new and less-exploited fishing grounds, and by 1922 pole-and-line fleets were well established in Micronesia. After 1927, when Micronesia became a Japanese mandated territory under the League of Nations, fishing operations in the area were strengthened. Shore facilities were established to service fleets, and by 1938 more than 7,600 Japanese tuna fishermen were operating in Micronesian waters.

World War II drastically reduced Japan's fishing capacity and operations. After the war, legislation to vigorously promote distant-water fishing as a reconstruction activity paved the way for the virtually unrestricted movement of Japanese distant-water tuna fleets into the Pacific islands region. Initially, tuna fishing was re-established in Micronesia, but by the early 1960s tuna bases had been established as far south and east as American Samoa, Fiji, French Polynesia, and Vanuatu to support Japanese distant-water longline fleets fishing for albacore. In most cases, these bases were established to supply tuna canneries in American Samoa and Hawaii and on the U.S. west coast.

The successful Japanese penetration of the Pacific islands region encouraged tuna longline fleets from Korea and Taiwan to follow suit. Japanese albacore fleets were gradually replaced by Korean and Taiwanese fleets as the Japanese longliners began to concentrate on more valuable sashimi tuna species.

The rising demand for canned tuna in the 1960s and 1970s, the desire to eliminate bait problems associated with pole-and-line fishing, and the demonstrated labor and fuel efficiency of purse seine fishing prompted the Japanese government to initiate exploratory purse seine fishing in the Pacific islands in 1974. Surveys produced encouraging results and led to the full-time deployment of commercial Japanese seiners in the region in 1976 for the first time.

FLEETS AND CATCHES

All foreign vessels operating in SPF member countries must be registered on the FFA's regional register of fishing vessels. As a matter of policy, SPF member countries do not license distant-water vessels under their respective access agreements with DWFNs if vessels do not have good standing on the regional register.

Longline fleets

Distant-water longline fleets from at least seven countries operate in the Pacific islands region. In 1985, 1,128 longliners were included on the regional register. However, this number does not accurately reflect the number of vessels operating in the region because (1) it includes Japanese longliners that operate in the Australia/New Zealand southern bluefin fishery, and (2) it probably excludes some or all of the Japanese and Korean longline vessels that operate in the French territories.

Longline vessels operating in the Pacific islands region range in size from 20 to 300 gross registered tonnes (GRT). In SPF member countries, the longline fleets predominantly fish in Micronesia, Papua New Guinea, and Solomon Islands, though larger-class vessels also operate in the southern portion of the region—for example, in the Cook Islands' EEZ. In the territories, the Japanese (113 vessels) and Korean (55 vessels) longline fleets are 150- to 300-GRT class vessels. Depending on flag and location of operations in the region, longline fleets concentrate on albacore or sashimi tuna species.

The Japanese longline fleet is the largest fleet fishing in the Pacific islands, with between 500 and 600 Japanese vessels currently in the fishery under close regulation by Japan's Fishery Agency. Vessels discharge their catches at three main ports in Japan—Yaizu, Misake, and Shimizu.

Korea and Taiwan also have sizable longline fleets operating in the Pacific islands region. Traditionally, these fleets have concentrated on albacore for the U.S. canning market, but more recently they have been directing their efforts toward producing fish for Japan's more lucrative sashimi market. This trend is expected to continue because (1) Korean and Taiwanese fishing costs are lower than those of Japanese fishermen, and (2) Japanese trading companies frequently contract and finance Korean and Taiwanese fleets.

Pole-and-line fleets

All distant-water pole-and-line tuna fishing in the Pacific islands region is undertaken by Japanese fishermen. Approximately 80 pole-and-line vessels currently operate in the region from ports in Japan carrying "hardened" live baitfish. The vessels fish in the EEZs of Micronesian countries (including Kiribati) and, despite being permitted by agreement, rarely fish in Papua New Guinea's EEZ.

The pole-and-line fleet concentrates on surface-swimming schools of tuna: skipjack and juvenile yellowfin. According to Japanese industry reports, the species composition of pole-and-line catches averages approximately 94 percent skipjack and 6 percent yellowfin.

Tuna landed by Japan's pole-and-line fleet normally goes to higher-quality uses than canning. In particular, pole-and-line-caught tuna is used to produce smoked and dried products as well as *tataki* (sashimi lightly grilled over straw). In this regard pole-and-line-caught fish—which is more costly per unit of output than purse-seine-caught tuna—is not competing with the purse seine product.

Distant-water pole-and-line fishing has declined from 500 vessels in 1973 to 120 in 1981. This reduction in the size of Japan's pole-and-line fleet has been due to technological advances in fishing (i.e., the rise of purse seining) and to increased fuel and labor costs, as well as the Japanese government's pole-and-line replacement program.

Japan's distant-water pole-and-line fleet is expected to decline further, despite efforts by industry and government to sustain its current level. The extent and rate of the decline of the fleet will be determined by both the demand for higher-quality skipjack in Japan and the ability of the pole-and-line fleet to deliver reasonably priced fish to meet that demand.

Purse seine fleets

In 1980 the Japanese single purse seine fleet in the Pacific islands, consisting of 14 vessels, was joined by Japanese group seiners (5 vessels) and single seiners from several other countries. The fleet expanded from 68 to 72 seiners in 1981, but by 1984 the number of single seiners exceeded 115. The number of seiners declined in 1985 and 1986, primarily because fewer U.S. vessels were deployed in the central and western Pacific.

The regional register of fishing vessels had 146 single purse seiners registered in 1985. Seiners ranged in size from 300-GRT vessels to superseiners of over 2,000 GRT. The fleet came from 11 countries, though 63.6 percent of the registered vessels were from Japan and the United States. As a result of the continuing reduction in the size of the U.S. fleet, the number of seiners operating in the region in the future probably will not exceed the 1984 figures.

The purse seine fleet operating in the Pacific islands region is characterized by a "core" fleet of vessels consisting of the entire Japanese fleet and about 15 seiners of other flags. This core group of vessels operates throughout the year. In addition, a "floating" fleet component is highly mobile and operates both within the Pacific islands region and outside.

The core purse seine fleet fishes almost exclusively in the EEZs of the Federated States of Micronesia, Palau, and Papua New Guinea, as well as the high-seas pocket between these zones. The floating component of the fleet not only fishes in the zones of these same three countries, but also operates quite extensively around Nauru, Kiribati, and elsewhere.

In contrast to the operations of distant-water longline and pole-and-line fleets in the Pacific islands region, a significant proportion of the tuna harvested by the purse seine fleet is transshipped or discharged in the region, but not within the SPF member countries. Purse-seine-caught tuna is transshipped at Guam and Tinian (Northern Mariana Islands) for transport to canneries in American Samoa and Puerto Rico or, alternatively, is discharged directly by the seiners at the American Samoan canneries.

Tuna caught by the Japanese purse seine fleet is usually taken to Japan at the completion of each fishing trip. The fleet is based at the ports of Yaizu and Shimizu, where catches are discharged. Some of the purse seine product is processed in Japan, but a significant proportion is exported to the United States and elsewhere (e.g., Thailand) for canning.

Purse seine vessels operating in the Pacific islands region concentrate on skipjack and associated yellowfin schools. With its 499-GRT vessels, the Japanese fleet typically averages 30 to 35 tonnes of tuna per day, while the average catches of larger vessels are 40 to 50 tonnes per day. The species composition of the purse seine catch for smaller class vessels is approximately 70 percent skipjack and 30 percent yellowfin, while the larger seiners average 60 percent skipjack and 40 percent yellowfin and other deep-sea pelagic species (e.g., bigeye [*Thunnus obesus*] and billfish). Larger-class vessels use deeper nets, which enable fishermen to harvest a higher percentage of deep-sea species.

Eleven group seiners and their support and carrier vessels are registered on the regional register of fishing vessels. Ten of these seiners operate seasonally in the Pacific islands region, usually for about four months each year (February to May). Nine of the group seiners are Japanese. The seiners confine their operations to the Federated States of Micronesia and Papua New Guinea, though one group—owned by Taiyo Gyogyo, the Japanese joint-venture partner in Solomon Islands—operates year-round in Solomon Islands' waters.

Group-seine-caught tuna is mainly sold to U.S. canners for processing.

1984 catches

Distant-water tuna fleet catches in the Pacific islands region in 1984 were about 600,000 tonnes. Catches were estimated to have a market value of $662.7 million (Table 2). Overall, Japan remained the leading DWFN in the Pacific islands region in 1984 in terms of both catches and catch values.

ACCESS CONSIDERATIONS

Prior to the introduction of extended jurisdiction in the Pacific islands region, the DWFN fleets fished virtually unhindered. When island countries declared EEZs, the DWFNs were required (1) to cooperate with Pacific island countries in exploiting the region's tuna resources and (2) to pay a fee for fish harvested. This situation led to the conclusion of fishing access agreements between DWFNs and island countries. In entering into these agreements, the DWFNs acknowledged (1) the sovereignty of island countries over tuna stocks within their respective EEZs and (2) the right of island countries to regulate and control the exploitation of their tuna stocks.

In licensing distant-water fleets, island countries essentially view their tuna stocks as a tradable commodity. They are prepared to sell harvesting rights to derive maximum financial returns. The French

Table 2. Ranking of estimated catches by distant-water fishing nations and estimated value of catches in Pacific islands region, 1984

	Catch		Value	
Country	Tonnes	%	US$m	%
Japan	295,600	49.4	385.1	58.1
USA	208,000	34.7	131.8	19.9
Korea/Taiwan	56,320	9.4	90.9	13.7
Other	38,800	6.5	54.9	8.3
Total	598,720	100.0	662.7	100.0

territories adopt a slightly different approach from the SPF member countries in licensing distant-water fleets in that their principal objectives are non-monetary. The territories seek the transfer of technology, training, research, and assistance in the development of domestic tuna industries rather than immediate financial returns.

In return for access and harvesting rights, the DWFNs pay predetermined access fees to island countries. Generally, Pacific island countries seek to put their access arrangements on a commercial footing and to relate access fee payments to the value of tuna harvested by each class and type of vessel. They strive to appropriate at least 5 percent of the value of tuna harvested by way of access fee payments.

In addition to access fee payments, the DWFNs frequently offer development assistance to island countries. This assistance represents a subsidy from the DWFN government to its fishing industry. The assistance is usually designated for fisheries development projects and includes the provision of goods and services, training, and research.

The purpose of the aid is to promote more lenient access terms for the DWFN fleets. However, Pacific island countries generally maintain that as a matter of principle the two issues—access and aid—should be de-linked. Island countries also have expressed dissatisfaction over the quality and type of fisheries aid proposed and supplied by the DWFNs. Often goods supplied—for example, fishing vessels—are inappropriate for either commercial fishing or research. Some training is unsuitable for fisheries officers and conditions in Pacific island countries, and in some cases it constitutes little more than a public relations exercise on the part of the DWFN supplying it.

Negotiated terms and conditions of access also form an integral component of all tuna access agreements in the Pacific islands region. These terms and conditions (1) govern the operations of distant-water fleets in the EEZs of licensing countries and (2) define other responsibilities that must be accepted and adhered to by vessel operators. Agreements specify terms and conditions relating to matters such as restricted fishing areas, license application procedures, vessel reporting requirements, submission of catch records, vessel identification, and the placement of observers on vessels. Failure to adhere to agreement terms and conditions can result in the imposition of penalties and, in the extreme, the removal of vessels from the regional register.

Licensing distant-water fishing fleets involves Pacific island countries in transaction (negotiation and administration) and surveillance costs. Island countries, in seeking to maximize net financial returns from their access arrangements, attempt to minimize these costs. Upward pressure on access fee levels can be expected if the returns to island countries are eroded by increased costs. It is therefore in the financial interests of the DWFNs to ensure that administration and surveillance problems for island countries are avoided. This can be best achieved by voluntary compliance by fishermen with access agreements. If costs exceed financial returns from agreements, countries would either terminate existing agreements or refuse to renew agreements in the future.

AGREEMENTS

In 1985 eight SPF member countries and three dependent Pacific territories had tuna access agreements with the DWFNs (Table 3). The agreements covered the three principal fishing methods: longlining, pole-and-lining, and purse seining.

All Pacific island countries listed in Table 3 have access agreements covering longline operations. It is this high-value fishery that brings the largest number of island countries into contact with DWFNs. Furthermore, it is this fishery that generates the largest amount of access fee payments for island countries.

The number of island countries with agreements providing for surface fisheries—pole-and-line and purse seine fishing—is considerably

Table 3. Distant-water fleet operations in Pacific islands region, 1985

Country	Longline	Pole-and-line	Purse seine
South Pacific Forum member countries			
Cook Islands	x		
Federated States of Micronesia	x	x	x
Kiribati	x	x	x
Marshall Islands	x	x	
Palau	x	x	x
Papua New Guinea	x	x	x
Solomon Islands	x	x	
Tuvalu	x		
Dependent territories			
French Polynesia	x		
New Caledonia	x		
Wallis and Futuna	x		

Source: Updated and extended from Clark 1985b and based on distant-water fishing nation agreements with Pacific island countries and information supplied by French territories.

Note: Table does not include foreign vessels specifically contracted to Star-Kist (American Samoa), PAFCO (Fiji), Solomon Taiyo (Solomon Islands), and South Pacific Fishing Co (Vanuatu).

smaller than that for longlining. All countries with active surface fisheries share two characteristics: (1) having a location on or near the equator, which has the least seasonal variation in surface tuna stocks, and (2) being members of the subregional fisheries grouping, the Nauru Group.

Six Pacific island countries had access agreements providing for distant-water pole-and-line vessels in 1985 (Table 3). The region's purse seine fishery was confined to the EEZs of four countries, though some DWFNs would welcome the opportunity to have the same geographic area of operation as the pole-and-line fleet. However, this change is not likely because (1) the Marshall Islands and Solomon Islands opt not to license purse seine vessels, and (2) in the case of the Japanese fleet, the Japanese government does not currently permit seiners to operate beyond 166°E.

Five DWFNs had access agreements with Pacific island countries in 1985 (Table 4). Notably absent was the United States, despite the fact that U.S. seiners are important operators in the region. The leading DWFN, Japan, had agreements with ten island countries, Taiwan with seven, Korea with three, and the USSR and Mexico with one country each. In recent years the region's agreements with Japan, Taiwan, and Korea have been fairly stable, though Mexico (1984) and the USSR

Table 4. Distant-water fishing nation tuna access agreements with Pacific island countries, 1985

Country	Japan	Taiwan	Korea	USSR	Mexico
South Pacific Forum member countries					
Cook Islands	x	x			
Federated States of Micronesia	x	x	x		x
Kiribati	x		x	x	
Marshall Islands	x				
Palau	x				
Papua New Guinea	x	x			
Solomon Islands	x				
Tuvalu		x			
Dependent territories					
French Polynesia	x	x	x		
New Caledonia	x	x			
Wallis and Futuna	x	x			

Sources: Compiled from distant-water fishing nation agreements with Pacific island countries and information supplied by French territories.

(1985) are newcomers to the region. Mexico is seeking to deploy part of its overcapitalized purse seine fleet in the Pacific islands and to enter into arrangements with island countries on more than a distant-water fishing basis. The USSR is also keen to broaden its tuna fishing operations in the Pacific islands region, and it is prepared to participate in shore-based fisheries development where feasible and possible.

Japanese agreements

In 1978 Papua New Guinea was the first Pacific island country to conclude a tuna access agreement with Japan (Table 5). Initially, the agreement was with the Japanese government, but following its break-off in 1979, an arrangement was established with industry in 1981. Shortly after the conclusion of the Papua New Guinea agreement, the Japanese government entered into access agreements with Solomon Islands and Kiribati. Between 1978 and 1981 Japan concluded bilateral agreements with nine Pacific island countries and territories.

At first, Japanese access fees in the Pacific islands region were based on a guaranteed lump-sum payment system. However, Papua New Guinea and Solomon Islands changed to a per-vessel and per-trip system in 1979. The demonstrated financial advantages of this system—despite increased administrative costs—stirred interest among Pacific island countries. The system's flexibility and "pay-as-you-go" approach also found favor with Japanese industry. In 1984 the Federated States of Micronesia, Kiribati, and the Marshall Islands moved to a vessel/trip system of determining access fee payments. The French territories are likely to change to a vessel/trip system in 1987, and Palau is also expected to follow the other SPF member countries.

All countries in the islands region except Papua New Guinea receive fisheries developmental assistance as part of their access agreements with Japan (Table 5). Papua New Guinea has refused aid in the past because it maintained that fisheries access should be negotiated on strictly commercial terms and that aid should be considered and granted independently of access considerations.

Apart from the Federated States of Micronesia and Papua New Guinea, Japan's access agreements with island countries are concluded on a government-to-government basis (Table 5). There are distinct advantages in having government-to-government agreements because diplomatic pressure can be brought to bear by island countries should problems arise with the administration and enforcement of agreements. Papua New Guinea has indicated a desire to conclude future agreements with the Japanese government—and where possible with the governments of other DWFNs—but this initiative has meet with some resistance from Japan's Fishery Agency.

Table 5. Japanese distant-water tuna fishing agreements with Pacific island countries

Country	Year commenced	Basis of lump sum	Calculation per vessel	Aid	Status	Vessels licensed	Vessel/catch restrictions	Duration[a]
South Pacific Forum member countries								
Federated States of Micronesia	1981	1981	1984	yes	industry	b,c,d	yes	1 year
Kiribati	1978	1978	1984	yes	government	b,c	yes	1 year
Marshall Islands	1981	1981	1984	yes	government	b,c	yes	1 year
Palau	1981	1981	—	yes	industry	b,c,d	yes	1 year
Papua New Guinea[e]	1978	1978	1979	no	industry	b,c,d	no	open-ended
Solomon Islands	1978	1978	1979	yes	government	b,c	yes	1 year
Dependent territories[f]								
French Polynesia	1979	1979	—	yes	government	b,c	yes	1 year
New Caledonia	1979	1979	—	yes	government	b,c	yes	1 year
Wallis and Futuna	1979	1979	—	yes	government	b,c	yes	1 year

Source: Compiled from Matsuda and Ouchi 1984 and various Japanese agreements with Pacific island countries.

[a]One-year renewable agreements are subject to annual negotiation.

[b]Longline.

[c]Pole-and-line.

[d]Purse seine.

[e]The 1978 agreement was with the Japanese government. From 1979 the agreement has been with Japanese industry.

[f]French territories worldwide are covered under a single agreement between the government of Japan and the government of France. While the agreement provides for the licensing of longline and pole-and-line vessels in the Pacific French territories, only longline vessels generally operate. However, Japanese distant-water pole-and-line vessels operated on a limited basis in New Caledonia during 1979–84, French Polynesia during 1981–83, and Wallis and Futuna in 1981.

Japan's access agreements with six Pacific island countries provide for the licensing of longline and pole-and-line vessels, while three agreements provide for longline, pole-and-line, and purse seine fleets (Table 5). Apart from Papua New Guinea's agreement, all other agreements provide for restrictions. Restrictions usually consist of either ceilings on the number of vessels that can be licensed and deployed or restrictions on the quantity of tuna that can be harvested. Japanese agreements with Pacific island countries are usually concluded for one year and are subject to renegotiation and renewal. However, Papua New Guinea has an open-ended agreement that continues in force until either the Japanese industry or the Papua New Guinea government gives three months' notice of its intention to terminate it. The agreement has relieved the pressure and cost of holding annual negotiations; for this reason, other countries in the region are expected to adopt an open-ended approach in future.

From 1978 to 1982 Japan paid a total of $49.5 million to all countries in access fees. Of this amount, Japan paid $27.3 million (55 percent) to Pacific island countries.

Since 1982 Japan's access fee payments to Pacific island countries have been progressively negotiated upward to more accurately reflect catch values at landing points in Japan. This adjustment has been possible because of increased flows of catch, landing, and market information to island countries. The increased information has enhanced the negotiating positions of island countries and directly affected the financial returns from licensing distant-water tuna fleets. Japan's total access fee payments to the 16 countries with which it had access agreements in 1985 were approximately $40 million, and between 30 and 40 percent was paid to Pacific island countries. Moreover, fisheries aid disbursements associated with access agreements in 1984 were approximately $1 million, of which 50 percent went to countries in the region.

U.S. agreements

A small number of U.S. purse seiners moved into the Pacific islands region in the late 1970s and commenced fishing operations on a trial basis in Micronesia and, to a lesser extent, in Papua New Guinea's waters. All U.S. fisheries agreements with Pacific island countries have been concluded between island governments and U.S. industry—the American Tunaboat Association (ATA). And apart from the 1982 agreement with Papua New Guinea, all agreements have been concluded on a multilateral basis. The first ATA agreement in the region was with the Federated States of Micronesia, the Marshall Islands, and Palau from July 1980 to June 1982 (Table 6). The agreement was extended on an interim basis to December 1982 in anticipation of a broader multilateral agreement to be concluded in 1983.

In March 1982—following the apprehension of the U.S. seiner *Danica* for illegal fishing in Papua New Guinea's EEZ—Papua New Guinea entered into a nine-month access agreement with the ATA. This was necessary in order to have a U.S. tuna embargo lifted on Papua New Guinea's tuna exports. Papua New Guinea did not renew its agreement with the ATA when it expired in December 1982, though it did participate in preliminary discussions with the Federated States of Micronesia, Kiribati, the Marshall Islands, and Palau for a broader multilateral agreement to be effective January 1983. The Marshall Islands withdrew because it did not accept the fee levels proposed by the ATA, and Papua New Guinea could not reach agreement with the participating Pacific island countries on how the revenue obtained under the agreement would be shared.

The second multilateral ATA agreement extended from January 1983 to December 1984 and involved the Federated States of Micronesia, Kiribati, and Palau. In December 1984, talks to extend the agreement became deadlocked and failed.

With a view to expanding the range of operations for the U.S. tuna fleet in the Pacific islands region, the ATA sought a multilateral

Table 6. American Tunaboat Association agreements with Pacific island countries, 1980–85

Country	1980	1981	1982	1983	1984	1985
South Pacific Forum member countries						
Cook Islands				x	x	
Federated States of Micronesia	x	x	x	x	x	
Kiribati				x	x	
Marshall Islands	x	x	x			
Niue				x	x	
Palau	x	x	x	x	x	
Papua New Guinea			x			
Tuvalu				x	x	
Western Samoa				x	x	
Dependent territories						
Tokelau			x	x		

Source: Compiled from American Tunaboat Association agreements with Pacific island countries.

agreement with five Polynesian countries—Cook Islands, Niue, Tuvalu, Western Samoa, and Tokelau (Table 6). The agreement extended from mid-1983 to December 1984. The conclusion of this agreement represented a gamble for the ATA because purse seiners had not previously fished extensively in the EEZs of these countries due to the highly seasonal nature of skipjack stocks in these waters. Nonetheless, U.S. fishermen were cautiously hopeful that new fishing grounds would be located closer to the American Samoan canneries.

Tactically, the ATA's principal reason in concluding the Polynesian agreement was to reduce its fleet's dependence on operations in Micronesia and Papua New Guinea and thereby to strengthen its negotiating position in dealing with these countries. However, because fishing conditions in Polynesia did not prove favorable, the ATA was unable to meet the agreement's minimum requirements. Consequently, when the agreement expired, the ATA made no move to renew it.

Fee payments for ATA agreements in the Pacific islands region were based on an annual per-vessel basis but related to the net registered tonnage (NRT) of vessels and not to the quantity of tuna harvested and its market value. Island countries proposed that per-trip fees be used, a condition that was rejected by the ATA negotiators. They maintained that the ATA membership would not entertain a change in determining access fee payments because the previous ATA agreements with Latin American countries had been predicated on a NRT approach. On an annual basis in 1983–84, fees paid by U.S. fishermen to operate in the region were lower than those paid by their Japanese counterparts.

Toward the end of 1983, U.S. vessel owners approached Papua New Guinea for permission to purchase fishing licenses. With no ATA access agreement in force, Papua New Guinea decided to license the U.S. seiners under the terms and conditions of its agreement with Japan. The U.S. fishermen agreed to abide by the terms and conditions contained in this agreement, and overall there was a high degree of compliance. In 1984 more than 50 percent of the U.S. purse seine fleet (65 vessels) was licensed by Papua New Guinea in this way.

No ATA agreements have been in force in the Pacific islands region since December 1984. However, since September 1984, SPF member countries have been meeting with the U.S. government to establish a tuna treaty that would provide for, as well as govern, the operation of the U.S. purse seine fleet in the islands region. Progress has been slow, despite a genuine commitment by countries in the region and the importance accorded to it by the U.S. government for fishing, political, and strategic reasons. More rapid progress in negotiations could have been made if the United States had approached the Nauru Group rather than the entire SPF membership because the Nauru Group (1) has proprietary interests in the purse seine fishery and (2) is a closely knit grouping whose members are accustomed to working together.

Other agreements

Access arrangements with other DWFNs in the islands region are similar in content and approach to the Japanese agreements. For most countries, arrangements with Japan have become benchmark agreements, and several countries have adopted the approach and format of the 1981 agreement between Papua New Guinea and Japan.

REGIONAL COOPERATION

The importance of tuna to the Pacific islands region—in terms of both the absolute size of the resource and its financial significance to island countries—prompted the SPF to examine the concept of a regional fisheries organization in the mid-1970s. The organization's primary task was to coordinate and manage the development and exploitation of the region's tuna fishery. After lengthy debate concerning the organization's membership, the members decided that the DWFNs and the region's metropolitan powers would be excluded. Most Pacific island countries adopted this position because they believed that the organization's purpose and integrity would be compromised through broadly based membership.

Consequently, the organization called the FFA was established in 1979. The agency's principal objectives are (1) to coordinate and harmonize fisheries policy so as to ensure maximum benefits for Pacific islanders, (2) to facilitate fisheries development and the harmonization of fisheries policies in the region, (3) to promote a coordinated approach in relations among member countries, and (4) to improve the capacity of island countries in the surveillance and policing of distant-water fishing activities in the region. The agency has a comprehensive work program that deals specifically with issues such as marketing fish, training, negotiating access, and compiling fisheries-related statistics. The FFA's work program and activities are directed and maintained by the Forum Fisheries Committee. Composed of officials from the SPF member countries, this committee reports annually to the SPF.

Since its inception—and particularly since 1982—the FFA has assisted member countries to achieve considerable financial gains in dealing with the DWFNs. The agency also has facilitated the standardization of tuna access agreements, including methods for determining access fee payments and terms and conditions of operation. In addition, it has instituted the regional register of fishing vessels as a means of controlling and regulating access to the region's tuna fishery, assisted member countries in standardizing and improving domestic fisheries legislation, provided training in negotiation skills and procedures for Pacific islanders, and conducted studies relating to the development and expansion of domestic tuna industries.

In the mid-1970s, Pacific island countries recognized that, despite the formal declaration of extended jurisdiction and their large tuna resources, they would reap few benefits from the new ocean regime if they did not join forces to cooperatively protect their interests. Countries realized that to avoid dissipation of benefits from the region's tuna fishery, they (1) must not compete with each other in dealing with the DWFNs and (2) must cooperate on fisheries matters in ways not previously envisaged or practiced. The dividends from this cooperation paid off, and while there is still scope for closer cooperation in some areas, the benefits gained by island countries attest to their success.

Nauru Group

The Nauru Group is a subregional grouping of Pacific island countries—the Federated States of Micronesia, Kiribati, the Marshall Islands, Nauru, Palau, Papua New Guinea, and Solomon Islands—that have agreed to consult on matters relating to their shared fisheries. The group has much in common and thus is a logical subgrouping within the SPF. The Nauru Group countries share common tuna stocks and distant-water fisheries and have contiguous EEZs.

The Nauru Group was set up following informal discussions at the 1980 SPF meeting in Kiribati between the Federated States of Micronesia and Papua New Guinea. These initial contacts led to a series of formal discussions and negotiations, and an agreement was subsequently signed in Nauru in February 1982.

In many respects the Nauru Group became a trendsetter for relations between the DWFNs and the SPF member countries. The Nauru Group countries have the largest vested interest in distant-water tuna fishing in the region, and policies adopted by the group tend to be subsequently adopted throughout the region.

The DWFNs reacted apprehensively to the formation of the Nauru Group. This feeling was particularly evident in 1982 when Papua New Guinea discussed the Nauru Agreement in consultation with Japanese industry for the first time. The matter of a regional licensing arrangement for the region's purse seine fishery was also tabled but met with a negative response from Japanese industry and government representatives. This reaction seemed to stem from a fear that a regional access agreement would result in increased bargaining power for participating island countries. The advantages of such an arrangement—fewer negotiations and reduced administrative costs—now appear to be recognized and accepted by both Japanese industry and government.

CONCLUSION

The Pacific islands region will remain an important area of operation for distant-water fishing fleets, and Japan will continue to dominate the fishery in numbers of both vessels and catches. The region also will remain an important area of operation for the U.S. fleet, but its relative importance in comparison with the eastern tropical Pacific will depend on the extent of further reductions in the size of the U.S. purse seine fleet. Total access fees paid to Pacific island countries by all DWFNs are approximately $15 million per year. However, future payments will depend on (1) changes in fish prices, (2) the amount of fishing effort applied in the region, and (3) the bargaining power of island countries in negotiations.

In addition to the payment of access fees, Pacific island countries will seek increased benefits from the operation of distant-water fleets in their EEZs. Until now, distant-water fleets have had limited interaction with the economies of island countries, and some countries are anxious that this situation not continue. Island countries are seeking more active involvement in the fishery, and if this does not eventuate, they possibly could invoke Law of the Sea Convention provisions and require vessels to off-load their catches at domestic ports. In this way, the benefits to island economies will be increased. Foreign fishing vessels based at ports in the region also could receive preferential access in licensing.

In addition, some Pacific island countries seek greater assistance and participation from the DWFNs in developing domestic tuna industries. Distant-water fishing activity is quite compatible with the development of domestic industries in the region because the access revenues received by countries can be used to finance domestic industries. Moreover, the DWFNs are able to transfer technology to these industries, as well as provide appropriate training and applied research.

Market access for fish and other products to distant-water nations also could be tied to future fisheries access. A precedent for this type of arrangement already exists with New Zealand. Japan's distant-water fleet access to New Zealand's EEZ is tied to Japan's market access for New Zealand's agricultural and forest products. Papua New Guinea, for example, could tie fisheries access to sugar exports, and Solomon Islands could do likewise with exports of palm oil.

Regional fisheries cooperation among Pacific island countries has made significant advances since 1976. This will continue at a high level, and fishing limitations—probably in terms of the number of vessels permitted to fish—are likely to be introduced in the region. Restrictions most probably will be introduced first in the purse seine fishery. In addition to a ceiling on the number of seiners permitted to operate at any one time, vessels and fleets most likely will be prioritized and licenses issued accordingly. Priorities are expected to be established with reference to the financial and other contributions that individual vessels and fleets make to the overall development goals of Pacific island countries.

3.
A Summary of Domestic and Distant-Water Tuna Fishing Industries in the Pacific Islands Region

Mary-Cath Togolo

ABSTRACT—This summary chapter discusses tuna fishing, which has always been an important economic activity in the Pacific islands region. The introduction of large-scale commercial tuna fishing in the last 20 years has brought many benefits to the region. Access payments from distant-water fishing nations have brought revenue while domestic-based fleets have bolstered Pacific island economies. The declaration of exclusive economic zones, increased information about the tuna resource, and regional cooperation have led island countries to become more directly involved in the exploitation of their tuna resource. Increased regional cooperation and improved tuna fisheries management are expected to bring additional benefits to the islands region.

INTRODUCTION AND BACKGROUND

The Pacific islands region is an internationally important tuna fishing area.[1] As the size and extent of the resource have become better understood, Pacific island countries have become more concerned about its exploitation, and they are becoming increasingly aware of the direct benefits of promoting domestic tuna industries. At the same time, the distant-water fishing nation (DWFN) fleets are competing for the resource. The fact that the impact of domestic and DWFN industrial fleets on the resource is not well understood demonstrates the need for greater regional cooperation and improved management of the tuna resource.

The Pacific islands region covers the central and western Pacific Ocean and includes Melanesia, Micronesia, and Polynesia. There are 14 independent countries and 8 territories in this region. Tuna stocks in the area are known to be extremely large and are generally considered underexploited. The economies in the Pacific islands region increasingly depend on the exploitation of tuna and related pelagic species.

Commercial exploitation of the region's tuna resource began in the 1920s with the arrival of Japanese fishermen. Since the 1960s the exploitation has increased as the United States, Korea, Taiwan, and other nations have become active in tuna fishing, which has brought opportunities for development to countries in the region. Governments have concluded agreements with foreign corporations and invested in vessels and shore facilities. Increasingly they have seen the benefits to be gained by promoting domestic tuna industries. By becoming involved in tuna industries, governments hope to broaden and strengthen their economies and to generate employment and foreign exchange. The benefits from domestic industry development are potentially greater than the benefits currently received from the DWFN access fee payments.

Countries and territories of the region declared their exclusive economic zones (EEZs) between 1977 and 1984, and many of the established tuna fishing grounds fall within the EEZs of these countries. The independent countries are members of the region's main political organization, the South Pacific Forum (SPF), and its Forum Fisheries Agency (FFA). All foreign vessels operating in SPF member countries are registered on the FFA's regional register of fishing vessels. SPF member countries do not license distant-water vessels under access arrangements unless they are registered and have good standing on the registry.

TUNA RESOURCES

Large and commercially proven resources exist in the Pacific islands region. Although ten species of tuna are known to occur in the region, four species make up almost the entire commercial catch. These are skipjack (*Katsuwonus pelamis*), yellowfin (*Thunnus albacares*), albacore (*Thunnus alalunga*), and bigeye (*Thunnus obesus*).

The most prolific species of tuna is skipjack, which accounts for approximately 40 percent of the world's total landings of tuna. Skipjack catches in the Pacific islands region between 1977 and 1982 accounted for an average of 64 percent of total annual catches. Skipjack is primarily used for canning, though it is also used for smoked tuna (*katsuobushi*) and lower-grade sashimi in Japan. It is estimated that the total resource of skipjack tuna in the region could sustain annual catches considerably more than the present total annual yield of between 200,000 and 300,000 tonnes.

Yellowfin tuna catches in the Pacific islands from 1977 to 1982 averaged 26 percent of the region's total catches. Resources of yellowfin should be able to support total annual catches of at least twice the present catch levels. Catches of this species are primarily destined for canning and sashimi markets.

Production of albacore and bigeye tuna in the Pacific islands is relatively small, and statistics on the resource are limited. But it is generally agreed that stocks of these species are underexploited at present. Albacore is canned, and bigeye tuna is sold to the sashimi market.

FISHING ACTIVITY

The tuna fishery is the region's most valuable commercial fishery. The unprocessed value of tuna caught in the region is estimated to be worth $600 million to $750 million annually. According to statistics of the Food and Agriculture Organization (FAO), an average of 20 percent of total world catches of the main tuna species was caught in the region between 1977 and 1982—an average of 326,662 tonnes annually. Island countries want some of the tuna harvested by DWFN fleets in their EEZs to be transshipped, if not processed, in the region.

Tuna is exploited by (1) artisanal fishermen for subsistence needs, (2) fishing fleets based in the region, and (3) fleets operating from ports in foreign countries.

Artisanal production

There is little information about total catches by artisanal fishermen. Although the artisanal fisheries have great social significance, their total catches are estimated at less than 10,000 tonnes annually.

Domestic fleets

Commercial fishing fleets using pole-and-line, longline, and purse seine techniques are based permanently in 12 countries and territories in the Pacific islands region. Tuna harvested by these fleets either is exported frozen or is processed at facilities in the region and then exported. Facilities for handling and processing tuna are found in 10 countries and territories, while transshipment facilities exist in only 2 countries. Several other countries are expanding or considering the establishment of new processing facilities.

The rising international demand for tuna for canning led to industrial fleets being based in American Samoa, Fiji, and Vanuatu beginning in 1950. In the 1960s, longline fleets were based in French Polynesia by the U.S. canning companies Star-Kist and Van Camp. Pole-and-line fleets were established in Fiji (1963), Papua New Guinea (1970), and

Solomon Islands (1973). Purse seiners have been based in American Samoa, Guam, Nauru, Papua New Guinea, and Vanuatu beginning in the 1980s.

In 1985 six longline fleets with a total of 161 vessels were based in the Pacific islands. Of these vessels, 86 were operating from American Samoa on charter to Star-Kist and Van Camp.

The pole-and-line fishing method is viewed by some Pacific island governments as a technologically appropriate fishing method. It represents an intermediate step between artisanal operations and sophisticated purse seining. However, pole-and-line fishing is more costly than purse seining. The average production cost is estimated to be at least $100 per tonne higher than that for tuna harvested by purse seiners. The pole-and-line product is usually of better quality than the purse seine product and thus generally sells for a premium.

Sixty pole-and-line tuna vessels are based in seven countries. The size and location of individual fleets are determined largely by the availability of reliable and good-quality baitfish supplies. The largest pole-and-line fleets are, or have been, found in Papua New Guinea and Solomon Islands, which have robust baitfish stocks.

In 1985, 29 purse seiners were based and/or registered in the Pacific islands region. Over 70 percent of these vessels, which were based in American Samoa and Guam, supply tuna to the Star-Kist and Van Camp canneries in American Samoa and Puerto Rico and occasionally to Thai canneries.

Two important points should be noted about the domestically based fleets. First, foreign aid has helped several countries to establish their fleets. For example, the Japanese aid programs have provided vessels to Solomon Islands, Tonga, and Tuvalu, while Britain has aided Kiribati. This aid was in addition to aid supplied for the development of the artisanal fisheries. Second, although fleets are based in the region, they are mostly foreign-owned. Approximately 90 percent of the total capital invested in the region's tuna fleets is estimated to be owned and controlled by individuals and corporations outside the region.

Distant-water fleets

Distant-water tuna fleet catches in the Pacific islands region in 1984 were about 600,000 tonnes, with a market value of $662.7 million. Japan was the leading DWFN in terms of catches and catch values.

In 1985 about 64 percent of the world's purse seine catches of tuna were harvested by purse seine fleets operating in the region. Vessels range from 300 gross registered tonnes (GRT) to superseiners of more than 2,000 GRT. In 1985 there were 146 single purse seiners on FFA's regional register. Although the fleet came from 11 countries, about

64 percent of the registered vessels came from Japan and the United States. A significant share of the tuna harvested by the fleet is transshipped at Guam and Tinian (Northern Mariana Islands) for transport to canneries within the region and outside. The tuna caught by the Japanese fleet is usually taken to Japan, but a portion of it is exported to the United States for canning. Of 11 group seiners on the regional register, 9 are Japanese. Most of the tuna caught by these vessels is sold to U.S. canneries.

The Japanese longline fleet is the largest fleet in the Pacific islands, with between 500 and 600 vessels. Vessels discharge their catches at several ports in Japan, but the most important ports are Yaizu, Misake, and Shimizu. Korean and Taiwanese longline fleets also operate in the region, often under financial arrangements with Japanese companies.

All distant-water pole-and-line tuna fishing is done by Japanese fishermen. About 80 pole-and-line vessels operate from ports in Japan. Pole-and-line-caught tuna is usually used for quality smoked and dried products. The size of the pole-and-line fleet has declined because its vessels are less able to compete with more efficient purse seiners.

DWFN agreements

The declaration of EEZs by Pacific island countries led to the conclusion of fishing access agreements between the DWFNs and island countries. The DWFNs acknowledge that island countries have sovereignty over their tuna stocks and the right to regulate and control fishing activity. In return for exploiting the resource, the DWFNs must pay access fees. Island countries try to relate the access fees to the value of the tuna harvested, or at least 5 percent of the value. The DWFNs also offer development assistance to island countries. However, these countries consider that aid and access are two separate issues. Because some of the aid supplied is inappropriate, it is often more a public relations exercise than practical assistance.

Agreements between DWFNs and island countries govern the operations of the DWFNs and define their responsibilities. It is clearly in the interest of the DWFNs to adhere to the conditions agreed upon in order to minimize demands by island countries to increase access fees in future.

In 1978 Papua New Guinea was the first Pacific island country to conclude a tuna access agreement with Japan. At first, access fees were paid on a lump-sum basis. But in 1979 Papua New Guinea and Solomon Islands changed to a per-vessel and per-trip system. Other countries also adopted this approach.

Japanese access agreements are concluded on a government-to-government basis with all island countries except Papua New Guinea and the Federated States of Micronesia. Most agreements are concluded for one year and are subject to renegotiation. Papua New Guinea has an open-ended agreement with Japan.

From 1978 to 1982 Japan paid $27.3 million to Pacific island countries in access fees, or 55 percent of the total paid to all countries with which it had agreements. Since 1982 fee payments have increased as more information has become available to island countries as a basis for negotiation of access fee levels.

In 1985 eight South Pacific Commission (SPC) countries and three territories had tuna access agreements with DWFNs covering the three main fishing methods. Five DWFNs had agreements with island countries in 1985. The United States has had no agreements since 1984, although U.S. seiners are important operators in the region. Japan had agreements with ten island countries, Taiwan with seven, Korea with three, and the USSR and Mexico with one each.

TUNA PROCESSING AND TRANSSHIPMENT

Both Japanese and U.S. multinational corporations (MNCs) are involved in tuna canning in the region. Until 1984 five canneries were operating in American Samoa, Fiji, Hawaii, and Solomon Islands. American Samoa has 74 percent of the total processing capacity (with the Star-Kist and Van Camp canneries). In recent years the Fiji cannery has not operated at full capacity due to inadequate supplies of longline- and pole-and-line-caught fish, while the cannery in Hawaii closed in 1984 but may re-open in 1987. In Solomon Islands, the Taiyo Fishery Company operates a tuna cannery in a joint venture with the government. Tuna exports from Solomon Islands accounted for 23 percent of total exports in 1978 and 41 percent in 1984.

Katsuobushi is also produced in the Pacific islands. The production process involves the boiling, smoking, and drying of tuna, usually skipjack. It is used for soup bases, etc., in Japan. The demand for *katsuobushi* outside Japan is limited. Currently there are plants in the Marshall Islands and Solomon Islands; until 1979 and 1981 there were plants in Papua New Guinea and Palau.

Shore-based transshipment facilities already exist in Palau (no longer in use) and Vanuatu (underutilized), while one base is being constructed in the Marshall Islands. Tuna is also being transshipped from purse seine vessels in Guam, Tinian (Northern Mariana Islands), and Papua New Guinea. The future of transshipment facilities in the region will depend on three factors: (1) the extent to which island countries enforce the provisions of the Law of the Sea Convention (Article 62) requiring the DWFN fleets to land tuna at domestic ports, (2) the capacity of island countries to provide and manage appropriate facilities, and (3) the size and type of the DWFN fleets operating in the region.

INVOLVEMENT OF MULTINATIONAL CORPORATIONS IN TUNA OPERATIONS

The operation of multinational corporations in the Pacific islands region has been important to fishing and processing as well as tuna trading. In the 1960s and 1970s the primary objective of these corporations was to obtain tuna for trade and processing. The long-term needs of Pacific island countries to establish industries for the benefit of their people were not generally taken into consideration by MNCs. In fact, the tuna development plans of some countries were disrupted by MNC activities. When market conditions changed or when cheaper raw materials became available, the multinational investors relocated their operations, forgetting their promises or abandoning their plants.

Although vessels from Korea, Taiwan, and other nations operate in the region, the main investment in the tuna fishing industry has come from Japanese and U.S. corporations. Three U.S. corporations have had a total of nine ventures in the region, but their main investment has always been in American Samoa as single investors. In contrast, Japanese corporations have been involved in consortiums, joint ventures, and single investments in Pacific island countries. In addition, the Japanese government has provided fisheries aid to several countries. Japanese investment in the region has tended to be more concerned with the long-term interests of Pacific island countries than has U.S. investment.

REGIONAL COOPERATION

The establishment of the FFA in 1979 was an important step for Pacific island countries. Established by the SPF, the agency is designed to coordinate fisheries policies and conduct research for the benefit of members. The work of the FFA includes training, compiling statistics, and standardizing access agreements.

The Nauru Group, formed in 1982, is a subregional grouping of Papua New Guinea, Nauru, Palau, Solomon Islands, Kiribati, the Marshall Islands, and the Federated States of Micronesia. Policies adopted by the Nauru Group tend to be followed by other countries in the region.

THE FUTURE

The tuna resource in the central and western Pacific is large, and the most plentiful species, skipjack tuna, appears immune to overexploitation. But although the resource is immense, it is not infinite, and those fisheries operating close to each other in both time and space could affect catch rates without affecting the total renewable resource.

Available scientific data do not enable precise determination of the degree of interaction among the various types of tuna fishing in the region. But the interaction is greatest in areas where artisanal and large-scale commercial fleets operate simultaneously. This has important implications for governments wishing to develop commercial tuna fishing in areas where artisanal fishermen exploit the same species. Governments must evaluate the social as well as economic implications at the artisanal, commercial, national, and international levels of increased competition for a finite resource.

Both the development of flexible agreements between DWFNs and island countries and the establishment of regional groupings indicate the determination of countries in the region to become more actively involved in the exploitation of their tuna resource. To this end, it is possible that provisions of the Law of the Sea Convention might be invoked as a means to bolster island economies. Increased assistance is being sought to establish domestic tuna industries. Limitations on the number of vessels permitted to fish are likely to be imposed, and licenses will be given with preference to the contributions made by individual vessels and fleets to the overall development goals of Pacific island countries.

NOTE

1. This chapter synthesizes the chapters by Doulman and Doulman/Kearney for PIDP's tuna project.

II. MARKETS

4.
The U.S. Tuna Market: A Pacific Islands Perspective

Dennis M. King

ABSTRACT—This chapter describes the U.S. tuna market and the major economic forces that affect this market. The emphasis is on general relationships in the market rather than on recent trends, but recent market shifts are used to illustrate general rules and the ever-present exceptions to the rules. An attempt is made to demonstrate how forces in the U.S. tuna market are transmitted to other tuna markets and how they can eventually influence those economic benefits that Pacific island nations can expect to derive from their tuna resources.

INTRODUCTION

World tuna and the U.S. market

The tuna harvest from the Pacific islands region has increased significantly in recent years and in 1985 accounted for about 40 percent of global tuna supplies. Most of the tuna taken from the region is harvested by purse seiners and is eventually processed into canned tuna products, which make up two-thirds of the international tuna trade. The United States accounts for over 50 percent of the global canned tuna market and dominates international trade in cannery-quality tuna.

Although the U.S. tuna industry has a high profile in the Pacific islands region, the U.S. tuna industry is small in comparison with other U.S. industries. Nonetheless, the size of the U.S. tuna market provides an attractive target for Pacific islands tuna development because even the value of the raw tuna used for the U.S. market is greater than the gross domestic production of most island nations. To the Pacific islands region, which is capable of producing high volumes of cannery-quality tuna, the U.S. market is important not only because of its size but also because of its reliance on imports of raw/frozen and canned tuna.

U.S. market and Pacific islands

Those Pacific island nations considering ventures in the international tuna trade should consider the U.S. market from several different perspectives. The first and least-risky option would be for them to participate in the U.S. market indirectly by licensing the foreign fishermen who operate in the Pacific islands region and who then sell tuna to the processors supplying the U.S. market. A second and more aggressive option would be to develop tuna harvesting, processing, or transshipping systems to supply those few large U.S. companies that have established food distribution and marketing networks in the United States. A third option, which would require a great deal of effort and entail significant financial risks, would be to attempt to enter the U.S. market independently (or with non-U.S. affiliates) and compete directly with established market leaders by selling processed tuna products in U.S. retail and institutional markets.

Whichever strategy is selected, island nations will not be able to avoid the strong competitive forces that prevail in the U.S. tuna market. Thus some familiarity with the U.S. market is essential if they are to evaluate the risks and understand the economic pressures that can affect the behavior of foreign competitors and joint venture partners.

U.S. tuna market in perspective

Some perspective on the influence of the U.S. tuna market can be gained by the following example: a 5 percent decline in U.S. tuna consumption, or a decline of less than 1/10 kg per capita in 1985, would not only reduce retail tuna sales in the U.S. by about $90 million but also release an extra 36,500 tonnes of raw/frozen tuna, or its equivalent of 2 million standard cases of canned tuna, onto the world market. A small decline in U.S. tuna consumption could dump enough extra tuna onto the world market to create disruptions in all national tuna markets and could result in idle capacity in fish harvesting and processing in many regional tuna industries.

In 1981–82 an unexpected 13 percent decline in U.S. tuna consumption (from 1.4 kg per capita to 1.2 kg per capita) contributed to a worldwide glut of tuna and a 7 percent decline in international tuna prices during 1981–83. This price drop, occurring at a time when fuel, labor, and finance costs were rising rapidly, caused financial turmoil in global tuna industries and helped force many traditional U.S. tuna harvesters out of business. Aided by an unusually strong U.S. dollar during this period, non-U.S. tuna producers were able to withstand the cost-price squeeze better than U.S. producers, and as a result they penetrated the U.S. raw/frozen and canned tuna markets. This, in turn, contributed to the geographical redistribution of tuna fishing and helped start the restructuring of the world tuna industry that is currently under way.

Any entity making investment, production, and trade decisions related to Pacific island tuna industries will need to become familiar with the U.S. tuna market.

What about Japan?

The celebrated Japanese markets for sashimi-grade tuna and dried, smoked, and fermented tuna products are larger and more exotic than the canned tuna market of the United States. Most foreign producers, including those in Pacific islands, do not produce the high-priced tuna products that are in demand in Japan and would have great difficulty developing the kinds of marketing connections required to penetrate the Japanese market. As a result, the Japanese tuna market, despite its size, exerts a relatively small influence on those aspects of the world tuna trade that are important to Pacific islands. Japan's most important role in the world tuna trade is as a supplier of tuna rather than as a tuna market; Japan is the world's largest tuna supplier, taking nearly 40 percent of the global tuna harvest in most years, and it is a major exporter of canned and raw/frozen tuna to the United States and elsewhere.

New entrants in the global tuna industry and those considering new tuna-related investments are intrigued by what appear to be lucrative Japanese markets. Many have already learned, however, that while large profits exist in these markets, they accrue to the traditional Japanese companies that have controlled the channels for marketing and distributing tuna in Japan for centuries. It is certainly in the interest of Pacific island nations to deal with Japan as a joint-venture partner and to monitor Japan as a competing supplier of tuna, but the United States dominates the world market for most of the tuna taken in the region, and there are few better leading indicators of change in the international tuna market than the fluctuations in U.S. tuna markets.

FUNDAMENTALS OF THE U.S. TUNA MARKET

U.S. food preferences

Americans prefer meat to fish; the average American eats ten times more red meat and five times more poultry than fish. In the United States, convenience is an important quality in all food products purchased for home preparation. Canned tuna is very popular because it is skinless, boneless, precooked, and convenient to store and serve. Because of these characteristics, canned tuna stands apart from most other seafood and is the only seafood staple in the American diet, accounting for about 50 percent of U.S. household expenditures on seafood.

The popularity of canned tuna in the United States can be seen across most socioeconomic and ethnic groupings. The results of a recent national food consumption survey indicate that over 70 percent of U.S. households include canned tuna as part of their regular supermarket purchases, which is three to four times the number reporting regular purchases of any other seafood product.

Types of tuna products

Canned tuna is the most important component in the U.S. consumer markets for whole tuna, tuna steaks, tuna fillets, frozen tuna nuggets, and prepackaged tuna casseroles.

There are two major product distinctions in the U.S. canned tuna market: whitemeat tuna made from albacore tuna, which is harvested in temperate waters of the world and has relatively firm white flesh, and lightmeat tuna, which refers to products made from yellowfin and skipjack tuna, which is harvested in tropical waters of the world and has darker, softer flesh. Because of its firm, white, "chickenlike" flesh, whitemeat tuna is more highly valued by U.S. consumers. Because lightmeat tuna is less expensive, it is more popular, and its sales account for over 70 percent of the U.S. market.

Besides the species distinction (whitemeat vs. lightmeat), canned tuna in the U.S. market is also labeled according to the type of meat (solid, chunk, and flake) and the packing fluid (water or oil). Because whitemeat tuna flesh is firmer and holds together better than lightmeat tuna, most of the solid or fancy pack tuna available in the U.S. is whitemeat. Most lightmeat is processed into a chunk product, and the two industry standards familiar to most Americans are "chunk lightmeat" and "solid whitemeat."

U.S. consumers have become more health-conscious in recent years, and a noteworthy shift has occurred in the preferences of U.S. tuna consumers toward products packed in water and away from tuna packed in oil.

The only other noteworthy change in the traditional U.S. canned tuna market in recent years has been consumer acceptance of slight changes in can size.

Major supply sources

There are dozens of different labels on the canned tuna that U.S. consumers find on supermarket shelves, and most U.S. consumers neither know nor care that only three major U.S. tuna companies supply 75 percent of this product, with the rest supplied by a handful of foreign processors. During the last ten years, the share of the domestic U.S. tuna pack made from domestic landings and the share from raw/frozen tuna imports have been remarkably constant, with each source supplying about 50 percent. The important change taking place is the rapid increase in the share of the U.S. market supplied by canned tuna imports, rising from 9 percent in 1975 to 28 percent in 1985. Over the past five years, the share of the U.S. market held by canned tuna imports has increased by 25 percent per year, and were this pace to continue, canned imports in another five years would take 95 percent of the U.S. tuna market.

The major exporters of canned tuna products to the United States are Thailand, the Philippines, Japan, and Taiwan, which together accounted for 95 percent of U.S. canned tuna imports in 1985.

Like many other sectors of the U.S. economy, the U.S. tuna processing industry has had trouble competing on a cost basis with operators in nations such as Thailand and the Philippines, where labor productivity is high and labor costs are low and where operators face less restrictive environmental, employee health, and social regulations than in the United States. The relocation of U.S. tuna processing operations to offshore U.S. territories (American Samoa in the Pacific and Puerto Rico in the Atlantic) was an attempt by U.S. processors to develop a more competitive cost structure by taking advantage of lower labor costs and a more liberal tax and regulatory environment in these areas.

Despite these attempts, however, a strong U.S. dollar during the early 1980s gave foreign producers a significant competitive advantage over U.S. processors. Now that foreign producers have successfully penetrated the U.S. market and have established themselves as reliable suppliers to the institutional and retail markets, they will remain in the U.S. market despite a weakening U.S. dollar. Nations like Thailand and Taiwan have demonstrated their ability to produce high-quality, low-cost products so convincingly that major U.S. tuna companies have begun to rely on them for canned products and are selling imported canned products under their own nationally advertised labels.

The critical question concerning the processing of tuna for the U.S. market is whether U.S. canners operating at their offshore processing sites will be able to compete with Asian processors or whether the move to these offshore sites will be only an interim step in the westward migration of tuna processing operations. U.S. tuna companies continue to increase their purchases of the imported canned product and may eventually stop processing altogether and become U.S. distributors of foreign-produced canned tuna products. Simple cost comparisons suggest that U.S. operations at offshore processing sites cannot compete on a cost basis with Asian processors, so such developments are possible. There are also indications that Asian processors are experiencing declines in productivity and efficiency and are facing labor problems and thus would have difficulty increasing production beyond current levels. The global tuna industry is in a particularly dynamic period, and the strong competition in the world market is eroding the efficiency of operators in the United States, Asia, Europe, and elsewhere—a trend that portends even more significant changes in the structure of the global tuna industry.

Market forces

Over the past ten years, the demand side of the U.S. tuna market has been relatively stable, even stagnant. Annual retail sales are approximately 30 million standard cases, which in 1985 had an approximate retail market value of $1.5 billion and a wholesale value (to the processor) of approximately $1 billion. The total amount of the canned product sold in the United States in 1985 required the processing of approximately 700,000 tonnes of raw tuna, which was purchased from U.S. and foreign fishermen for about $500 million.

Demand-side shifts in the U.S. tuna market can cause significant changes in tuna prices in the United States and elsewhere. However, supply-side shifts (caused by natural changes in the abundance and availability of tuna in ocean fisheries or by changes in the size, composition, or disposition of the world's tuna fleets) are usually the more important determinants of price.

The recent slump in world tuna prices was brought on by a combination of both supply and demand changes. The supply-side effects include the emergence of productive central/western Pacific tuna purse seine fisheries, a growing international tuna fleet, a strong U.S. dollar, and a few large year-classes of skipjack. The more subtle demand-side shifts were caused by recessionary economic conditions, low relative beef and poultry prices, and the pricing of canned tuna products beyond an apparent "psychological threshold" of $1 per can (for chunk light). These factors can be considered separately, but many demand-side and supply-side forces influence the U.S. tuna market at the same time, and the impact of individual market forces cannot always be isolated.

It is the timing of market relationships, the lag between cause and effect, and the strength of market relationships that are important, and it is the constant blending of the offsetting and complementary market forces that makes tuna markets so dynamic.

STRUCTURE AND PERFORMANCE OF U.S. TUNA MARKETS

Overview

The U.S. tuna industry is composed of three main industrial sectors (harvesting, processing, and distribution/marketing), which are distinctly separate from one another. The competitive forces influencing supply-and-demand conditions and tuna prices at different levels of the U.S. market (ex-vessel, wholesale, and retail) are not identical, but changes at one market level are reflected through all levels. A demand decrease that lowers retail tuna prices, for instance, will cause lower wholesale and ex-vessel tuna prices. Similarly, a worldwide supply decrease that causes higher raw/frozen tuna prices on the international market will force an increase in U.S. ex-vessel tuna prices that will be reflected at the wholesale and retail market levels.

The difference between tuna prices at each market level provides some idea of the value added to tuna as it passes between market levels. Ex-vessel prices are slightly more volatile than wholesale prices, and wholesale prices are slightly more volatile than retail prices.

Value added

Tuna increases in economic value as it passes through each stage of production from fishermen through processing, distribution, and marketing.

The value added at each stage is the difference between the market value of tuna entering each stage and the market value of tuna passing on to the next stage.

Actual profits at each stage, of course, will also reflect the operator's market power and negotiating skills (the ability to receive high selling prices and to pay low input prices) and the operator's management skills (the ability to control production costs). Because each stage of production is linked with every other stage, and because market power and negotiating skills are not distributed evenly, the potential profit at one level of production may well be realized by operators at some other stage due to their greater market power or better negotiating skills.

U.S. harvesting sector

The U.S. tropical tuna fleet consists of purse seiners and bait boats. Purse seiners account for over 95 percent of domestic U.S. tuna landings and are the only U.S. vessels ranging as far as the Pacific islands region. The size and capacity of the U.S. tuna purse seine fleet have declined drastically over the past ten years from a peak of 138 vessels in 1976 to only 68 vessels in 1986. This attrition in the U.S. fleet reflects the transfer of some vessels to foreign flags and the idling of other vessels because of deteriorating economic conditions and the relocation of U.S. canners to offshore sites. Many of the U.S. tuna purse seiners either were sold to foreign interests or are being operated under foreign flags to take advantage of benefits such as favorable taxes, fuel subsidies, and off-loading privileges, but are still managed by U.S. fishing companies.

The purse seiners still operating under the U.S. flag tend to be highly productive; even though the hold capacity of the fleet declined by 34 percent between 1976 and 1985, tuna landings by the U.S. tuna purse seine fleet declined by only 22 percent. Since 1984 about one-half of the U.S. fleet operates in the traditional tuna fishing areas in the eastern tropical Pacific, and about one-half is operating in the new tuna fishing areas of the central/western Pacific. There is virtually no U.S. tuna fishing in the Indian Ocean, and the U.S. tuna harvest from the Atlantic is fairly small.

The productivity of an individual tuna purse seiner depends on fishing skills and the amount of time that the vessel spends fishing. A vessel's daily catch rate (adjusted for the value of fish) is a reliable indicator of fishing skills, and the vessel's fishing time (days spent fishing or days at sea) is a reliable indicator of shoreside vessel management skills. In competitive international tuna fisheries, both kinds of skills are required to maintain a profitable operation, and as tuna prices have gone down, the productivity and efficiency of vessel operations have become more critical. Ironically, as tuna prices go down, many vessel operators do attempt to increase productivity to improve their own financial performance. This process contributes to the downward pressure on tuna prices and exacerbates the downward spiral.

Ex-vessel tuna markets

U.S. ex-vessel tuna prices (the price paid at the point of vessel off-loading) have been established for many years through an auction administered by the American Tuna Sales Association (ATSA). ATSA prices are published by the U.S. Department of Commerce through the National Marine Fisheries Service (NMFS) Market News Reports, which are known as "yellowsheets."

When most U.S. canners were located on the U.S. mainland, the published U.S. west coast tuna price or the yellowsheet price was the major determinant of world tuna prices. The price of cannery-quality tuna at many foreign ports was pegged at the yellowsheet price, less shipping costs from the foreign port to the U.S. west coast. Now that the location of the U.S. ex-vessel tuna market has shifted from the U.S.

west coast to offshore processing locations, the role of ATSA and the U.S. ex-vessel market has weakened, but the price paid at U.S. canneries to U.S. fishermen is still an important reference price for the world trade in cannery-quality tuna.

Ex-vessel price trends

Many different ex-vessel tuna prices are based not only on the species but also on fish size and fish quality. U.S. tuna buyers, like buyers in most of the world, pay higher prices for larger fish because they yield more edible meat and require less labor to clean than smaller fish. In 1982 U.S. canners introduced a complex schedule of ex-vessel tuna prices with many more distinctions based on fish size and additional distinctions based on fish quality and salt content.

Some U.S. fishermen claim that this new complicated pricing system provides canners with too much opportunity to contrive lower tuna prices by misrepresenting the characteristics of the fish being delivered. U.S. tuna processors, however, claim that the international tuna industry has become so competitive that they must pay for tuna on the basis of its real economic value, which depends on the yield of edible meat (standard cases per tonne) and processing costs.

Tuna processing sector

In the early years of the U.S. tuna industry and up to the mid-1970s, supplies of tuna were barely keeping pace with the growing consumer demand for canned tuna. To increase fleet production and to ensure adequate supplies of raw fish, U.S. canners began to integrate by investing in tuna vessels and transshipping operations, as well as in processing facilities and distribution and marketing networks.

As new tuna fisheries and new foreign fishing fleets developed during the late 1970s and early 1980s, the structure of the U.S. tuna industry began to change. Because of increasing competition among U.S. and foreign tuna suppliers, U.S. canners found in many cases that they could buy frozen tuna on the international market at a lower cost than their own vessels could produce it. They thus divested themselves of the boats they owned, entered into fewer long-term financial and contractual arrangements with individual U.S. fishermen, and began to rely more on market purchases of tuna from independent domestic and foreign producers. The U.S. tuna fishermen who had been tied closely to U.S. tuna processors were now exposed to direct competition from a growing international tuna fleet, and many of them have not survived. As a result of these changes, the relationship between conditions in the U.S. ex-vessel tuna market and the international tuna market and the distinction between the harvesting sector and processing sector of the U.S. tuna industry have become more apparent.

A similar pattern of restructuring seems to be taking place in the U.S. canned tuna market. Due to the development of reliable tuna processing companies outside the United States during the early 1980s, the major U.S. tuna companies now have opportunities to purchase canned tuna from foreign processors. This development has further distinguished the processing sector of the industry from the distribution/retailing sector. U.S. tuna processors now compete with foreign tuna processors in U.S. wholesale tuna markets. The large U.S. food companies maintain their market power in the modern U.S. tuna industry by regulating the food distribution/marketing networks and by supporting their nationally advertised brands; they have no real interest in competing with the rest of the world to harvest and process tuna.

Wholesale tuna market

The wholesale tuna market involves sales of canned products by processors and wholesalers to retailers and institutional buyers. Canned tuna products destined for retail trade are marketed under nationally advertised labels, private labels, or supermarket housebrands. Although the products sold under these different brands are nearly identical, there are some quality differences. The large U.S. tuna companies try to reserve the highest-quality products for their own nationally advertised labels. The higher prices charged for nationally advertised brands also reflect the cost of advertising and maintaining brand preference. Within the industry, the price premium is viewed as a return to the long-term investments required to keep brands popular.

Distribution/marketing sector

Competition among U.S. tuna producers at the wholesale market level takes place in four different ways: (1) canners offer price discounts and promotional allowances on nationally advertised brands; (2) canners engage in their own national and regional promotions to develop and nurture consumer allegiance; (3) canners offer off-brands or private-label packs at prices substantially lower than their major brand price; and (4) canners offer canned tuna to be sold under housebrand labels by major supermarket chains.

Domestic canners normally maintain canned inventories near major market areas so that they can react quickly to market changes, while importers of canned tuna products (who sell primarily in the housebrand and private-label market) usually do not maintain warehouse inventories and sometimes have imported products shipped directly to the buyers. As a result, the U.S. companies tend to be more responsive to the U.S. market and are therefore more competitive.

Retail tuna markets

Most Americans shop in large supermarkets, which carry all of the major nationally advertised brands of canned tuna and may stock one or more lesser known private labels and perhaps even their own housebrand.

The large U.S. tuna companies are subsidiaries of conglomerates that produce and sell many other foods besides canned tuna, and they advertise extensively to stimulate name recognition for their products.

In typical years, advertising expenditures amount to 1 percent of retail tuna sales. Housebrands and private-label canned tuna are usually advertised as part of regional supermarket promotion campaigns, and they are used frequently by supermarkets as loss leaders—products advertised at prices at or near cost to attract shoppers.

The level of retail tuna prices usually reflects the level of wholesale prices and the discounts and promotional allowances being offered by tuna producers; however, they also reflect decisions made by retailers regarding profit margins and the use of shelf space. Not all discounts provided by processors to retailers are passed on to consumers in the form of lower retail prices, and U.S. tuna processors are not in control of retail tuna prices. As a result, they are not always in a position to stimulate consumer sales when supplies are abundant or inventories are building up. Significant retail price fluctuations occur mainly in housebrand labels when U.S. tuna producers offer their lowest prices (greatest discounts) to increase the volume of sales when canned inventories become excessive.

U.S. retail tuna prices increased dramatically during 1980–81 and, combined with an overall deterioration in U.S. economic conditions and relatively low beef and poultry prices, caused a significant decline in the volume of U.S. canned tuna sales. During this same period, rising costs of raw fish, labor, energy, and financing increased cannery costs to a level where cutting wholesale canned tuna prices to stimulate the market would have caused substantial financial losses to U.S. processors. As a result, wholesale and retail prices remained high during almost a full year of slack demand until late 1982, when canners (1) reduced raw fish costs by offering lower ex-vessel tuna prices, (2) renegotiated labor contracts, and (3) began passing cost savings on to retailers who eventually passed some of the savings on to consumers in the form of lower prices. During this same period, low-cost canned tuna imports put further downward pressure on retail prices, creating even more competition and generating even more financial problems for U.S. harvesters and processors.

ANALYSIS OF U.S. TUNA DEMAND

Overview

The overall size of the U.S. market for tuna is influenced by two factors: changes in consumer food preferences and changes in the size and composition of the U.S. population. For purposes of analysis, it is useful to disregard the influence of population changes and to focus only on changes in food preferences, which are best reflected by changes in per capita tuna consumption. This focus is particularly useful in evaluating the future of the U.S. tuna market because the U.S. population is no longer growing very rapidly (< 1.0 percent annual growth in 1985) and because changes in the size of the U.S. tuna market primarily reflect changes in per capita tuna consumption.

Most published U.S. tuna consumption statistics are computed using tuna production and import data rather than sales data because production and import data are compiled routinely by the government and are easy to obtain, whereas sales data are not.

Factors affecting demand

Changes in U.S. demand for tuna, as reflected by changes in per capita tuna consumption figures, are influenced primarily by shifts in disposable household income and by changes in canned tuna prices and the price of substitute goods such as chicken and beef. Tuna demand can also be affected by non-market phenomena such as fear of product contamination (e.g., botulism or mercury scares) and consumer reaction to bad publicity (e.g., porpoise kills by tuna fishermen) or good publicity (e.g., health benefits of Omega-3 acids found in fish).

Changes in disposable income and in tuna and substitute food prices explain 86 percent of the annual changes in tuna demand, and disposable income is a significantly more important factor than tuna or substitute prices in determining tuna demand. Surprisingly, statistical results indicate that changes in substitute prices (the prices of chicken and other meat products) have more impact on tuna demand than changes in tuna prices. This may reflect the fact that canned tuna is a staple in the U.S. diet, and consequently consumers not only maintain a fairly steady level of tuna consumption across a wide range of tuna prices but also buy more tuna when the prices of substitute products go up.

Although the tuna demand does not seem to respond significantly to small price changes of any kind, many market observers believe that consumers do respond to psychological thresholds and that when canned lightmeat was priced near $1 per can in the early 1980s, the decline in tuna demand was more abrupt than would be indicated by the historical price-demand relationships. Threshold impacts aside, the elasticity of tuna demand with respect to changes in key economic indicators based on statistical analysis can be summarized as follows:

- Response to tuna price changes. In retail tuna prices, a 1 percent decline results in a 0.2 percent increase in the amount of canned tuna sold in the U.S. market.
- Response to substitute price changes. In the price of meat and poultry substitutes for tuna, a 1 percent decline results in a 0.3 percent decline in the amount of canned tuna sold in the U.S. market.
- Response to changes in household income. In disposable household income in the United States, a 1 percent decline results in a 1 percent decline in the amount of canned tuna sold in the U.S. market.

While the overall U.S. demand for tuna does not seem to be very responsive to changes in tuna prices, the demand for any specific brand may be very price-elastic. Brand allegiance among U.S. tuna consumers is not strong, and when the price of an individual brand changes, that brand's market share will change correspondingly with a more-or-less offsetting change in the market share of a competing brand.

Seasonality

The peak tuna marketing periods in the United States are during Lent (March/April), when some Americans eat more fish and less meat for religious reasons, and during the peak summer months (July/August), when vacationing Americans prefer the convenience of prepared foods that require no cooking. During the fall and winter seasons, especially around Christmas, more Americans cook and consume big holiday meals and leftovers, and thus tuna sales tend to be low.

The seasonal pattern of tuna demand is reflected in wholesale and retail price trends, with more promotional discounts offered during the slack fall/winter seasons and with fewer discounts offered during the spring/summer seasons. Changes in cannery demand for raw fish, of course, precede changes in wholesale/retail demand. Traditionally, U.S. canners have been short of fish during January/February (after the traditional Christmas tie-up of the U.S. fleet) and during May/June, when inventories have been depleted during Lent and the peak summer period is beginning. In recent years, U.S. canners have had no shortages of fish from domestic or foreign sources, even during periods of high demand, and have been able to meet production targets without increasing ex-vessel prices during peak canning periods. The supply of fish from U.S. vessels has been more consistent throughout the year as financially troubled fishermen choose to work through the traditional Christmas break and postpone drydock and maintenance schedules.

INSTITUTIONAL CONSIDERATIONS

Imports and embargo policy

Under U.S. law the seizure of a U.S. vessel by a foreign nation almost always results in a prohibition against the importation into the United States of some or all fishery products from that nation. In the case of seized tuna vessels, embargoes have been placed only on tuna products, although the United States has occasionally threatened to embargo all seafood products from the offending nation.

Over the past five years, the United States has prohibited imports of tuna from Canada, Peru, Ecuador, Costa Rica, Mexico, Papua New Guinea, and Solomon Islands. An embargo usually remains in effect until the U.S. State Department determines that the cause for implementing the embargo no longer exists. As a practical matter, this usually means two years from the date of the vessel seizure if there are no other incidents, or until U.S. industry leaders recommend that the embargo be lifted.

U.S. tariffs and quotas

In general, imported canned tuna in oil is subject to a 35 percent ad valorem tariff, and imports of canned tuna in water are subject to a 12.5 percent ad valorem tax. There has been a shift in U.S. consumer preference toward canned tuna in water and away from canned tuna in oil, which provides some tariff relief to those countries exporting canned tuna to the United States.

The U.S. territories (such as Puerto Rico) and the U.S. insular possessions (such as American Samoa, Guam, and the Virgin Islands) can ship canned tuna free of duties to the U.S. market. The shift of U.S. tuna operations to American Samoa also affects the tariff paid by those nations that export canned tuna to the United States because the current tariff schedule for canned tuna packed in water states that "imports not in excess of 20 percent of the preceding year's U.S. pack are dutiable at 6 percent ad valorem" and that those imports "in excess of the quota are dutiable at a rate of 12.5 percent ad valorem." For the purposes of establishing the quota on canned tuna in water, tuna canned in American Samoa is not counted as part of the U.S. pack. The quota allowing canned tuna imports at a duty of 6 percent ad valorem therefore shrinks automatically with the decline in U.S. mainland canning operations, and more imports will be dutiable at the higher rate.

Other regulations for imported canned tuna

Regulations for canned tuna imports are very specific. For example, the species *Sardi chilensis*, commonly called east Pacific bonito, is not regarded as tuna and must be marketed in the United States as bonito or bonita, even though it yields a product that is similar to the product made from skipjack. Similarly, the species *Seriola dorsalis* or yellowtail cannot be labeled as tuna in the United States, even though it also has flesh similar to that of skipjack.

Aside from basic health and labeling requirements, quality standards for products entering the United States are generally applied by the U.S. buyers themselves and not by the government. Failure to comply with guidelines set by the major U.S. tuna companies with regard to color, salt content, texture, and other properties can result in either a declassification of the product or its outright rejection. Any fish with a mercury content greater than 0.5 parts per million is always rejected, and U.S. tuna companies are also attentive to the appearance and smell of the meat and the can. A Tuna Certificate for exportation to the U.S. is also necessary because tuna and tunalike species may be caught with porpoises, which fall under the jurisdiction of the Marine Mammals Protection Act (1972).

Industry/trade organizations

The major U.S. food conglomerates, H.J. Heinz (owner of Star-Kist) and Ralston Purina (owner of Van Camp), have an extremely large stake in the U.S. tuna industry and are powerful institutions in the United States. There are several tuna-related trade organizations, most notably the United States Tuna Foundation (USTF) and the American Tunaboat Association (ATA), but the direct influence of the major food conglomerates on international fisheries and trade policies exceeds the influence of these tuna-oriented organizations. In fact, the position of the major U.S. tuna companies on tuna issues (e.g., tariffs, quotas, and the 200-mile zone) is frequently established on the basis of the broader interests of the parent corporation and will not always reflect what is specifically in the best interests of U.S. tuna operations. As a result of all these conflicting allegiances among tuna companies and trade organizations, organized labor and local or regional government organizations are frequently the most active in protecting the interests of domestic tuna operations.

CONCLUSION

The tuna market in the United States is so large that it has a major impact on conditions in other tuna markets around the world. This market also is relatively accessible to foreign suppliers. Whether island nations choose to deal directly with U.S. tuna companies in the U.S. market or to focus their efforts elsewhere, the financial performance of tuna-related ventures will be influenced by conditions in the U.S. market.

While the U.S. tuna market is primarily a market for canned products, there may be other markets for specialty tuna products that are too small to attract the major tuna companies, and these markets may offer a niche for small island tuna producers. U.S. consumers prefer meat to fish, and although the traditional canned tuna product is popular in the United States, it is the only seafood staple in the diet of most Americans. In fact, attempts to market other kinds of processed tuna products have never been successful.

Despite the fact that only a few firms market tuna in the United States, the competitive pressures at the wholesale and retail market levels are sufficiently strong so as to prevent U.S. buyers of raw/frozen or canned tuna from offering a price to any foreign supplier that exceeds the prevailing international price. Opportunities exist in the U.S. market for new foreign tuna harvesters and processors who can provide quality products on a regular basis at prices that are competitive with those of other international suppliers.

While the development of a competitive domestic tuna industry is essential for success in penetrating the U.S. tuna market, of equal importance is the distribution and marketing of tuna products once they reach the United States. The major U.S. food companies operate elaborate marketing/distribution networks and are in a position to buy raw and processed tuna from many different domestic and foreign sources. Before Pacific island nations make any investments in harvesting and processing operations, they should consider how these activities will fit into a complete tuna production/distribution system that takes tuna from the sea and provides it to U.S. consumers. It is possible to operate exclusively at one or two stages in the production/distribution process, but if risks are to be minimized, long-term relationships should be established within a known production/distribution network.

If the developing tuna industries in Pacific island countries expect to compete with the established tuna producers on a cost/price basis, they will need to attract overseas financial assistance through government channels and should not expect to receive any advantageous supply or market arrangements from commercial U.S. tuna interests. The only exception is where the U.S. government chooses to pursue its strategic interests in the Pacific islands region by providing incentives to U.S. corporations to enter into financial, contractual, or trade arrangements that are advantageous to Pacific island nations. When dealing with U.S. corporations, Pacific island leaders should not mistake a willingness to discuss cooperative ventures with an interest in making investments, and they must guard against pursuing prolonged negotiations that have little chance of resulting in economic benefits and that may even inhibit the adoption of less attractive but more realistic strategies.

Most individual Pacific island nations are not in a position to develop a competitive national tuna industry or command enough market power to deal effectively with the major tuna companies. Thus they do not have any significant impact on the international tuna trade. On a regional basis, however, Pacific island nations could control 40 percent of the global tuna supply, which is a large enough share to make a major impact on all world markets.

If Pacific island nations view the large U.S. market from a regional rather than a national perspective, they can develop strategies that will give them more market control and allow them to negotiate more effectively with the companies that control the established product distribution and marketing networks in the United States. If Pacific island nations act collectively, they can take advantage of economies of scale by centralizing the information-gathering, surveillance and enforcement, and marketing functions. Large U.S. food companies have developed their strength in the tuna industry by centralizing control over the tuna production/distribution systems. Pacific island countries must seek a similar type of strength and control if they are to compete and negotiate successfully with the U.S. food companies.

5.
U.S. Tuna Import Regulations

Jesse M. Floyd

ABSTRACT—This chapter describes U.S. import regulations for fish and fishery products with particular emphasis on tuna imports. Following a review of technical and legal aspects of import requirements, the changes in the U.S. tuna canning industry are discussed in terms of how they might affect import regulations and policy. Court actions relating to the U.S. tuna industry are outlined and conclusions drawn.

INTRODUCTION

The United States has a free-trade policy on all categories of fish and fishery products except canned tuna. Import deposits and licenses are not required by U.S. fish importers, and foreign exchange is freely obtainable to finance fish imports. Exporters of fish and fishery products to the United States, however, must pay close attention to several regulations governing imports to the United States.

U.S. IMPORT REGULATIONS

The United States imposes strict standards on the import of fish and fishery products, which are enforced by the Food and Drug Administration (FDA). Two major pieces of legislation cover raw and canned tuna imports: the Food, Drug, and Cosmetic Act (FDCA) covers quality of all imported fish products, and the Fair Packaging and Labeling Act (FPLA) specifies labeling requirements for imported products arriving in consumer-size packages (e.g., retail-size cans). The intent of these regulations is to ensure that all food product imports, including fish, are free from dangerous diseases and pests and that they conform to the same standards of wholesomeness, sanitation, and labeling required of products processed and packed in the United States. The FDA commissioner has the power to periodically revise these standards to ensure maximum protection for the consumer.

According to the FDCA, all fish and fishery product imports must comply with the same product specification requirements as domestically produced and processed fish—a principle generally referred to as "good manufacturing practices." The act prescribes standards in eight areas:

- Misbranding
- Definitions and standards of identification
- Food additives
- Adulteration
- Tolerances for poisonous and deleterious substances
- Pesticide residues
- Defect action levels
- Good manufacturing practices

Products that contain any poisonous or deleterious substance are in violation of the FDCA unless (1) such substance is required in the production of the food or cannot be avoided by good manufacturing practices or (2) a tolerance has been established for a particular poison in a particular food, and the residual amount does not exceed the established tolerance. Those products exceeding the prescribed levels are denied access to the U.S. market.

Packaging standards for imported fish and fishery products are covered by the FPLA. These standards pertain to the size and weight of allowable packages, preservation of the product within the package, and acceptable packaging materials. Where fish is packed in a medium (e.g., tuna in oil), the container should hold as much fish and as little oil as possible. Failure to declare the presence of added salt or the kinds of oil used as the packing medium for canned fish has resulted in the detention of fish product imports to the United States.

Regulations governing marking and identification standards specify that label information must be conspicuously displayed and presented in terms that the consumer is likely to read and understand under ordinary conditions of purchase and use. The label must be in English and the country of origin identified. All text written in foreign languages must be translated into English. The name, address, city, and state of manufacturer, packer, and distributor must appear on the label. The net amount of the fish must be listed in the avoirdupois weights and measures system.[1] The common name of the product must be displayed. Ingredients used in the product (as well as any artificial flavoring, artificial coloring, or chemical preservative) must be listed by common name in order of their predominance by weight. Foods that are adulterated in any way or contain any inferior or impure materials or additives are strictly prohibited. Imitations must be labeled as such. If the product is not manufactured by the person or company whose name appears on the label, the name must be qualified by "manufactured for," "distributed by," or a similar expression.

Imported food products may be examined on entry by FDA officials under the FDCA because the FDA has no authority to inspect food processing operations in other countries. Foods that do not comply with existing regulations are impounded. Importers have up to 30 days to appeal the notice of detention by FDA authorities and to justify how the product complies with the regulation in question. In the event that the appeal is denied, the product is either destroyed or re-exported at the exporter's expense. In addition to FDA examination, U.S. customs officials inspect all imported products to ensure that labeling and packaging requirements are met.

Proper documentation to import goods to the United States is required by law. All imported goods in excess of $500 must be accompanied by Special Customs Invoice No. 5515 and a bill of lading from the exporting country.[2] No more than one shipment from one consignor to one consignee by one vessel may be included on the same customs invoice. No special requirements apply to bills of lading except that two copies are required by the consignee. A packing list and pro forma invoice designating the value of the product are not compulsory, but these documents facilitate customs clearance.

Quotations indicating the value of the product should be specified in U.S. dollars at the port of arrival. FOB (free on board) and CIF (cost, insurance, and freight) trade quotations are generally used, but other trade arrangements are common. Payment is usually made by letter of credit, but under appropriate circumstances other commercial financial arrangements may be made, such as dollar drafts, open account, or consignment.

Fish and fishery products imported to the United States may be subject to duties. The value for duty is almost always quoted ad valorem and is usually determined on an FOB basis.[3] Sales tax, ranging up to a maximum of 12 percent of the retail value, may also be levied by state and municipal bodies in the United States.

The rate of duty imposed by the United States varies according to the exporting country's trade status with the United States. Most favored nations (MFNs) and less developed countries pay the lowest import duties. Other countries fall under general trade and usually pay the highest rates of duty. All Pacific island countries and territories have MFN trade status with the United States.

U.S. IMPORT REGULATIONS ON TUNA COMMODITIES

In addition to the general regulations and requirements applicable to fish and fishery product imports to the United States, other regulations apply specifically to tuna commodities.

Frozen tuna

To import frozen and unprocessed tuna to the United States, an importer must first obtain a tuna importation certificate from the National Marine Fisheries Service (NMFS). The purpose of this certificate is to ensure that the imported tuna has been caught in accordance with U.S. law, which prohibits catching tuna with certain types of gear (especially purse seines) that could endanger porpoises. Porpoises are protected under the U.S. Marine Mammal Protection Act of 1972. The import certificate must carry the signature of a responsible person in the country of origin who certifies that the porpoise protection requirements have been met. The certification does not relate to quality of the tuna. Quality assessments are between buyer and seller within the conditions of the food and drug legislation.

There is no duty on frozen and unprocessed tuna irrespective of country of origin. On prepared and preserved tuna such as cooked loins, the U.S. International Trade Commission (USITC) established a tariff on 1 January 1972 of 0.5 cents per pound excluding the weight of the container.

Although frozen and unprocessed fish is allowed to enter the United States duty free, the Nicholson Act prohibits foreign flag fishing vessels from landing their fish, including tuna, on U.S. soil. The product must be off-loaded outside the United States and imported into the United States. Countries exempted under the Nicholson Act in the Pacific islands region, where foreign flag fishing vessels are permitted to off-load their catch, are American Samoa and the Commonwealth of the Northern Mariana Islands (Tinian) and Guam. Foreign flag fishing vessels are also permitted to off-load in Puerto Rico.

Canned tuna

The United States imposes tariffs on canned tuna imports in two categories: tuna in oil and tuna not in oil. The tariffs on canned tuna were introduced in 1930, but they have been changed several times since then. The rates of duty currently in effect were established on 1 January 1972. At that time, U.S. consumers generally preferred lightmeat tuna packed in oil. As a result of increasing concern among Americans about the amount of cholesterol and sodium in their diets, U.S.

consumption has shifted significantly in favor of lightmeat tuna packed in water.

The tariffs on tuna canned in oil and not in oil differ substantially. Under Tariff Code 112.90, tuna prepared or preserved in oil in airtight containers weighing 15 pounds each or less is subject to 35 percent duty (ad valorem) from MFN countries and 45 percent duty (ad valorem) from all other countries. Canned tuna not in oil is subject to a rate of 6 percent from MFN countries and 25 percent from all others (Tariff Code 112.30). Because all Pacific island countries and territories have MFN trade status with the United States, they pay 25 percent duty on canned tuna in oil and 6 percent on canned tuna not in oil. As a result of the comparatively high duty on canned tuna in oil, imports to the United States are negligible.

American Samoa is the notable exception in the Pacific islands region. Under Headnote 3(a) of the Tariff Schedules of the United States, products from territories such as American Samoa are accorded duty-free entry to the United States if the value-added component in American Samoa is at least 50 percent of the product's value. The Headnote 3(a) exemption is easily met by canned tuna, regardless of the origin of the raw fish.

In addition to the 6 percent ad valorem duty, a tariff rate quota is imposed on canned tuna not in oil. The quota was imposed on 14 April 1956 and is recalculated annually on the basis of 20 percent of the previous year's domestic pack. Those imports exceeding the annual quota are dutiable at a rate of 12.5 percent ad valorem, which is approximately twice the regular rate and translates into roughly $1 per case. Domestic production in the United States—including Hawaii and Puerto Rico but not American Samoa—reached its highest level in 1978. That year domestic production was 629.1 million pounds (286,000 tonnes) or 31.5 million standard cases.[4] In 1979 the quota was established at 125.8 million pounds (55,000 tonnes), 20 percent of production in 1978.

Imports of canned tuna have exceeded the quota in several years since 1978. Imports of canned tuna not in oil in 1983 were more than 122 million pounds (58,000 tonnes) while the established quota was 92 million pounds (42,000 tonnes) (Table 1). Those imports arriving after the quota closed on 16 July 1983 were dutiable at 12.5 percent ad valorem. The quota was exceeded in 1985 by 116.8 million pounds (53,091 tonnes). By 30 September 1986, the quota had been surpassed by 121.4 million pounds (55,182 tonnes). The quota for 1986 was 81.8 million pounds (37,182 tonnes) and was filled by March 1986.

The quota is filled on a first-come, first-served basis. Imported tuna is counted against the yearly quota when it is released for consumption from warehouses. Some firms store tuna in bonded custom warehouses late in December for release after the first of the year in order to assure sales at the lower tariff.

A significant event, which occurred in 1983, was the exclusion of canned tuna not in oil packed in American Samoa from being counted against the quota. Prior to 1983, an estimated one-third of the annual quota was filled by canned tuna processed in American Samoa, even though these products entered the United States duty-free. The exclusion of American Samoa's tuna products in 1983, therefore, allowed foreign countries to export much more canned tuna not in oil to the United States at the lower tariff duty rate.

CHANGES IN THE U.S. TUNA CANNING INDUSTRY

The rapid growth of imports of canned tuna not in oil to the United States has been accompanied by a sharp decline in canning capacity on the U.S. mainland. At the end of 1982 the NMFS reported that 20 plants were canning tuna in the United States: 12 on the U.S. mainland, 5 in Puerto Rico, 2 in American Samoa, and 1 in Hawaii. This reflects a decrease in the total number of plants canning tuna in the United States since 1977 but not a substantial change in total canning capacity. As of September 1986 there were 5 operating plants in Puerto Rico and 1 on the U.S. mainland (Pan Pacific in Terminal Island, California). The plant in Hawaii closed in 1984. All other mainland plants have closed, largely as a result of high labor costs, increases in the value of the U.S. dollar, and rising imports from overseas producers. In addition, prices for canned tuna on the U.S. market have been depressed since 1982 and have shown little sign of recovery.

Table 1. Quota and imports of canned tuna not in oil to the United States, 1979–85

Year	Quota (000 pounds)	Under quota (000 pounds)	Over quota (000 pounds)	Total imports (000 pounds)
1979	125,813	53,703	—	53,703
1980	109,074	63,553	—	63,553
1981	104,355	70,851	—	70,851
1982	109,742	87,579	—	87,579
1983	91,904	91,904	30,425	122,329
1984	89,699	89,699	116,800	206,499

Source: USITC 1984, A-7 and A-22.

At the same time that canning capacities have declined on the U.S. mainland, offshore operations in Puerto Rico and American Samoa have increased. In Puerto Rico, Star-Kist in 1985 added a third work shift to increase production to 660 tonnes a day. Ralston Purina's National Packing has reopened after renovations and a temporary closure in 1983, and Sun Harbor, formerly an independent company, has changed ownership. Both canneries will presumably intensify their production. Neptune Packing (a subsidiary of Mitsui) is also canning tuna in Puerto Rico. The Bumble Bee canneries in Puerto Rico and Hawaii closed because the parent company, Castle & Cooke, was having financial difficulty. Bumble Bee has since been sold and has reopened canneries in Puerto Rico.

Canning operations in American Samoa have expanded even more rapidly than the operations in Puerto Rico. Both Star-Kist/Samoa and Van Camp's Samoa Packing are renovating their plants in Pago Pago and plan to increase their production to 88,000 and 66,000 tonnes a year, respectively. According to industry sources, the shift of production to American Samoa is due to its proximity to the western Pacific tuna fishery, reasonable wage costs, and tax incentives offered by the American Samoan government for offshore investment. American Samoa's exemption from the Nicholson Act and its duty-free access to the United States under the provisions of Headnote 3(a) of the U.S. Tariff Schedules are other advantages that make American Samoa an attractive site for tuna canning operations.

Despite the advantages enjoyed by Puerto Rico and American Samoa, imports of canned tuna not in oil to the United States from other countries have increased steadily from 54 million pounds (25,000 tonnes) in 1979 to over 206 million pounds (73,000 tonnes) in 1984 (Table 1). As shown in Table 2, Thailand was the largest supplier of canned tuna not in oil, accounting for 33 percent of the imports in 1983, followed by the Philippines, which accounted for 26 percent. Prior to 1981 Japan was the largest exporter of canned tuna to the United States, but in 1983 Japanese tuna exports accounted for only 17 percent of the total import volume. Taiwan accounted for 15 percent and has steadily increased its market share over the last five years.[5]

Preliminary reports indicate that Thailand increased its market share in 1984 and accounted for more than half of U.S. canned tuna imports not in oil. According to industry sources, U.S. imports of canned tuna not in oil were 162 million pounds (74,000 tonnes) in 1984. Thailand exported 89.7 million pounds (41,000 tonnes) to the United States, followed by Japan (26.9 million pounds—12,000 tonnes), the Philippines (22.2 million pounds—10,000 tonnes), Taiwan (17.9 million pounds—8,000 tonnes), and all others (5.3 million pounds—2,400 tonnes).

COURT ACTIONS

In addition to intensifying and consolidating their operations in Puerto Rico and American Samoa, several U.S. tuna operators have reacted to the increasing volume of imports by filing grievances in court. On 15 February 1984, the U.S. Tuna Foundation, C.H.B. Foods, the American Tunaboat Association, the United Industrial Workers, the Fishermen's Union of America, and the Fishermen's Union filed a petition for import relief with the U.S. International Trade Commission (USITC) in Washington, D.C. The petition claimed that canned tuna was being "imported into the United States in such increased quantities as to be a substantial cause of serious injury, or threat thereof, to the domestic industry producing articles like or directly competitive with the imported article." After an investigation, the commission dismissed the petition on 15 August 1984 by a 4–1 vote, ruling that imports were not the major cause of the U.S. industry's current difficulties. The only panel member voting in favor of a higher import tariff on canned tuna not in oil indicated that she would have recommended federal help for boat owners and laid-off cannery workers before recommending a tariff increase to protect the U.S. tuna canning industry.

In a similar action the Tuna Research Foundation, on behalf of the U.S. tuna canning industry, filed a countervailing duty petition on

Table 2. Imports of canned tuna not in oil to the United States, 1983

Country	Quantity (1,000 pounds)	Share of total (%)	Value ($000)	Unit value (per pound)
Thailand	39,930	32.7	43,259	1.08
Philippines	32,108	26.2	32,291	1.01
Japan	20,387	16.7	24,643	1.21
Taiwan	18,707	15.3	22,767	1.22
Malaysia	3,083	2.5	4,068	1.32
Australia	2,799	2.3	3,684	1.32
Indonesia	2,634	2.2	2,679	1.02
Canada	2,104	1.7	2,982	1.42
Singapore	332	0.3	386	1.16
Republic of Korea	68	0.1	69	1.02
All others	71	0.1	77	1.08
Total[a]	122,132	100.0	136,906	1.12

Source: USITC 1984, A-23.

[a]Columns may not sum to totals because of rounding.

canned tuna from the Philippines with the International Trade Commission in 1982. The subsequent investigation resulted in a countervailing duty of 0.72 percent being levied against Philippine imports to the United States. According to U.S. government sources, the purpose of the countervailing duty was to offset subsidies received by tuna packers based in the Philippines. This duty has been re-evaluated several times and is subject to revision in the future.

In February 1985 a group of U.S. independent tuna boat owners in San Diego filed a $1 billion antitrust suit in the San Diego Federal Court against the nation's three largest tuna processors. The suit alleged that Star-Kist, Ralston Purina (Van Camp), and Castle & Cooke conspired to restrain trade, fix prices, monopolize trade, and destroy the independently owned tuna fleet. The suit was settled out of court in 1986.

CONCLUSION

The changes and conflicts occurring in the U.S. tuna industry raise several questions about its future. One important question of interest to the Pacific islands region, as well as to the international tuna industry in general, concerns the U.S. quota. Between 1979 and 1984, the annual quota for tuna canned in oil declined by 29 percent (Table 1). In 1985 the quota was expected to decline further because of plant closures on the mainland and in Hawaii and Puerto Rico. Although this will result in a higher proportion of canned tuna imported at the 12.5 percent ad valorem rate (because 20 percent of U.S. production in 1984 will result in a lower quota), imports from overseas producers are expected to continue to increase. In response, the U.S. government or the USITC could raise the import duty on canned tuna not in oil. The current composition of the USITC is considered anti-protectionist, and final decisions on tariffs in the future could be more sensitive to trade impacts of the tariff than to industry impacts.[6] The U.S. tuna industry could bring political pressure to enforce U.S. import regulations on tuna with increased scrutiny in an attempt to limit foreign canned tuna imports. One thing, however, is certain. The continued success of overseas producers depends on their ability to meet U.S. processing and packaging standards and administrative requirements for entry to the U.S. market.

NOTES

1. The short ton of 2,000 pounds is used in the United States. A short ton is equivalent to 909 kilograms. To avoid confusion, short tons should be used in the import documents.

2. Commercial invoices are acceptable for shipments of $500 or less.
3. The notable exception is canned clams, which are valued for duty at the U.S. selling price.
4. A standard case holds 48 cans weighing 185 g (6.54 oz) each. The case weighs approximately 9 kg (20 lb).
5. Each of these countries has MFN trade status with the United States.
6. The USITC is composed of political appointees of the U.S. president and professional specialist staff members who undertake research related to trade issues.

6.
The European and Middle East Tuna Market: A View from the Pacific Islands

Richard Elsy

ABSTRACT—This chapter analyzes the current status of the European and Middle East tuna market from the vantage of Pacific island countries. An overview is provided of the western European market, including an analysis of markets by country. Middle East markets are reviewed, and the prospects are discussed for those island countries selling tuna in these markets. The next section reviews marketing standards and other regulations imposed on canned tuna imports in Europe. The chapter then addresses implications for Pacific island countries wanting to penetrate the European market or to expand existing sales.

INTRODUCTION

Overview

By 1985 an estimated 51 percent growth had occurred in demand for canned tuna imports in west European markets since 1975. From an estimated 5.6 million case equivalents in 1975–76, total imports reached 11.5 million in 1985. The total level of imports in 1981 was 8 million cases.

With minor exceptions in the late 1970s, this upward trend has been sustained throughout 1975–85 in all European markets. The annual average growth rate between 1975–76 and 1981 was 5 or 6 percent, increasing to 7.6 percent between 1981 and 1985.

For the purpose of this chapter, the various European markets are as follows:

European Economic Community (EEC): France, United Kingdom, West Germany, Netherlands, Italy, Belgium/Luxembourg, Denmark, Greece, Spain, Portugal, and the Republic of Ireland.

Other Europe: Switzerland, Sweden, Finland, and Norway.

Unlike the United States, western Europe consists of a diverse set of markets that have evolved in different ways.

These markets are classified for the purpose of comparison as "mature" markets and "non-mature" markets. Mature markets are France, Italy, Spain, Portugal, Greece, and Belgium/Luxembourg. With the exception of Belgium/Luxembourg, these countries have domestic catching and canning industries and long-standing traditions of tuna consumption. Consequently, local and national pack preferences have evolved. The inclusion of Belgium/Luxembourg is the result of strong French influences in the retail food market. Non-mature tuna markets are exclusively importing nations. It is in these markets that the main growth in European tuna demand has occurred in recent years.

Three markets account for an estimated 74 percent of Europe's canned imports: France (34 percent), United Kingdom (22 percent), and West Germany (18 percent). Canned imports are shown in Table 1.

Insofar as generalizations can be made about such a diverse range of markets, a trend has occurred in European markets toward trading down—that is, increasing the share of market of the lower unit-value product. The growth in market shares that the Thai producers have enjoyed is both a result and a cause of this trend, which is discussed in more detail in the individual country sections. European tuna canneries processed approximately 170,000 tonnes of tuna in 1985 compared with imports of around 70,000 tonnes and exports of about 5,000 tonnes. Thus the market size within the European Economic Community (EEC) is 235,000 tonnes, of which 30 percent is supplied by imports.

Western Europe's production is concentrated in Spain (74,000 tonnes), Italy (64,500 tonnes), France (25,000 tonnes), and Portugal (4,500 tonnes).

The product of Pacific islands origin seems to enjoy a high level of acceptance in those European markets that are effectively open to it. This fact is reflected in the high prices that the Pacific product commands.

Estimated per capita consumption of tuna in western European countries is shown in Table 2. Per capita consumption ranges from 1.1 kg in France to 0.14 kg in the Netherlands. The average per capita consumption is approximately 0.63 kg.

MARKET DESCRIPTIONS

France

Current and past status. Of the total 7 million cases recorded for 1985, 3.9 million (56 percent) were imported, and 3.8 million (54 percent)

Table 1. Imports of canned tuna to western Europe, 1980–85

	(48 x 198-g cases)					
	1980	1981	1982	1983	1984	1985
France[a]	2,586,076	2,913,700	3,140,333	3,575,400	3,421,207	3,879,684
United Kingdom[b]	1,158,977	2,020,376	1,400,242	1,850,200	2,530,000	2,523,448
West Germany	1,604,237	1,430,290	1,596,272	1,695,094	2,011,364	2,086,805
Belgium/Luxembourg	444,891	457,750	427,014	506,224	523,080	552,539
Italy[c]	320,000	191,530	264,691	353,327	370,557	524,438
Netherlands	176,738	148,780	149,079	146,027	228,009	212,284
Denmark	117,000	136,890	131,196	180,532	229,076	237,218
Other EEC[d]	11,784	17,000	101,824	100,551	114,767	182,705
Total EEC	6,419,703	7,316,316	7,183,651	8,408,155	9,428,060	10,199,121
Sweden	238,800	219,120	246,571	261,603	287,257	268,548
Estimated total Western Europe[e]	7.3m	8.4m	8.0m	9.5m	10.7m	11.5m

Source: Foodnews. Various years.

[a]Total French market including domestic production (3.4 million cases) less exports (490,000 cases) is estimated at 6.8 million cases for 1985.

[b]United Kingdom imports for January–November 1986 reached 3,432,647 cases.

[c]Italy exported 215,243 cases in 1985.

[d]Includes Republic of Ireland and Greece from 1981.

[e]Includes Finland, Norway, Switzerland, and Austria. No breakdown is given in the national statistics for these last two countries, but the estimate is 500,000 cases and 250,000 cases, respectively. Cases in millions (m).

originated in the west African countries of Senegal and Ivory Coast. This compares with a total for 1979 of 6.2 million cases; 2.5 million (40 percent) were imported, of which 2.3 million (37 percent) came from these two countries. Since 1976 a minimum of 97 percent of imports to France has originated in Senegal and Ivory Coast, where French industry has significant investment in canneries. These investments are effectively wholly owned French subsidiaries, supplying lightmeat chunk and solid packs to the French market.

The market is a mature European market, showing less than 2 percent growth annually; it has continued dependence on two supplying countries, with little change in the overall balance between domestic canning and imports.

The tuna catching and canning industries in France have a strong political lobby, and no serious consideration is given to the idea that there will ever be a day when either one or both of these enterprises will be abandoned. The French fleet has moved farther away from its traditional west African fisheries in recent years. Costs of canning in France have risen, and despite the decline of the U.S. dollar, the French industries' commitment remains firm. Clearly, this commitment is endorsed by government. Questions periodically arise over the matter of hidden subsidies, such as those that are alleged to be allowing French canners to export the product at present against a weak U.S. dollar. It is also reported that some infringement of EEC regulations may be occurring in this matter, though no action is being taken against France.

The French market is overwhelmingly a lightmeat chunk-style pack market with a small demand for the whitemeat pack. This latter market

Table 2. Estimated per capita consumption of tuna in western Europe

Country	Tuna consumption (kg/head)[a]
France	1.10
United Kingdom	0.54
West Germany	0.29
Belgium/Luxembourg	0.47
Italy	1.50
Netherlands	0.14
Switzerland	0.69
Sweden	0.29
Average	0.63

[a]Based on author's estimate.

has shrunk with limited consumer spending power over the past ten years, but it appears to be maintaining about 5 percent of the market.

The flake/hors d'oeuvre pack grew in the late 1970s to attain about 30 percent of the market by 1980. However, it appears to have stabilized around that level since 1980, despite a relatively high degree of product innovation. The majority of the market, over 50 percent, is held by lightmeat brine packs, with oil packs having slipped to below 10 percent.

Distribution of canned tuna in France is relatively straightforward. A high degree of concentration exists in the market for both producers' and retail outlets. Companies supply both housebrand packs to the major retailers and their own nationally marketed packs. The industry as a whole also supports a generic advertising and public relations campaign.

Accessibility. By common consent and historical evidence, this market is virtually inaccessible to non-traditional suppliers. The French industry has considerable investment in all phases of the industry in the supplying countries (Senegal and Ivory Coast), and it is able to wield sufficient authority to ensure that other suppliers will simply not be allowed to ship the product in. The legality of this action has been questioned on many occasions, and some buyers in France have from time to time sought the product from other suppliers. The halting of shipments of tuna by French customs is reported. The consensus in the trade is that while on paper the French market is a potentially good market in which to develop a small niche, it is effectively inaccessible. While this state of affairs appears to be in contravention of both the General Agreement on Tariffs and Trade (GATT) provisions and the Treaty of Rome, no action has been taken to challenge or counter the French position.

Forecast. No basis exists for supposing that any of the market parameters described above will change in the foreseeable future. Growth rates are expected to remain the same, and despite the problems facing the industry, there is no likelihood of access being allowed to other than the traditional suppliers. Pacific island producers would not be advised to direct any efforts to additional research or development of this market.

United Kingdom

Current and past status. The United Kingdom has proved to be the most dynamic of the European markets in recent years and, indeed, promises to be so in the future. Between 1975 and 1979 the total imports rose from 0.62 million cases to 1.1 million, an average annual growth of

11 percent. Between 1979 and 1985 growth averaged 9.3 percent to reach 2.5 million cases, a slightly lower total than in the previous year. In 1986, however, based upon the January–November import statistics, it appears that the total will have reached 3.6 million cases, an increase of over 30 percent from the previous year. The slight drop in 1985 was probably due to large carryover inventories and cautious buying in 1985, which led to a marked increase in 1986. However, this is a remarkable increase in volume, and sources in the trade are confident of further significant growth.

The single biggest change in the supply side of the market has been the growth of Thai imports since 1979. Thai imports were 5 percent of the market in 1984, but by 1986 they had achieved a 50 percent share. Fiji imports peaked at 18 percent of the market share in 1983 and have since fallen to around 10 percent. Solomon Islands' imports peaked at 7.6 percent of the market in 1980, then, after running steadily at around 100,000 cases, dropped to about 3 percent of the market.

While solid oil packs dominated the early days of the market, a marked trend has occurred toward the chunk brine pack. Between 1982 and 1985 the share of chunk packs grew from 12 percent to 24 percent and is estimated to have reached 30 percent in 1986 with growth continuing. The brine pack share reached 21 percent in 1984 and 35 percent in 1986 and is continuing to grow.

These changes reflect consumer responses both to price rises and to increased awareness of the health implications of oil packs, which have combined to push up sales of the chunk brine pack. The market is highly competitive, with John West Foods as the acknowledged brand leader, having an estimated 35 percent of the market. Housebrands have developed the market share, notably Sainsbury, which now dominates the retail food market and continues to expand with an aggressive store-building program outside its traditional stronghold in the south of England.

Buying practices by U.K. outlets have altered to accommodate the increasing high level of competition. Some buyers are deviating from the norm of buying only through Japanese trading companies. Some are obtaining the product from new sources and buying directly from canneries. Buyers (either by the major firsthand distributors, including John West and Princes) or retail companies have traditionally preferred to purchase tuna from Japanese trading companies. This was because they saw this as a form of insurance for the product that they were selling in the market.

Accessibility. The United Kingdom has traditionally been the most accessible of European markets, particularly for Pacific island producers.

Tariff-free entry is common to all EEC markets, but this fact—combined with the lack of any existing domestic tuna fishing or canning industry and a common language—makes the market more accessible than most others.

Forecast. The 1986 figures are supported by trade views that argue a continued high rate of growth in the U.K. market leading to a doubling of market size by 1991. The slower growth rate witnessed during the early to mid-1980s encouraged many observers to believe that the "steam had gone out of the market" and that sustained steady growth with alteration of the market structure was the best that could be hoped for. However, product innovation in the U.K. market will play a major part in the development of share gains, with obvious possibilities for sauce and low-salt packs. Pack size is also coming under scrutiny with attention being focused on the need to develop a market share for the 400-gm pack, as opposed to the 200-gm pack. This reflects the growing acceptance of tuna in general in the United Kingdom and the increased scope for major promotion of the product in the larger family-size pack.

West Germany

Current and past status. Between 1975 and 1985 the West German market grew at an annual average rate of 2.7 percent to reach a total volume of 2.01 million cases. The growth rate was somewhat higher between 1980 and 1985 (4.7 percent) and close to the average for the European countries surveyed in this chapter.

The West German market is quite different from other European markets and is particularly characterized by a heavy reliance on price, which has determined the entire evolution of the market. West German buyers do not recognize any latent consumer brand loyalty, neither are any efforts being made to develop this loyalty other than through constant pressure on price. This has necessarily led to frequent and large-scale shifts in sourcing the product over recent years, with practically every supplying country having sold tuna at one time or another.

Starting in 1971 Japan supplied about 81 percent of all tuna to the West German market, though by 1980 Taiwan was supplying 64 percent. By 1984 both of these supplying countries had been superseded by Thailand, which accounted for 55 percent, and the Philippines, which supplied 19 percent. Of all buyers in Europe, the West Germans are most susceptible to offers from new suppliers who are able to compete over price. This contrasts markedly with other markets, notably neighboring Switzerland, which has a firm policy on quality and consistency.

The West German market was dominated by a low-value flake and vegetable pack up to the early 1980s, and it still accounts for around

40 percent of the total market, though its share has declined slightly in recent years in favor of low-cost solid and chunk packs. The low-value flake/vegetable pack has been commonly used as a loss leader in store promotions and has helped to sustain the pressure on price. The decline in the market share of the dressing pack is believed to be a direct result of consumer resistance to the ever-inferior packs, which in turn were the result of intense price competition among Thai packers.

In this market, brand shares are not perceived as a significant means of assessing the development of the market. Housebrands and brokers' own label brands share the market, competing entirely on price.

Accessibility. In principle, the West German market is highly accessible. In practice, access is gained only by extreme competition on price, which results in a downward spiral in quality.

Forecast. There are no indications of any changes either in the volume growth trend or in the structure of the market. Until such time as the structure improves to the extent of allowing a higher unit-value product to be marketed, Pacific island producers have little reason to commit any resources to developing the market.

Belgium and Luxembourg

Current and past status. Belgium and Luxembourg are usually combined because of their economic union. Volume growth in this market has been slightly below the average for all countries considered in this chapter. Growth averaged only 3.6 percent per year between 1975 and 1985. In 1985 the total tuna imports reached just over 0.5 million cases. Only slight variations around that average growth rate occurred in recent years with the exception of 1982, when a drop was recorded. This decline was thought to be the result of a high stock carryover.

The composition of the tuna-supplying countries has been relatively stable, with even the Thai suppliers reaching only 8 percent of the market in 1985.

France and Italy remain the major suppliers with 14 percent and 12 percent of the market, respectively.

The Belgium/Luxembourg market is generally a high-value market, with solid-pack yellowfin only recently giving way to a chunk pack. The brine pack is also increasing its market share for reasons of both health and price, though the difference in price is narrowing.

Housebrands dominate the market as a result of the highly concentrated retail food market.

Accessibility. The market is considered to be readily accessible, given the wide variety of supplying countries. However, buyers have a tendency to be conservative to the extent of buying high-value packs from

European canneries. Given the right quality of pack, Belgian buyers can be described as more consistent price takers than those found elsewhere in Europe.

Forecast. Little change is forecasted either in the growth trend or the structure of the market. The market will continue to offer prospects for the Pacific island producers that are reported to have been in contact with the prominent retail chain buyers.

Italy

Current and past status. As a mature tuna market with an established tuna canning industry, Italy has tended to be discounted as a target market for exporting nations in the past. Imports were traditionally in the form of catering packs, while the retail market was entirely satisfied by the domestic canning industry, which was supplied from imported frozen tuna. Current figures on production levels in the domestic industry are not available, though production is estimated to be about 1 million cases per year. Imports have risen annually at an average rate of 6.5 percent, while exports have grown at a lower rate of 3.6 percent. Total canned tuna imports reached 0.52 million cases in 1985. Exports of tuna were 0.21 million cases in 1985.

The Italian market is generally diverse with respect to regional brand loyalties, though this factor is becoming less of a constraint to access. The traditional pack is lightmeat (yellowfin in olive oil), which accounts for about 90 percent of all products sold; some skipjack also finds its way onto the market.

Accessibility. The main barrier to entry in this market has usually been the high costs associated with penetrating such a diffuse market combined with strong brand loyalties (both national and regional). These loyalties are being tested by the ever-increasing costs of canning domestically, and the retail industry seems to be moving toward a greater degree of homogeneity. Consumers are accustomed to paying a high price for a good-quality pack, and they may now be better disposed to trying an imported product.

One concern is that the Italian canning industry, much of which is state-owned, will respond to increased imports in the same way as their counterparts in France. The answer to this problem depends upon the government's willingness to subsidize the industry, albeit covertly, and to implement illegal non-tariff barriers. The general feeling is that this is unlikely, and past evidence shows that the only major difficulties imposed on exporters to the Italian market have been in the form of rather complex currency exchange regulations.

Forecast. Continued growth is expected in imports, with emphasis on retail packs.

Other European markets

Netherlands. While import volume in 1985 was only slightly above volume for 1976, marked fluctuations have occurred in the intervening years with heavy declines in the early 1980s despite a 36 percent increase between 1983 and 1984. The fluctuations indicate that this is a young and unstable tuna market where canned salmon continues to outsell tuna by an average ratio of 2.5:1.

The import volume is limited (212,000 cases in 1985), and there was no assurance that the volume would rise in 1986. Moreover, the Netherlands is not a high-value market, and while highly accessible in terms of structure, it tends to be extremely price-competitive. Growth is possible, though somewhat dependent on the future of canned salmon sales.

Sweden. Total imports of 268,000 cases in 1985 represented a decline of nearly 7 percent over the previous year, as well as the first drop in what had been a steady if slow growth trend. The annual average growth has been around 1.8 percent since the mid-1970s, with a slightly higher rate being recorded in more recent years.

The market has been trading down, and over 67 percent of all imports now originate in Thailand and the Philippines. The market has been overwhelmingly a chunk light-pack market, with the water pack steadily gaining popularity over the oil pack. A small whitemeat market, accounting for around 20 percent of volume in 1980, is thought to have been seriously eroded in recent years.

The market is readily accessible and, because of the concentration in the retail food sector, is relatively inexpensive to penetrate. An agency of the Swedish government (Impod) has been established with the objective of assisting developing countries to introduce their products to the Swedish market.

The prospects for real growth in the near future are limited, and the downward trend in quality gives little cause for optimism among Pacific island suppliers.

Switzerland. Regarded as the top-quality market in Europe, Switzerland has seen no growth in volume since 1980. Import data on canned tuna are not disaggregated in the official statistics, though estimates confirmed by trade sources show that the volume has remained in the region of 0.45–0.50 million cases since 1980.

The only change has been in the structure of supplies, with a large fall in the Japanese share of the market from about 275,000 cases per year between 1980 and 1984 to about 157,000 in 1985. The Japanese share dropped from over 50 percent of the market to only 30 percent. The Japanese decline has been offset by a corresponding rise in imports from Thailand. However, it is not accurate to deduce from this switch in supply that there has been a strong move toward trading down in this market, even though the small market share held by skipjack chunks has grown. The albacore solid pack appears to remain the market leader and accounts for over 50 percent of all volume. Top prices continue to go for whitemeat solid packs in olive oil.

The market is easily accessible provided the pack is of the highest quality.

No significant change is forecasted either in total volume or in the structure of the market.

MIDDLE EAST MARKETS

Little information is available on these markets, and the existing information does not greatly encourage potential suppliers.

The total market size is estimated at no more than 2 million cases, with the largest portion imported by Jordan and Saudi Arabia—the dominant markets. Egypt is known to be a significant market but is commonly regarded as being difficult to trade in because of (1) the general difficulties of a complex bureaucracy and (2) the highly fragmented nature of the retail food sector.

The largest share of the Middle East market is supplied by Safcol, though the Indonesian product is reported to be making headway in Jordan. The product is purchased in different ways, often through a government tendering system. It is understood that the large foreign trading companies in the region have handled a large quantity of product over the years, particularly the Korean trading houses (many of which are directly tied to construction companies).

In general terms, the majority of tuna consumers are expatriate workers, and the fall in oil revenues has hit this market heavily. The market may be classified as middle- to high-value. For a variety of reasons common to many trading sectors in the region, market penetration can be difficult to achieve, which is one of the reasons why the international trading companies have found themselves in the business; because they have gained access to one area of trading, they are better placed than a new competitor.

Clearly, these markets need closer examination, even though there appear to be few worthwhile incentives for Pacific island canners to commit resources to developing markets in the region. Despite the existence of middle- to high-value markets, they are not readily accessible and are in decline as a result of falling oil revenues.

GOVERNMENT/EEC POLICY

Health and packing regulations

Buyers generally dictate terms for product standards if the basic health criteria have been satisfied. Each EEC country has its own regulatory body that lays down the minimum specifications, and these should be referred to by suppliers at the outset. Most countries have fairly complex legislation on food regulations because they have to cover a wide range of eventualities.

Mercury levels and histamine levels should be verified by country as these vary slightly.

Both Sweden and Switzerland have strict codes involving registration of the exporting company with their own ministries of agriculture. Both countries have resolved to introduce minimum atomic count levels, though it is not yet known whether these levels have been set for canned tuna. However, several importing nations of both the frozen and canned product were concerned over the possible effects of the continued nuclear testing at Mururoa atoll in French Polynesia.

Generally, each importing nation requires health certificates and certificates of origin. Other specific conditions that need to be satisfied are described at the outset by the importer.

Rules of origin

The EEC commission is charged with policing the special relationship between the Asian, Caribbean, and Pacific (ACP) nations and EEC countries, particularly regarding preferential access to the markets. Effectively, the ACP exporter is allowed tariff-free access to the market, thereby avoiding a duty of 24 percent, which is levied on the product of non-ACP origin.

The tariff-free entry for the ACP product is granted only if the product satisfies certain criteria described in the Rules of Origin. These rules involve questions of percentages of added value within the ACP country and various other considerations. These issues regularly require that the commission find workable compromises between the exporting nation and itself and EEC members.

Insofar as tuna and tuna products are concerned, the critical question of origin is determined by where the product was caught and by whom. Within the national 12-mile limit (territorial seas) of the ACP country, the product may be caught by vessels of any flag. Once beyond the 12-mile limit, the product must be landed either by the ACP country undertaking the export to the EEC or by a vessel flying the flag of another ACP country. The product caught by non-ACP vessels outside the 12-mile limit may nevertheless be accepted tariff-free into EEC markets, subject to the successful granting of a derogation to the the Rules of Origin. The ACP government must apply for this derogation through the ACP Council, which in turn submits the application to the joint ACP/EEC Customs Cooperation Committee. The purpose of the Rules of Origin is to offer protection to the European canning industry, and applications for derogations are really of concern only to France, Italy, Spain, and Portugal.

In general, the business of seeking and negotiating derogations has been made more difficult with the recent accession of Spain and Portugal, two more tuna-producing countries, to the EEC. Any derogations that are granted are not being offered assurances of renewal, and failure to fully utilize the specified quotas may prejudice the renewal of a derogation.

IMPLICATIONS FOR THE PACIFIC ISLANDS REGION

Constraints and opportunities

The physical remoteness from the marketplace and the consequent problems of arranging shipping schedules convenient for buyers are obvious constraints, yet these are restated here based on comments from buyers. Partially associated with the physical remoteness (yet more a function of the limited volumes) is the matter of representation in the marketplace itself. Serious consideration has been given to this question in the period up to the withdrawal of C. Itoh from the operation of the Pacific Fishing Company (PAFCO). At this time, the entire issue of the cost of representation in the market was critically examined. In the past the producers have been exclusively represented by the Japanese trading company representative in the market at the agreed percentage charge. This arrangement, while costly in a low-profit-margin period for the industry, has met with the full approval of the buyers, who view the Japanese trading houses as adequate insurance against the ever-present risk of a health breakdown such as happened in the canned salmon trade. The small canneries are not considered to be sufficiently substantial to offer this sort of insurance to buyers.

The costs of maintaining permanent representatives of the companies in the European market are high, and it is difficult to see how this could be justified at the present volume levels, though the future may allow for this to be seriously considered.

The scale of production, the costs of production, and, in some cases, the availability of adequate supplies of the right quality of raw materials have placed Pacific island producers at a substantial disadvantage in comparison with the Thai canners in any price competition. Pressure from buyers to reduce prices inevitably results in a gradual decline in quality, and this is nowhere more clearly demonstrated than in the Thai industry. Plans for production growth in Indonesia, where labor costs are lower than in Thailand, could put even more pressure on the lower-value part of the market.

Clearly, the existing canneries in the Pacific islands cannot now or in the future compete at this level of the market. They appear to be limited to that part of the market that was traditionally held by the Japanese, and attempts to trade down in order to achieve higher growth rates lead inevitably to financial ruin. Trading up by means such as developing the yellowfin solid pack and selling an albacore solid pack in Europe could achieve growth, though this may prove difficult in the non-mature markets where the consumers' perception of product quality is not yet as well developed as it is in the mature markets. Trade sources indicate, however, that this perception is changing and that buyers will soon need to recognize a more educated consumer. This consumer will be less responsive to the price competition and increasingly disenchanted with the lowering of quality. The long-term results of such a trend would clearly be to the advantage of the Pacific canneries and the type of pack currently being produced.

CONCLUSION

While the overall growth across all European markets rose to a 7.6 percent annual average in the period 1981–85, it was not evenly distributed. The main growth in the market took place in the exclusively importing nations, or non-mature markets, notably the United Kingdom. The level of imports did nevertheless rise in the mature tuna canning countries. Although the barriers to entry for non-traditional suppliers are considerable in most of these markets, a breakthrough may be possible in Italy. The situation in France and Spain remains unchanged insofar as access is concerned, and no efforts at this time should be diverted to developing either market.

The key market for growth for Pacific island producers must be the United Kingdom, which offers a substantial premium-market segment despite the tendency to trade down.

The other European markets offering good prospects for premium Pacific pack are Belgium and Switzerland, the latter in the form of a whitemeat olive oil pack. A yellowfin olive oil pack should be costed and tested with Italian buyers.

There is a feeling in the trade that downward trading, which has seen the Thai packers make enormous gains in the markets, cannot continue indefinitely and that during the medium term buyers will detect a raised level of consumer perception of the product in the non-mature markets. Clearly, there are limits below which consumers will not go; this was evidenced in the West German market in recent years when resistance built up to the ever-lower quality dressing pack. There are good prospects for a limited amount of trading up in these markets as consumers gradually become more discriminating in their purchasing. This is not to say that price considerations will significantly diminish, and in fact the common complaint about buyers who demand top quality at middle-range prices will become even more prevalent. However, this trend could work to the advantage of the Pacific island canneries.

Pacific island producers should be aware of the implications of the accession on 1 January 1987 of Spain and Portugal, which means that there are now four tuna catching and canning countries in the EEC. Given the Rules of Origin, which are designed to protect EEC domestic industries, increased opposition can be expected to derogations from the Rules of Origin.

In general, Pacific island producers should continue the practice of targeting premium packs in these markets and should resist the pressure to trade down. New premium packs such as an olive oil pack should be developed and tested in both the Swiss and Italian markets.

7.
The Japan Tuna Market

Geoffrey P. Ashenden and Graham W. Kitson

ABSTRACT—This chapter examines and evaluates the Japanese tuna market and the demand for tuna. Government policies and regulations with respect to the tuna industry are reviewed. The chapter focuses on implications for those Pacific island countries that want to penetrate the Japanese tuna market.

INDUSTRY OVERVIEW

Introduction

The Japanese tuna market is based primarily on consumption of tuna in raw form—sashimi. The sashimi market demands large tuna species: bluefin, bigeye, and, to a lesser extent, yellowfin. The sashimi buyer is extremely sensitive to quality, which is affected by many factors but especially by fish size, fat level, and methods of handling. Larger sizes and higher fat levels are preferred. Because high fat levels are encouraged by cold water temperatures, the Japanese tuna industry does not believe that tropical waters can supply the highest quality of sashimi.

The Japanese canned tuna industry produces both for export and for domestic demand. The domestic market, which is presently limited, is expected to expand as younger Japanese eat more western-style dishes. The preferred species for Japan's canned market is yellowfin, which is growing in sales; the demand for canned skipjack and albacore, however, is limited.

A significant but static demand exists in Japan for smoked and cured skipjack loins, a product known as *katsuobushi* (known in sliced form as *kezuribushi*). A small demand also exists for other distinctive and traditional products processed from skipjack.

Several significant developments have recently occurred in the Japanese market for both canned and sashimi tuna. The major influence, which is expected to be long term, is the upward movement of the yen against most world currencies. The Japanese canned tuna producers can no longer compete against other suppliers on the international market; furthermore, Japan itself has started to import canned tuna from Thailand. Thus, while the Japanese canned market may be expanding, the Japanese canned tuna industry is declining, and its viability is at risk.

Competitive forces have not been so severe for the domestic sashimi tuna industry because the catching and processing technologies are highly specialized, but competition from Korean and Taiwanese longliners has had a marked influence, especially for the lower-quality sashimi species—yellowfin and bigeye from tropical waters. Despite voluntary quotas on Korean imports, stocks of lower-grade sashimi have increased markedly (especially in 1984 and 1985), thereby depressing prices. Market distinctions between high- and low-quality sashimi tuna are now more rigid.

A complementary development has been the growth in chilled sashimi tuna imports by air—bluefin from the United States and the Mediterranean; bluefin, yellowfin, and bigeye from Australia; and yellowfin from Taiwan and the Philippines. Chilled, air-freighted sashimi tuna imports, while currently only a small proportion of all tuna imports, are expected to increase by two or three times due to increasing market demand, coupled both with lower freight costs through the stronger yen and with the use of simplified fishing techniques (such as purse seining and set nets) to lower fishing costs while retaining quality levels. Growth has also occurred for air-freighted loins from the large Mediterranean tuna species.

The distribution system within Japan for tuna products is complex; it is dominated by specialist tuna trading companies—called *maguro shosha*—and the large trading companies or fishing companies with which many of them are affiliated. In addition, new trade channels have evolved outside the traditional wholesale auction market system.

Overview of products

Skipjack. Among all tuna species, the catches and landings are highest for skipjack. The Japanese catch over the last ten years has ranged from 259,000 to 446,000 tonnes, showing considerable year-to-year variations. Exports of the catch from Japanese fishing vessels while at sea (or in foreign ports) and exports of frozen skipjack from Japan have reduced the supply to the domestic market annually by between 20,000 and 90,000 tonnes.

The annual production of canned skipjack has consumed between 60,000 and 110,000 tonnes of whole-fish weight in recent years; this level is expected to drop as canned exports decline. Imports of canned skipjack fell to zero in 1985.

Traditional Japanese dried/smoked products (*fushi*) have used an estimated 160,000 tonnes of skipjack per year. The production volume of *fushi* products is not expected to increase significantly in the future. Imports of *fushi* products (skipjack and mackerel) amounted to 1,566 tonnes in 1985, down from nearly 2,000 tonnes in 1980. The main suppliers in 1985 were Taiwan, Thailand, and Solomon Islands. Imports from Solomon Islands peaked in 1983 at 308 tonnes.

Other traditional skipjack processed products take smaller volumes of fish. Fish products include heat-seared (*tataki*), soy-pickled (*tsukudani*), and fermented/salted (*shiokara*).

Consumption of skipjack in fresh or frozen form has ranged from 40,000 to 100,000 tonnes per year since 1975, depending on the excess of supply over processing and export demands.

Bluefin. The most important tuna species is bluefin, although the volumes are much lower than those for bigeye or yellowfin. Bluefin is the highest grade of sashimi species.

Catches of bluefin have declined by one-half between 1980 and 1985 to 30,000 tonnes. Catches of small bluefin (*meji*) have ranged from 17,000 to 20,000 tonnes. Imports have added 3,000 to 5,000 tonnes, mostly of frozen fish. Total supply to the market from all sources has ranged from 50,000 to 65,000 tonnes in 1980–85. The share of the market held by the species is limited by availability of supply; more fish could be sold if it were available. There are no exports.

Bluefin not consumed fresh is frozen at supercold temperatures to preserve quality. Freezing takes place at sea. A small volume of fresh catches is frozen onshore in Japan. The carcass is processed into major cuts while frozen or fresh. All bluefin is consumed as small consumer cuts without any processing by way of flavoring or heating.

Bigeye. Bigeye is the major tuna species in terms of catch. Volumes caught by Japanese vessels since the mid-1970s have ranged from 111,000 to 149,000 tonnes, with a common annual level of around 130,000 tonnes. Exports transshipped at sea from Japanese fishing vessels have been small volumes.

Imports of bigeye are significant; frozen volumes have ranged from 42,000 to 52,000 tonnes annually since 1981, while 1,000 to 2,000 tonnes arrive chilled each year. There are no exports.

Total supply to the Japanese market this decade has ranged from 154,000 tonnes in 1980 to 202,000 tonnes in 1985. Bigeye is important for its quality, and it is consumed raw as sashimi in the same manner as bluefin. Like bluefin, bigeye is processed only by cutting. Only a very small volume of fresh bigeye is frozen onshore.

Yellowfin. The yellowfin catch is second largest of the tuna species. The Japanese fleet's catch volumes since 1980 have been more than 110,000 tonnes, with its 1985 catch increasing to 134,000 tonnes. Exports at sea have been only a few thousand tonnes annually, and exports of frozen yellowfin from catch landed in Japan have ranged from 4,000 to 9,000 tonnes.

The supply of yellowfin is increased substantially by imports, with the frozen product ranging from 25,000 tonnes in 1980 to a record 62,000 tonnes in 1985 and with chilled imports adding between 10,000 and 13,000 tonnes per year. Total annual supply of yellowfin on the Japanese market was 150,000 to 170,000 tonnes between 1980 and 1984 and reached 209,000 tonnes in 1985, when both the Japanese fleet catch and imports increased.

Most yellowfin is consumed fresh (or thawed) in the form of cuts as sashimi (115,000 to 120,000 tonnes per year). Volumes going to the canned product increased from about 15,000 tonnes whole weight in 1980 to about 56,000 tonnes in 1984.

Albacore. Albacore catches by the Japanese fleet have been between 50,000 and 90,000 tonnes per year. Exports at sea from Japanese fishing vessels have been minimal; exports from catch landed in Japan have varied widely from 1,000 to 15,000 tonnes since 1975. Imports of frozen albacore have ranged from 2,000 to 4,000 tonnes, except for about 8,000 tonnes in 1983, when the Japanese catch was low. Total supply of albacore since 1980 has been close to 70,000 tonnes, except for 1983 and 1985, when supply was 60,000 tonnes or less.

Whole-fish usage for canning has fallen from a level of around 60,000 tonnes in the early 1980s to around 50,000 tonnes in 1983–84. Volume is likely to decline further, and the yen's strengthening will probably affect canned exports and encourage imports from lower-cost countries.

The residual volume of albacore consumed in fresh (or thawed) form is estimated to be about 5,000 or 6,000 tonnes per year.

Patterns of processing sashimi cuts

The processing of tuna species for sashimi involves the progressive cutting of the flesh into cuts and portions.

The pattern of processing sashimi has changed in Japan over recent years. Formerly, the breaking down of whole tuna into smaller cuts was undertaken mainly by subwholesalers operating within the major fish wholesale markets. But now the *maguro shosha* companies break down a large and increasing proportion of sashimi tuna for direct distribution to supermarkets or restaurants, usually in the form of a frozen loin or chunks, which have been cut by a band saw. Subsequent breaking down into retail cuts is done in the outlets after the loins have been thawed.

Canned tuna and skipjack production

Production of canned tuna has shown consistent growth from a 60,000- to 66,000-tonne level in 1978–80 to 86,000 tonnes in 1985.

Domestic canning operations in 1984 took 25 percent of skipjack supplies (110,000 tonnes of whole-fish weight), 32 percent of yellowfin (56,000 tonnes), and 74 percent of albacore (49,000 tonnes). Exports of fresh or frozen tuna as canning raw materials in 1984 consumed 72,500 tonnes of skipjack and lesser quantities of albacore (12,100 tonnes) and yellowfin (4,100 tonnes).

Tuna in oil represented about 77 percent of the canned tuna production in 1985; this percentage has increased every year since 1977. Flavored canned tuna has declined consistently as a proportion of all canned tuna to about 12 percent of production in 1985.

Total production of canned skipjack generally remained at a level of 28,000 to 36,000 tonnes during 1975–85. Production reached exceptionally high totals—42,000 tonnes in 1983 and 51,000 tonnes in 1984. Since 1980 the production of skipjack in oil has increased.

The decline in Japanese canned tuna exports since 1985 has resulted in pressure for sales in the domestic market, and the increased local competition has caused large layoffs in the Japanese canning plants and severe financial difficulties for the industry in general. The domestic canned market could be affected further by competition from Thai imports.

Major sellers

For fresh tuna and skipjack, the major sellers in the domestic Japan market, apart from the fishing companies, are wholesalers and subwholesalers operating in the landing markets and in the consumption area fish markets. These fish market dealers sell to restaurants, institutions, and retail shops.

In addition to the fish market dealers, the specialist tuna traders or *maguro shosha* and the major general fishing companies and trading companies are involved in the frozen tuna trade. This interest has spread because trading in frozen tuna does not need to occur within the market system used for fresh fish. The fishing and trading companies generally work through special, affiliated *maguro shosha* and/or fish market wholesalers.

Some of the landing market subwholesalers have developed businesses as *maguro shosha*, engaged in transactions in and distribution of frozen tuna outside the fish markets.

Toyo Reizo (a Mitsubishi affiliate operating freezer warehouses) has by far the largest share of the sashimi business held by any one company. Nippon Suisan (a fishing company) is a major operator, as are Hassui and Yashima, both *maguro shosha*.

For canned tuna, the specialist producer, Hagoromo, holds the largest share of the domestic market (50 percent); another specialist company, Inaba, ranks second with a 12.7 percent share. These companies do not export. The company producing the fifth-ranked brand (SSK) recently went into receivership.

The Japanese general trading companies and the six largest fishing companies (known as the 6 majors) are not involved in the direct production of canned tuna within Japan. A number of these companies do commission the production of canned tuna, which they sell under their own labels. Significant shares of the domestic market are held by brands of Nippon Suisan, Taiyo Gyogyo, and Nichiro—all large fishing companies. Although none of these companies is canning tuna in Japan, some are involved in overseas canning ventures.

Promotion and market development

Promotion of sashimi tuna tends to be generic for the industry as a whole rather than for individual brands. Recently, this promotion has been undertaken by the Federation of Japan Tuna Fisheries Co-operative Associations (Nikkatsuren), which in 1986 budgeted $1.25 million for advertising and promoting sashimi tuna as a health food.

For companies involved in the sashimi tuna market, the securing and nurturing of distribution outlets can effectively build the market share. The key to distribution has been control of supercold storage facilities; other important factors are (1) consistent supplies of a superior quality product, (2) the provision of extended trade finance, and (3) regular personal follow-up visits with customers and the creation of a network of personal obligations.

Canned tuna lends itself to brand discrimination more readily than sashimi tuna. Some canned brand promotion is undertaken by major companies, including the fishing companies Taiyo, Nippon Suisan, and Nichiro, as well as the large specialist canned tuna manufacturers Hagoromo, Inaba, and Shimizu. This promotion is primarily via magazine and in-store advertising.

Promotional effort for canned tuna is frequently associated with product innovation or modification. With canned tuna now being imported from Thailand, the Japanese producers are expected to respond by concentrating both on promoting the quality of the Japanese product and on packaging innovations.

Tuna distribution

Structure and control of the system. Sales of tuna and skipjack catches by Japanese fishermen can occur through a primary wholesaler in a landing market, to a *maguro shosha,* to a major fishing company, or to Nikkatsuren.

Distribution of fresh tuna is through the wholesale market systems at the landing ports and then onto the wholesale markets or major outlets in a city's consumption markets. Because most fresh tuna is caught locally, a significant proportion is sold and consumed in the port area at which it is landed. These ports are much more numerous than ports for frozen sashimi tuna.

A unique market structure has emerged for the distribution of frozen tuna, dominated by companies with supercold (−55°C to −60°C) storage facilities, which are located mainly at the principal landing ports of Shimizu, Yaizu, and Misaki. These specialist tuna trading companies (*maguro shosha*) generally are affiliated with either a major Japanese fishing company or a general trading company (*sogo shosha*). *Maguro shosha* also can be described as processing companies because of their role in both storing and breaking down tuna for subsequent distribution. Their linkages with other levels of the distribution chain are shown in Figure 1.

The distribution of frozen sashimi tuna from the landing ports by the *maguro shosha,* landing-market dealers, or major fishing companies can be direct to supermarket chains and large restaurants or restaurant chains, to central wholesale markets in major consumption areas, or to wholesale markets in other regions. Distribution also can occur directly from boat owners to wholesale markets in major consumption areas.

The major effort in winning a market share in frozen sashimi tuna relates to changes in distribution. Recently, rapid growth has occurred outside the wholesale market system through direct trade with supermarkets and restaurants. This business was developed initially by general trading companies through their *maguro shosha;* it has been emulated by those fishing companies that have a traditional strength in fish distribution through their financial linkages with primary wholesalers in the major markets. In addition, by virtue of their past

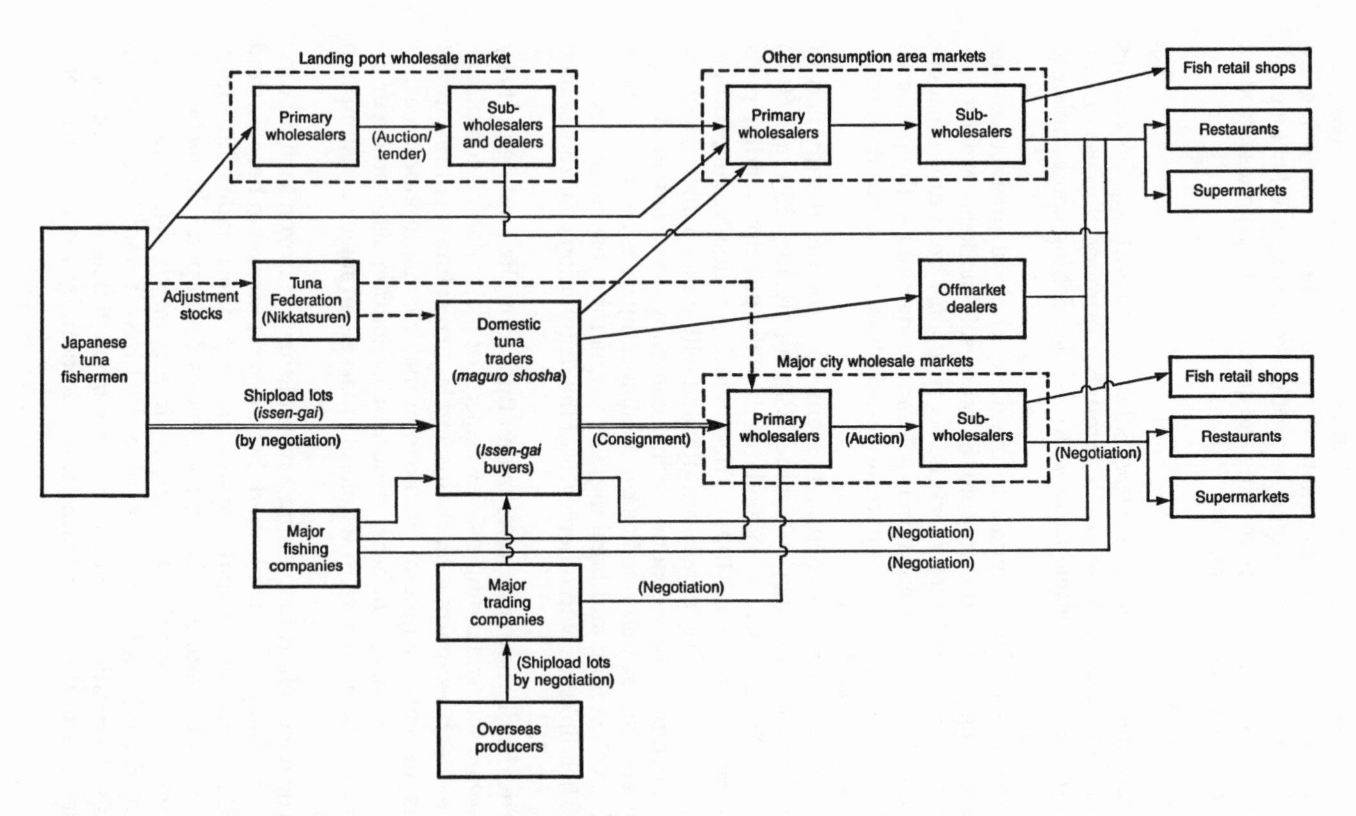

Figure 1. Distribution flows in Japan for frozen tuna

experience in conventional cold-storage ownership, the fishing companies have become major frozen food companies in Japan and have developed strong distribution links with supermarkets and other retail organizations.

Despite the growth in direct sales to retail outlets by the tuna traders, the wholesale markets in major cities remain important in the overall structure of sashimi distribution. This importance is reinforced because the major fishing companies work closely with financially affiliated wholesale companies that operate in the major wholesale markets.

Imported chilled tuna is usually sold through wholesale markets in the major consumption areas, but the more innovative importers distribute directly to supermarkets, restaurants, and regional wholesale markets.

For frozen skipjack, three-fourths of catch volumes are landed at the port of Yaizu, which is also the home of most tuna canneries and *katsuobushi* plants in Japan. At Yaizu about 60 percent of frozen skipjack sales are auctioned through the primary wholesalers, with about 10 percent by written tender and the balance as shipload sales. The Yaizu auction market is operated by the Yaizu Fisheries Cooperative Association, which takes a 2 percent commission on frozen skipjack sales. About 280 subwholesalers operate in this market, seven of which are large organizations acting for the ultimate buyers, mainly the processors.

The distribution system for fresh skipjack is similar to the system for fresh tuna.

Consumption area markets for wholesale fish operate like the landing-area fish markets at two levels. The primary wholesalers sell to secondary wholesalers (subwholesalers), which buy tuna that they break down into loins, chunks, and small cuts for sale to retailers and restaurants. The Tsukiji market (Tokyo) is the largest fish wholesale market in Japan and accounts for close to 25 percent of total tuna and nearly 40 percent of all southern bluefin traded in Japan. In the Tsukiji market, there are 5 primary wholesalers, 1,200 subwholesalers, and 300 large-scale "users"—processors, institutions, or supermarkets. These users can buy at auctions in direct competition with the subwholesalers, but they are not permitted to resell within the wholesale market. About 80 percent of total tuna sales made by primary wholesalers are to secondary wholesalers.

Both canned tuna and *katsuobushi* (or the *katsuobushi* shavings called *kezuribushi*) have a durable product, and distribution is through conventional wholesale/retail linkages.

Canned tuna distribution can be direct from processors to major retailers, or it can be through the established distribution channels of the major fisheries companies that have tuna processed under their own labels. Tuna canneries are specialist in their activity; that is, they produce only canned tuna. These companies rarely have outside capital affiliations.

The *katsuobushi* industry also is small in scale, and the large fishing or trading companies have little involvement. Sales of *katsuobushi* are by auction in consumption regions such as Tokyo, where the product is subsequently sold (often after being sliced) by small brokers located near the auction market. A growing proportion of *katsuobushi* is sold by the processors directly to other food manufacturers as input for other processed foods.

Transportation and storage of frozen sashimi tuna

On unloading in Japan, frozen sashimi tuna is taken by refrigerated trucks to nearby supercold storage warehouses and graded by species into five weight grades. Because grades are separated by species, boat, and size, frequently more than 1,000 different lines must be stored, allowing for only about a 30 percent storage utilization. Transportation within Japan is mainly by refrigerated trucks.

Inventories. Inventory levels for frozen sashimi tuna are cyclical and vary with price and resource availability.

Catches and stock levels of higher-quality bluefin have declined because of exclusive economic zone (EEZ) limitations, the increased costs of longlining, and the Japanese government's reduction of transshipment quotas. The 1986 bluefin catches were expected to be only about 15,000 tonnes against the normal 20,000 tonnes. Loss of access to traditional distant-water bluefin fishing grounds has forced Japanese longliners into catching more yellowfin and bigeye tuna, with a resulting accumulation of stocks of these species.

Because chilled tuna has a limited shelf life, distribution must be rapid; thus there is a limited need for holding an inventory.

DEMAND OVERVIEW

Demand for tuna and skipjack products

End-use sectors. Distinctions can be made between sashimi tuna species by the type of end use. The highest-quality northern bluefin is consumed exclusively in high-class restaurants; the lower-quality product is marketed as household sashimi. Southern bluefin also is used exclusively by restaurants. About 50 percent of bigeye is used in restaurants and 50 percent as household sashimi, while most yellowfin sashimi is consumed in households.

Skipjack is eaten as sashimi in households mainly when it is caught from domestic waters during the springtime in Japan. Other end uses for skipjack outside of canning and *fushi* are limited.

Regional supply and demand for tuna and skipjack

Tuna landing markets. The ports of Shimizu, Yaizu, and Misaki (all close to Tokyo) accounted for 90 percent of all the Japanese fleet's frozen tuna landings in 1984. Shimizu with 39 percent of landing volumes is the most important. Three-fourths of all tuna landings during 1984 were frozen tuna.

From the total frozen tuna volumes landed, about 73 percent went into cold storage near the landing port, about 17 percent went directly to canners and other processors in the region, 2 percent was thawed and consumed fresh in the region, and 7 percent was trucked to other markets in Japan.

The major ports for fresh tuna landings are located north of Tokyo, and together Katsuura and Shiogama have 36 percent of landings. Yaizu has 12 percent and Shimizu 4 percent. Landing ports are more numerous for fresh tuna than for frozen tuna. For fresh tuna 61 percent is transported immediately to other markets beyond the landing port, with about 13 percent consumed fresh at the landing region and the balance frozen or processed locally in the region.

Skipjack landing markets. Yaizu was the landing port for 75 percent of frozen skipjack in 1984, followed by Shimizu with 8 percent. Of the total product landed at all ports, about 12 percent went to cold storage, 25 percent to canning plants, and 44 percent to other processors, mainly *katsuobushi* plants. Only 19 percent was transported immediately outside the landing port regions.

The major landing ports for fresh skipjack in 1984 were Kessenuma, Choshi, Numazu, and Nakaminato.

Yaizu is the most significant port in the Japanese tuna industry, taking 75 percent of frozen skipjack, 7 percent of fresh skipjack volumes landed, and 38 percent of the frozen tuna caught by the Japanese in 1984. Yaizu is the home port of the majority of the 35 large Japanese purse seiners. Most of Japan's skipjack storage and processing facilities are located in the Yaizu region. Processing facilities include canneries and *katsuobushi, tataki,* tuna pickle, and *shiokara* producing plants. Yaizu is the major distribution point for these products.

Final consumption markets. Of the main regions, the greatest per capita demand for tuna species is found in the region centered on Nagoya, followed closely by the Tokyo region. Regional differences do occur, with the most notable being the bias in consumption in the Nagoya and Osaka regions toward yellowfin rather than bluefin or bigeye (particularly in comparison with Tokyo).

The per capita consumption statistics do not directly show the importance of various regions in total volume consumption because of differences in population concentrations. The general region of Tokyo is the largest consumption region in total for all tuna, followed by the Osaka and Nagoya regions. The Tokyo and Sendai regions dominate the total skipjack consumption.

Based on volumes passing through the central markets during 1984—albeit a record year—the Tsukiji market accounted for 52 percent of all tuna handled by the top ten central wholesale markets in consumption areas or major cities throughout Japan. The next-highest central markets in 1984 were Osaka with 14 percent and Nagoya with 12 percent. Yokohama, which is also in the Tokyo general region, accounted for an additional 8 percent.

Since 1976, household consumption of skipjack has leveled off at between 1,350 and 1,400 g per household. Household consumption of *katsuobushi* is on the decline, having fallen from 4,696 g per household in 1975 to 2,870 g in 1984. In total consumption, however, this decline has been compensated for by an increase in institutional and food manufacturing use. Trends in consumption of *katsuobushi* have moved strongly toward production of sliced unseasoned *fushi,* sold either to the household market or to the institutional market.

Exports. Exports of tuna and skipjack consist almost entirely of the canned product. Export volumes in the five years prior to 1983 consistently ranged between 34,000 and 38,000 tonnes per year. In 1984 these volumes reached 45,582 tonnes but in 1985 retreated back to pre-1984 levels—34,000 tonnes, of which some 19,000 tonnes were canned skipjack. During the first five months of 1986, an additional decline of 19 percent occurred over the same period in 1985. The prime reasons for the decline have been the increased competition from Thailand and a subsequent strengthening of the yen against most of the world's currencies.

Between 1978 and 1985 canned tuna exports as a proportion of canned tuna production fell from 47 percent to below 30 percent, except in the 1984 boom year, when exports represented 35 percent of Japanese production.

Exports of frozen tuna by Japanese longliners through sales at sea include only the albacore tuna species.

Demand and price determinants

The total tuna consumption increases sharply in December, with a post-New Year decline; other months have a more stable market. Consumption of sashimi tuna tends to be seasonally concentrated with celebrations of the New Year, *obon* in July, and school graduations or company promotions in April. For skipjack, most of which is eaten fresh, consumption tends to be a function of availability, with peak months in April, May, and June. During these months the quality is also superior (especially the fat content).

In an examination of the demand factors, it is important to note that distinctions exist not only between species but also between individual fish of the same species, giving rise to different tuna products with different price relativities. The value of each product is determined on a case-by-case basis by buyers and sellers.

For sashimi tuna, the various cuts have different market values, with the major distinctions being drawn between the different portions of the fish, each with differing fat levels. Market values are also set for similar cuts by the variations in fat levels between species. Sashimi tuna prices also vary according to fish size, catch region and season, freshness, and other quality features, including moisture content, flesh texture, visual appearance, and, of course, taste.

Prices for sashimi tuna are influenced by the way the product is handled. Buyers prefer to deal with the experienced tuna producers who have a good reputation for preparing tuna for the market and who have facilities on board to adequately handle tuna.

Demand elasticity and outlook

One important factor in future demand for tuna is a general move in consumption away from rice with fish and toward bread and livestock products.

In official estimates of future demand, however, the overall Japanese fish and shellfish consumption is projected to increase by 40 percent between 1979 and 2000, a consequence partly of population growth and partly of income growth.

The pattern of fish consumption does vary between species, and a general decline is occurring in consumer interest in species at the lower end of the quality spectrum—particularly mackerel and cod.

The tuna species, however, are at the higher end of the quality spectrum, and demand is expected to show continued growth, with a bias toward the upper end of this market and with frozen bluefin and fresh chilled bluefin, bigeye, and yellowfin being the major recipients of demand pull. Non-sashimi species at the lower end of the quality spectrum may be expected to decline. The recent buildup in stocks of frozen yellowfin and bigeye appears to have lowered the general market appreciation of these species in frozen form.

Consumption and apparent demand volumes are influenced by supply volumes. Large volumes of lower-priced, less-preferred species are caught locally in Japan and landed fresh; if supply volumes increase considerably, the sellers must lower their prices in order to clear volumes. Although statistics can show an increased volume of fish as having been demanded and sold, the reality may be a considerable sacrifice in price to achieve sales.

Many traditional processed products are experiencing low or no growth in Japan. Demand for the tuna processed products *katsuobushi*, *namaribushi*, tuna pickle, and *shiokara* is unlikely to increase significantly. Demand for domestic canned tuna is likely to increase as salads become more popular and also as the perception grows that canned fish has health-related benefits.

Price histories for tuna and skipjack

Price variations. The main price differentials within species occur for different size ranges and for quality assessments based on factors such as catch region, season, catch vessel, and handling methods, as well as individual fish quality attributes. Higher premiums in general are paid for larger weight ranges, for fresh rather than frozen sashimi tuna (except for bluefin), and in major consumption areas as compared with landing ports. Higher sashimi tuna prices also result from factors such as enhanced color preservation and higher fat content.

The result of these influences is such that average price statistics tend to obscure more significant price variations, which may occur at any time, so that the data are significant only with respect to general trends. The effective monitoring of price trends can be achieved by selecting one or more key species for Pacific islands interests and by monitoring the monthly price-range data for the product from the leading vessels operating in selected fishing regions, as well as for major landing ports and central market outlets for that particular species.

Landing-port prices

Prices for the three prime sashimi tuna species (two bluefin species and bigeye) peaked in 1983–84 and then declined in annual average price in 1985. (This decline reversed during June–September 1986.) The differential has been growing between the price for southern bluefin and the prices for other species, including Japanese bluefin. Yellowfin prices peaked in the mid-1970s and have declined since 1980. Skipjack prices also declined from 1980 to 1984, showed a slight upward movement in 1985, and then fell again in 1986.

Consumption area prices

Price movements in the consumption regions generally reflect price movements in the landing regions, although the reverse is also true. Prices in consumption regions for fresh bluefin are lower on average than for frozen bluefin, but the differential is less than in markets at the landing ports. Prices for frozen bluefin in the landing regions follow the major city wholesale market prices quite closely. For the other species the prices in the landing regions are markedly lower on average than in the major city markets.

Sashimi tuna prices in Tsukiji (the major city wholesale market for sashimi species), as with the landing-port price trends, rose to a peak in November 1984. Prices since then have registered a decline of about $3 per kg so that they were lower in September 1986 than in August 1982. This decline affected all species over the 1984–85 period except the minor species of albacore, small bluefin, and small yellowfin. Fresh skipjack prices also increased markedly between 1984 and 1985. Substantial premiums exist in Tsukiji for fresh sashimi tuna over frozen sashimi tuna, except for southern bluefin. The average differential per kg in Tsukiji in 1985 was bluefin $4.80, bigeye $6.15, and yellowfin $1.45. For frozen southern bluefin the price was actually higher in 1985 by $6.50 per kg, a reflection primarily of the difference between domestic catches (Japanese supercold longliners) and imported fresh/chilled bluefin.

Import prices. In 1985, the cost, insurance, and freight (CIF) prices for imported frozen bluefin, bigeye, yellowfin, and albacore declined from 1984 prices, reflecting overall market conditions; prices for frozen skipjack increased in 1985.

Fresh/chilled sashimi tuna imports consist generally of yellowfin. Average CIF prices for yellowfin have been fairly stable over the 1980–85 period. Fresh/chilled bluefin CIF prices have increased each year since 1981. Fresh/chilled bigeye import prices have declined slightly from the 1983 peak in total, but the U.S. product has experienced increasing prices since 1981.

Retail prices. A large difference in retail price exists between the best bluefin cuts and yellowfin sashimi. Resource valuations are affected accordingly.

Among the processed products, *katsuobushi* retail prices increased through the 1970s until 1983. The 1975 average price was $0.75 per 100 g as compared with $0.98 in 1983. The retail prices for canned tuna have increased marginally in recent years. For tuna in oil (No. 2 can), for example, the price for one brand in January 1983 ranged from $1.86 to $2.18 and from $1.97 to $2.15 in May 1985. Considerable variation does occur between brands, and prices have been substantially discounted.

Recent developments in sashimi tuna imports

Air-freighted, chilled whole tuna. Substantial growth is expected in volumes of chilled sashimi tuna packed in ice and air-freighted from distant fishing grounds. This expectation is based partly on the bias in demand toward the higher-quality end of the market and partly on the market's acceptance of simplified catching methods; in addition, improved transportation has resulted in lower costs.

This combined method of fishing and transporting tuna does not require expensive supercold on-board freezer facilities, and because the vessels are always within one or two days of a port, there is little non-fishing time. Fisheries of this type operate from New York (bluefin), the Mediterranean (bluefin), eastern Australia (bluefin, bigeye, and yellowfin), Philippines (yellowfin), and Taiwan (yellowfin and bigeye).

Chilled, air-freighted sashimi tuna imports currently are about 3,000 tonnes per annum for yellowfin, about 2,000 tonnes for bigeye, and about 1,000 tonnes for bluefin. Volumes are expected to triple in the near future. Growth in this trade is expected in all species, with yellowfin forecast to reach 30,000 tonnes and ultimately to be the largest individual species traded. Chilled bluefin tuna imports will continue from other existing sources: Spain, Norway, Italy, Turkey, and the eastern United States. Growth is also expected from New Zealand. The best months for shipment for all species are from December to April.

Prices for air-freighted, chilled sashimi tuna vary considerably. Yellowfin, for example, varies from about $1.88 to $15.60 per kg, but the Philippine yellowfin varies between $1.88 and $9.40 per kg. Margins for the Philippine yellowfin for Japanese importers are too narrow to warrant outright purchase, and Philippine exporters ship on a consignment basis. Bigeye from the eastern United States varies from $18.75 to $31.25 per kg, with fish with lower fat level fetching from $3.15 to $12.50 per kg. Bluefin from the Mediterranean and the eastern United States varies from $12.50 to $62.50 per kg. Smaller Australian bluefin cost only $3.15 per kg in 1986.

Air-freight and other distribution costs to Japan also can vary; for example, the costs from Australia in late 1986 were in the range of $2.50 to $3.00 per kg. When the air-freight charges were paid in yen in advance (using prepaid bills of lading), the increasing strength of the yen meant a substantial reduction in air-freight costs in 1986.

A 5 percent duty payable for chilled tuna imports can be avoided by removal of the fish head. The reception for large U.S. bluefin marketed in this way has been satisfactory.

Chilled and frozen sashimi tuna loins. In some cases, rather than ship the whole fish, only the loin portions are shipped, thus reducing transportation costs. Sashimi tuna loins (chilled and frozen) currently come only from large bluefin from the Mediterranean. Attempts to import loins from smaller species have failed. Shipments of frozen yellowfin loins from South Korea also failed because of color changes during processing. Distribution in Japan of sashimi tuna in loin form is difficult. Where tuna is sold through auction, the system is geared to the whole tuna.

Frozen sashimi tuna imports. Consistent volumes of yellowfin and bigeye have been imported from South Korea and Taiwan. Yellowfin imports from Solomon Islands have increased to over 5,000 tonnes per year. The United States in 1985 supplied nearly 10,000 tonnes of yellowfin. Imports of frozen bluefin tuna from Australia have increased to over 3,000 tonnes per annum. Frozen sashimi whole-tuna imports have also begun from other countries.

Imports of frozen skipjack from Solomon Islands increased from 3,000 tonnes in 1981 to 10,000 tonnes in 1982, but trade has since declined and virtually ceased by late 1985.

GOVERNMENT POLICY AND REGULATIONS

Policy aspects

The principal objective of the Japanese government's tuna policy is to allow adjustment to occur within the market and industry in such a way as to minimize damage to various sectors of the Japanese industry. This policy involves reducing the fishing effort in many regions while ensuring that the reduced effort does not cause a collapse of any associated sectors. The policy has meant that adjustment to a reduced scale of industry has been slow.

The level of transshipments has been regulated by the Japanese government, and volumes have been reduced in recent years to slow the fishing effort; the government concurrently has taken steps to ensure that reduced Japanese capacity will not be immediately replaced by increased foreign capacity and has restricted sales of secondhand Japanese longliners to vessels over ten years old. Similarly, vessels of different types are licensed to operate in specified regions, again to reduce the fishing effort.

The Japanese government is subject to considerable pressure from interest groups in the tuna industry. It is lobbied by various tuna fishing groups for favorable treatment (at the expense of other groups) with respect to numbers of boat licenses, access to fishing grounds, and periods of access; it is petitioned to subsidize sales of canned tuna to developing countries; it is lobbied by groups, including seamen's unions, to reduce or stop tuna imports or replace existing voluntary Korean import quotas with formal quotas.

In general, the Japanese tuna industry and the Japanese fishing industry do not have a great deal of influence with the central government as compared with the agricultural industry.

Government food regulations

Regulations for food standards in Japan are administered by the Ministry of Health and Welfare. The ministry has established standards for food additives, for bacteria counts, and for packing materials. In the case of exported canned tuna, the ministry's regulations also apply to packaging and labeling.

Japan Agricultural Standards (JAS), established by the Ministry of Agriculture, Forestry and Fisheries (MAFF), also apply to tuna products. These standards of quality provide higher levels of consumer protection. JAS is voluntary and relies on consumer preference for its level of acceptance by food processors. Only plants and products that reach certain standards can use the JAS seal of quality. Eligibility involves product sampling and grading by an organization approved by MAFF, plant inspection, and the achievement of quality control standards. Foreign producers can apply for listing as approved JAS plants.

The critical factors of product quality that affect prices and demand in the marketplace are not related directly to government-imposed standards. The demands of buyers and the pressure from competition give rise to the fine differentiations in product quality and prices.

IMPLICATIONS FOR THE PACIFIC ISLANDS REGION

Canned tuna

The ability of Pacific island countries to participate in the Japanese canned tuna market will depend on two factors. The first is whether imports from Thailand will continue to be a commercial success. Indications are that they will be, and if so, Japanese distributors will become interested in canned tuna imports, and volumes will quickly grow. Distributors state that preference will be given to imports from Thailand until alternative suppliers are proven.

The second significant factor is the ability of the Pacific island countries to compete with Thailand in production cost and product quality; some Pacific island canneries already have difficulties in competing on other worldwide markets.

Apart from the distributors, the Japanese canned tuna producers and importers also express doubt over Pacific island countries' ability to compete with Thailand and show little interest in investing in the region's cannery operations. The Japanese specialist canning companies do not have the financial strength to make Pacific island cannery investments even if they did find such operations attractive; their own future is already in doubt.

Sashimi tuna

The opportunities for Pacific island countries to participate in the chilled tuna business will depend primarily on their sashimi tuna resource and access to air transportation to Japan. Tropical Pacific island countries face a major difficulty in overcoming two perceptions of the Japanese market: (1) that tropical regions produce only tuna with low fat levels, which is thus suitable only for low-grade sashimi, and (2) that the tuna caught by Pacific island operators would be inadequately handled because the operators lack the skills and equipment to handle sashimi tuna. While these perceptions persist, little support can be expected from the Japanese to develop this business, except for technical assistance. This assistance will lack commercial commitment.

The claim that the quality of Pacific island sashimi tuna is low could vary with location of the catch. For example, the deeper-water yellowfin in Solomon Islands' region is thought to have higher fat levels than other resources. The ability of a Pacific island country to participate in the sashimi market could be enhanced by access to bigeye rather than lower-grade yellowfin.

Without a commercial commitment from Japan, the Pacific islands will have to develop trade on a consignment basis in the way that yellowfin is handled from the Philippines. The principal risk is that the returns will not cover the costs of catching and shipping the tuna. The risk is likely to be greater in the initial phases, when returns will be influenced by the initial lack of credibility for shippers, with prices varying greatly according to product quality, and with a greater possibility of technical mistakes. An important requirement in developing a business in air-freighted, chilled tuna, therefore, is the establishment of a structure that can accept the risks involved during the development phase.

For frozen sashimi tuna, risks are also involved in developing trade with Japan. The major element at risk is the capital expenditure on small secondhand supercold longliners. These vessels currently cost about $150,000 each. Technical assistance and management input from Japanese sources also are required. As with chilled sashimi tuna, those countries that are endowed with higher-quality sashimi tuna resources will have the advantage.

Skipjack

Additional opportunities may exist for those Pacific island countries with skipjack resources to participate in Japan's *katsuobushi* market. Although this market is not growing, Japanese producers are expected to find increased difficulty in competing with foreign competitors in their own market. Successful operations already exist in Pacific island countries, and more could be established. The Japanese health regulations regarding *katsuobushi* processing are stringent, and initially the product probably would need to be the semifinished form, *arabushi*, rather than *katsuobushi*. A subsequent option could be *kezuribushi*, the sliced form of *katsuobushi*.

Participation in the market

Because of the complexities of the Japanese tuna distribution system and the dominant position of the major operators, Pacific island exporters are unlikely to initiate distribution operations within the Japanese domestic market or to encourage new operators in competition with the established distributors. Entry and participation in the Japanese market should be through an association with Japanese marketers of tuna products who already have the know-how and networks necessary for effective marketing. Participation should be in the form of supplying the tuna and skipjack product to reputable Japanese marketers who are already established in the business.

8.
The Canadian Tuna Market

Geoffrey Waugh

ABSTRACT—Over 95 percent of Canada's domestic production of tuna is canned. There is only one major producer of canned tuna in Canada (Star-Kist), and production figures are confidential. Furthermore, at least 95 percent of the country's domestic tuna consumption is in the form of canned tuna. Moreover, almost all of Canada's tuna consumption is dependent on imported supplies, either in the form of raw fish for domestic processing or canned product. Import statistics are therefore used to provide a reasonably accurate description of Canada's tuna market.

INTRODUCTION

A large fishing nation with only a small tuna fishery, Canada relies extensively on tuna imports to supply its requirements. For reasons cited below, little direct information is available on the Canadian tuna market. Nonetheless, the nature of the tuna market can be gauged by examining import data because nearly all frozen tuna imports are used for canning. This information, together with the data on the catching sector, provides a reasonably complete description of the Canadian tuna market.

This chapter examines three aspects of the Canadian tuna market: (1) domestic landings, (2) imports of whole or dressed tuna, and (3) imports of canned product. Collectively, these data provide a description of the country's tuna market. However, such a description is necessarily brief and in some ways oversimplified. Important considerations, such as the changing nature of the canned product itself, oil vs. water packs, the effects of differing can sizes on costs of production and sales, the changing level of per capita consumption as income changes, and the degree to which tuna has created its own niche, all must await a more complete analysis of the Canadian market.

LANDINGS AND PRODUCTION

Canada does not have a major tuna catching industry. Tuna is caught using vessels that are primarily engaged in other fishing activities, and relatively little tuna is caught in domestic waters. Landings data between 1980 and 1984 are shown in Table 1. In the Atlantic Ocean the main catch is northern bluefin tuna (*Thunnus thynnus*), while in the Pacific Ocean the catch is predominantly albacore tuna (*Thunnus alalunga*). Because tuna is not shown separately in Canada's export statistics, the quantity and value of exports are believed to be insignificant.

Canada's domestic production of fresh and frozen tuna is relatively small (Table 1). Information relating to domestic tuna processing, both canned and meal, is confidential because there is only one major producer in Canada (Star-Kist in St. Andrews, New Brunswick). However, over 95 percent of Canada's domestic consumption of tuna is in the form of canned tuna. Almost all consumption is dependent on imports of either raw tuna for domestic processing or canned

Table 1. Canadian tuna landings and production, 1980–84

		1980	1981	1982	1983	1984
Landings						
Atlantic[a]	Q[b]	375	615	298	417	254
	V[c, g]	749	883	536	872	916
Pacific[d]	Q	212	103	92	242	47
	V	289	218	211	344	86
Production[e]						
Fresh or frozen						
Atlantic	Q	n.a.	76[f]	210[f]	358	93
	V	n.a.	152[f]	511[f]	1,098	391
Pacific	Q	152	66	36	150	34
	V	277	175	144	345	83

Source: Department of Fisheries and Oceans 1986.

Note: Q = quantity, V = value.

[a]Mainly northern bluefin.

[b]Tonnes.

[c]$000.

[d]Mainly albacore.

[e]Canned tuna and fishmeal production data are confidential.

[f]Fresh tuna only; frozen tuna data are confidential.

[g]International Monetary Fund exchange rates are used for currency conversion from Canadian dollars to U.S. dollars.

product. Therefore, Canada's import statistics provide a reliable guide to the nature and extent of the domestic tuna market.

IMPORTS

Unprocessed tuna

Table 2 shows total Canadian imports of tuna, including whole tuna and tuna fillets (fresh, chilled, or frozen). There has been no discernible trend in total imports, the average level (excluding 1983, when the figures appear to be inconsistent) being 12,656 tonnes. The peak year appears to occur between 1980 and 1985, with tuna imports being in excess of 15,000 tonnes.

Table 2. Volume of Canadian frozen tuna imports, 1980–85

	Tonnes				
	1980	1981	1982	1984	1985
Brazil					544
Costa Rica		778			
Ecuador		1,299		18	
El Salvador			273		
France				3,456	2,503
Ghana	2,106		3,559	408	
Ivory Coast		3,218			
Jamaica		301			
Japan	1,610	151	127	3,479	2,692
South Korea		1		181	56
Mexico	12	2,294	545		4,375
Panama	3,607				
Papua New Guinea		2,062			
Philippines					20
Puerto Rico		499	363		
Senegal	1,061				
Spain		816	3,306	484	
Taiwan		272	635	21	
United States	5,178	2,184	3,202	2,135	608
Venezuela		429		873	1,542
Total	13,574	11,086	15,228	11,055	12,340

Source: Department of Fisheries and Oceans 1986.

Note: The 1983 data are excluded because they appear to be inconsistent with other data presented.

In terms of country of origin, Japan, France, the United States, and the Latin American countries have been the major suppliers. In 1984 and 1985 Japan contributed 26 percent of imports, France 26 percent, and the United States 11 percent. The remaining imports came primarily from Latin America. Mexico has been an irregular supplier but in 1985 contributed 36 percent.

Of the Pacific island countries, only Papua New Guinea exported unprocessed tuna in any significant amount to Canada between 1980 and 1985. In 1981 Papua New Guinea (probably Star-Kist, Papua New Guinea) sold approximately 2,000 tonnes of tuna to Star-Kist, Canada (Table 2). This amount was valued at $2.6 million (Table 3). Since 1980 no other significant imports of tuna from the Pacific islands have been

Table 3. Value of Canadian frozen tuna imports, 1980–85

	$000				
	1980	1981	1982	1984	1985
Brazil					387
Costa Rica		1,030			
Ecuador		1,746		5	
El Salvador			415		
France				2,363	1,522
Ghana	2,298		3,231	373	
Ivory Coast			3,971		
Jamaica		396			
Japan	2,296	267	231	3,337	2,328
South Korea		3		151	16
Mexico	22	2,922	678		3,578
Panama	3,500				
Papua New Guinea		2,560			
Philippines					10
Portugal				2	
Puerto Rico		1,345	708		
Senegal	1,235				
Spain		1,264	2,825	408	
Taiwan		723	1,477	43	
United States	6,873	2,945	3,895	2,862	1,103
Venezuela		659		986	1,282
Total	16,224	15,860	17,431	10,530	10,226

Source: Department of Fisheries and Oceans 1986.

Note: The 1983 data are excluded because they appear to be inconsistent with other data presented.

recorded. It seems that Canada is an untapped market for sales of unprocessed tuna from the Pacific islands.

The value of Canada's unprocessed tuna imports declined by 37 percent between 1980 and 1985, from $16.2 million to $10.2 milion (Table 3). This decline reflects the falling price of tuna over that period. Because the world tuna price appears to have steadied, tuna imports are also expected to remain constant during 1986–87.

Canned tuna

Table 4 shows Canada's canned tuna imports by country of origin. Some fluctuation has occurred in the level of imports, with perhaps a slight

Table 4. Volume of Canadian canned tuna imports, 1980–85

	Tonnes				
	1980	1981	1982	1984	1985
Brazil				14	
Cuba	133	115			
Ecuador				18	
Egypt					13
Fiji	1,595	1,559	551	1,014	560
France				1	1
Hong Kong			13		4
Italy	10	11	4	57	43
Indonesia				53	
Japan	3,453	3,069	3,135	4,826	4,423
Malaysia	543	368	326	599	752
Mexico			429	61	
Netherlands					1
New Zealand					13
Peru	109	175			
Philippines		1,604	1,450	2,181	2,470
Portugal	2	42	11	12	12
Puerto Rico				16	31
Singapore	113	153	52		
South Africa					13
Spain	2		5	5	8
Sri Lanka	149		13		
Taiwan	149	459	256	145	134
Thailand	55	89	253	2,378	2,469
United States	2,575	2,079	377	210	72
Total	8,755	9,741	6,990	11,590	11,019

Source: Department of Fisheries and Oceans 1986.

Note: The 1983 data are excluded because they appear to be inconsistent with other data presented.

upward trend. The average annual level of imports for the period 1980–85 (excluding 1983) was 9.6 million tonnes. Although Canada's canned imports have remained relatively constant, there has been a marked shift in the pattern of country of origin. In 1980 Japan and the United States were Canada's major suppliers; Japan contributed 39 percent and the United States 29 percent. Fiji was also a significant supplier in 1980 with 18 percent.

Internationally, the 1980s proved to be a period of transition for the tuna industry. As a consequence of the world tuna oversupply and the collapse of the tuna market (Waugh 1986), both the international catching and canning sectors responded to the crisis. In particular, the canneries on the west coast of the United States all experienced initial slowdowns, and consequent shutdowns (Waugh 1987a); by 1986 only one cannery was operating in the traditional canning areas of California. These events were caused by, as well as affected by, events in the international tuna market.

Despite the continued increase in volume of canned sales, the Canadian market responded in several ways. Most important, new sources of supply were sought for canned products. By 1985 the Canadian market, although still influenced by Japanese supply, was rapidly becoming dominated by supplies from other Asian countries. In 1985 Thailand and the Philippines had captured 45 percent of the Canadian market, rising from less than 1 percent in 1980. Over the same period Japan's share of the market remained relatively steady at about 40 percent. Although figures are not yet available, indications are that Thailand was Canada's main market supplier in 1986.

The domination of the Canadian market by Asian canneries is directly attributable to their cheaper costs of production (in 1986 labor costs were $0.40 per hour in Thailand). The main losers in the canned tuna stakes were the United States and Fiji. The U.S. share fell from 29 percent in 1980 to less than 1 percent in 1985. Fiji's share dropped from 18 percent in 1980 to 5 percent in 1985.

In value terms, the Canadian tuna market has declined along with the fall in price of raw tuna, despite the increase in quantity imported (Table 5). In 1980 total imports of canned tuna were valued at $38.2 million, whereas in 1985 the value was $29.7 million; this represents a decline of about 22 percent in value during a period when import volumes rose by 26 percent.

TUNA MARKET

As noted above, 95 percent of tuna consumption in Canada is in canned form, which in turn is dependent on the imported frozen tuna or canned product. Therefore, Canada's import statistics are thought to provide a fairly accurate description of the extent of the Canadian tuna market. Total tuna imports—frozen and canned—are presented in Table 6.

In volume terms, the Canadian market remained stable over the period 1980–85. Consumption of tuna rose only slightly from 22,329 tonnes in 1980 to 23,360 tonnes in 1985—an increase of approximately 5 percent. In value terms, however, consumption (at imported prices) fell from $54.4 million to just under $39.9 million over the same period—a decline of about 27 percent.

Table 5. Value of Canadian canned tuna imports, 1980–85

	$000				
	1980	1981	1982	1984	1985
Brazil				32	
Cuba		573	427		
Ecuador				33	
Egypt					27
Fiji	6,357	7,291	2,460	3,605	1,802
France	2	3		2	2
Hong Kong			37		10
Italy	37	53	21	247	168
Indonesia				136	
Japan	13,869	14,233	12,354	15,704	14,395
Malaysia	1,701	1,557	928	1,576	1,772
Mexico			1,031	130	
Netherlands					1
New Zealand					29
Peru	234	374			
Philippines		4,827	4,120	5,138	5,420
Portugal	3	189	42	39	34
Puerto Rico				58	81
Singapore	372	579	163		
South Africa					48
Spain	6		20	19	19
Sri Lanka	462		38		
Taiwan	527	1,873	926	421	326
Thailand	188	315	743	5,925	5,382
United States	14,451	10,585	1,380	1,005	185
Total	38,209	42,452	24,690	34,070	29,701

Source: Department of Fisheries and Oceans 1986.

Note: The 1983 data are excluded because they appear to be inconsistent with other data presented.

Table 6. Canadian tuna market, 1980–85

	Tonnes				
	1980	1981	1982	1984	1985
Canned tuna	8,755	9,741	6,990	11,589	11,020
Tuna whole or dressed	13,574	11,086	15,228	11,055	12,340
Total	22,329	20,827	22,218	22,644	23,360
	$000				
Canned tuna	38,209	42,452	24,690	34,070	29,701
Tuna whole	16,224	15,860	17,431	10,530	10,226
Total	54,433	58,312	42,121	44,600	39,927

Source: Tables 2, 3, 4, and 5.

Note: The 1983 data are excluded because they appear to be inconsistent with other data presented.

CONCLUSION

The Canadian market appears to provide a stable outlet for tuna supplies. The market, initially dominated by Japanese and U.S. suppliers, now relies on Asia as its main source of supply. Fiji made initial inroads into the Canadian market but has recently lost ground due to competition from Asian canners. However, opportunities may exist for Fiji to re-enter the Canadian market and for other Pacific island countries to export frozen tuna to Canada.

9.
A Summary of International Tuna Markets: Characteristics and Accessibility for Pacific Island Countries

Linda Fernandez and Linda Lucas Hudgins

ABSTRACT—This summary chapter discusses the Australian tuna market, the Japanese tuna market, the European and Middle East tuna market, the import regulations in the United States, and the U.S. tuna market. The chapter lists the characteristics of these markets and examines the potential for Pacific island participation in these tuna markets.

INTRODUCTION

Approximately 65 percent of the world's tuna catch is canned. This canned tuna is sold in the United States (51.5 percent), Japan (12.5 percent), and western Europe (27.7 percent) with the remainder going to eastern Europe, the Middle East, and Latin America.[1] Some countries import their total canned tuna supplies (e.g., Belgium, Luxembourg, Sweden, and United Kingdom). Others import large quantities of fresh/frozen tuna to supply domestic canning industries (e.g., United States, Japan, and Italy). Still others consume little tuna domestically yet harvest or process significant amounts of tuna for sale on world markets (e.g., Fiji and Thailand).

This chapter summarizes the characteristics of the wholesale and import markets, retail markets, and institutional considerations for each of the major tuna-consuming countries or regions: Australia, Belgium/Luxembourg, Canada, France, Italy, Japan, the Middle East, the Netherlands, Sweden, Switzerland, the United Kingdom, the United States, and West Germany. The concluding section offers suggestions for potential participation by Pacific island countries in these markets.

AUSTRALIA

Wholesale and import markets

The Australian canned tuna market demand is about 7,700 tonnes per year, valued at about $27 million. This demand is met 80 percent from domestic sources and 20 percent from imports. Competition for domestically caught southern bluefin tuna (*Thunnus maccoyii*) is keen among the Australian canners.

Australian canned tuna import shares have shifted over time since the 1970s, when Japan supplied 60 percent of the market and Thailand and the Philippines each supplied 20 percent. In the 1980s Thailand dominates imports with 60 percent, and Japan has 24 percent, while the Philippines has only 5 percent (Tables 1 and 2). Australia's canned tuna imports from Thailand and the Philippines come from canneries owned by the Australian food processing company, Safcol Holdings Ltd. Eighty percent of canned tuna distribution in Australia is handled by four food distribution conglomerates: Safcol, John West Foods, Seakist, and H.J. Heinz (Australia) Ltd. H.J. Heinz is the market leader with 21 percent market share followed by Seakist (18 percent), John West Foods (16 percent), and Safcol (13 percent). There is also a large tuna market in Australia for pet food production.

Retail market

The major determinants of Australian consumer demand for canned tuna are price and quality.

The four large distributors (Safcol, Seakist, John West Foods, and H.J. Heinz) provide an enormous variety of seafood products to the retail market. Heinz's products of different qualities of canned tuna allow the company to compete on different price levels in the market. Safcol's canned tuna in oil is its major tuna product.

Institutional factors

Resource problems with southern bluefin tuna have resulted in government regulation of the tuna industry. Australia, New Zealand, and Japan have concluded a trilateral management arrangement that is designed to monitor exploitation and to set quotas for the three countries.

BELGIUM/LUXEMBOURG

Wholesale and import markets

These two countries are combined because of the integration of their economies. They import 0.5 million cases of canned tuna per year. They have no domestic harvesting or processing industry; supplies from

France (14 percent), Italy (12 percent), and, to a lesser extent, Thailand (8 percent), meet the tuna market demand (Table 1). Distributors purchase these imports and sell under their own labels in the retail market.

Retail market

The major determinants of retail demand for the Belgium/Luxembourg market are price and health consciousness. Tuna products sold in Belgium/Luxembourg are primarily yellowfin chunk and brine packs. Retail market buyers are conservative in that they prefer high-value packs from European canneries.

Table 1. Canned tuna: Percentage distribution of exports and imports by exporting and importing country, 1980–85

	1980	1981	1982	1983	1984	1985
			Exports			
Thailand	0.9	5.8	11.0	16.2	28.6	38.0
Japan	36.2	25.3	26.0	21.4	22.5	16.0
Philippines	10.4	13.0	13.6	13.5	11.0	9.8
Ivory Coast	17.2	12.3	13.6	13.5	11.0	10.2
Senegal	11.2	11.0	11.7	11.5	9.7	—
Taiwan	0.9	9.7	7.8	8.9	6.2	5.3
Spain	7.8	8.4	1.3	2.1	1.8	1.6
All others	15.4	14.5	15.0	12.9	9.2	—
Total	100.0	100.0	100.0	100.0	100.0	—
			Imports			
United States	25.2	25.6	31.7	35.3	40.8	44.8
France	22.1	22.6	23.7	21.4	17.9	17.6
United Kingdom	9.5	15.3	10.1	11.6	12.9	10.5
West Germany	13.4	11.0	12.2	10.4	10.5	9.2
Canada	7.9	8.0	5.8	6.9	6.5	—
Sweden	2.4	2.2	2.2	1.7	2.0	—
Italy	2.4	1.5	2.2	1.7	2.0	2.9
Australia	2.4	1.5	0.7	0.6	1.5	—
Belgium	3.2	4.4	2.9	3.5	1.0	—
Demark	0.8	0.7	0.7	1.2	1.0	—
Netherlands	1.6	0.7	0.7	0.6	1.0	—
All others	9.1	6.5	7.1	5.1	2.9	—
Total	100.0	100.0	100.0	100.0	100.0	—

Source: Calculated from USITC 1986, 201.

Institutional factors

As members of the European Economic Community (EEC), Belgium and Luxembourg promote EEC policies of encouraging economic benefits for developing countries by offering financial assistance for export endeavors so long as these endeavors do not threaten the markets of the EEC tuna-producing countries: France, Italy, Portugal, and Spain. To protect European canning industries, there is a 24 percent duty on imports. Imports are defined as products coming from any region outside the territorial waters (i.e., 12 miles) of EEC countries that are caught by a foreign (non-EEC) vessel. Each EEC nation requires health certificates and certificates of origin to comply with food regulations.

Table 2. Raw/frozen tuna: Percentage distribution of exports and imports by principal countries, 1980–84

	1980	1981	1982	1983	1984
			Exports		
Japan	19.3	9.3	9.4	10.1	19.7
Korea	25.1	21.5	18.6	33.8	19.1
France	3.7	4.5	8.6	8.3	8.2
Solomon Islands	4.6	4.6	3.5	5.1	7.2
Mexico	—	—	3.5	2.1	7.0
Spain	4.5	9.2	12.1	8.2	5.8
Philippines	10.4	7.3	4.2	3.4	3.0
All others	—	—	40.1	29.0	30.0
Total	—	—	100.0	100.00	100.0
			Imports		
United States	47.9	46.2	37.9	34.5	29.7
Japan	16.3	17.0	21.6	24.8	17.9
Thailand	—	—	—	4.6	17.8
Italy	13.7	11.9	13.1	12.9	13.4
Ivory Coast	2.4	2.6	4.3	4.9	4.2
Singapore	2.4	2.6	1.5	3.1	3.2
France	1.9	2.0	3.4	3.3	3.1
Senegal	2.4	3.4	2.3	3.9	2.6
Spain	2.1	4.7	7.3	3.8	2.5
Ghana	5.0	4.6	2.9	3.5	2.5
All others	5.6	5.0	5.6	0.7	3.1
Total	—	—	—	100.0	100.0

Source: Calculated from USITC 1986, 200.

CANADA

Wholesale and import markets

The average annual consumption of canned tuna in Canada is 10 million cases. The market is valued around $40 million and is supplied predominantly by imports. While Canada has a large fishing industry, domestic tuna catches are incidental on board those vessels targeting other fish species such as salmon. Star-Kist (Canada), the only domestic tuna canning company, which supplied approximately 40 percent of the domestic market, closed its plant in 1986. Imports of raw/frozen tuna to supply Star-Kist came from Japan (26 percent), France (26 percent), the United States (11 percent), and several Latin American countries. Small quantities came from other sources, including Papua New Guinea (Table 2).

Imports of canned tuna to Canada in the early 1980s came from Japan (39 percent), the United States (29 percent), and Fiji (18 percent). By the mid-1980s, Japan still held a significant share (40 percent), but Thailand with 30 percent and the Philippines with 15 percent had displaced the U.S. share (Table 1). The closure of the Star-Kist plant presents an opportunity for additional supplies of canned tuna to be imported.

Retail market and institutional factors

Data regarding retail market characteristics and institutional factors are unavailable but are assumed to parallel the U.S. structure on a smaller scale.

FRANCE

Wholesale and import markets

France maintains harvesting fleets in Senegal and Ivory Coast that supply French-owned canneries in these countries. Of the total 7 million cases of tuna consumed in France in 1985, over 54 percent was imported from Senegal and Ivory Coast (Table 1). Canned tuna products from Senegal and Ivory Coast also supply other EEC countries.

The highly developed French tuna distribution system is virtually inaccessible to new suppliers. Distributors buy direct from the producer or work through an agent and resell under their own labels. The retail market outlets, in which tuna companies supply both housebrand packs and nationally marketed packs, are concentrated. The entire industry supports a generic advertising and public relations campaign.

Retail market

The major determinants of consumer tuna demand in France are real income, brand allegiance, and, more recently, health consciousness. Declining incomes have caused a shift to increased consumption of light-meat chunk-style packs, which account for over 50 percent of the market. Flake/hors d'oeuvre packs are 30 percent of the market. Increased health consciousness has led to a decline in sales of oil packs to 10 percent of the market.

Institutional factors

National interests and policies override EEC policies in France. The harvesting and cannery lobbies have strong political and financial support throughout all stages of production and distribution. In fact, shipments from other than the traditional suppliers, Senegal and Ivory Coast, are not allowed entry. While this violates the General Agreement on Tariffs and Trade (GATT) and the Treaty of Rome, the EEC has not taken action to cite this violation.

ITALY

Wholesale and import markets

Italy maintains both a marginal tuna harvesting sector and a large tuna canning industry that supplies 85 percent of the 1.2 million cases of canned tuna consumed annually. Imports of raw/frozen tuna continue to rise at an annual rate of 6.5 percent (Tables 1 and 2). These imports supply domestic canners who resell in retail markets. In 1985 over 0.5 million cases of canned tuna were imported to Italy, mainly from Portugal and Spain. The Italian retail market distribution system for canned tuna operates through well-established brokers. Generally, importers face high costs in penetrating this diffuse, state-subsidized market.

Retail market

The determinants of Italian consumer demand for canned tuna are quality and brand consciousness. Italian consumers prefer lightmeat yellowfin tuna in olive oil and are willing to pay a premium for quality packing. Aside from this product, which accounts for 90 percent of the market, skipjack packed in oil sometimes sells in Italian tuna markets.

Institutional factors

In addition to EEC policies, import suppliers must contend with the complex currency exchange regulations maintained by the Italian government.

JAPAN

Wholesale and import markets

Japan is the world's most developed fishing nation, and each sector of the Japanese industry is extremely productive. Japanese fisheries harvest a large variety of tuna to supply the domestic sashimi, canned tuna, and *katsuobushi* markets as well as the processors' needs for canned tuna exports. Annual tuna demand in Japan is estimated to be 900,000 tonnes, of which 70 percent is domestically supplied, between 18 and 20 percent is supplied by Korean fleets, and between 6 and 7 percent by Taiwanese fleets.

The Japanese canned tuna industry produces for both the domestic and export markets (Table 1). The highly structured processing and distribution/marketing sectors for frozen, canned, and *katsuobushi* industries are made up of wholesalers and *maguro shosha* (specialist tuna traders) that concentrate on supplying domestic markets. In fact, Japan's two largest canners control 70 percent of the market and do not export at all. Recently, canned tuna from Thailand has become more competitive in export markets due to relatively high costs of Japanese production and the appreciation of the yen. Continued appreciation of the yen may generate opportunities for new suppliers to the Japanese market.

Retail market

While Japanese consumers favor sashimi, tastes are becoming more westernized. This change in tastes has two important market implications: (1) at the same income level, consumers will increase their demand for canned tuna, preferably yellowfin in oil; and (2) with an increase in income, consumers will switch to livestock and shellfish products. Japanese canned tuna consumers are brand-conscious, favoring domestic and well-known brands.

Institutional factors

The fishing industry does not have a great deal of influence within the central government. However, tuna interest groups persist in their efforts to promote government policies favoring subsidized sales of canned tuna to developing countries and policies for formal import quotas.

MIDDLE EAST

Wholesale and import markets

Approximately 2 million cases of canned tuna from Safcol (Australia), Indonesia, and Korea constitute the entire Middle East market (primarily

Jordan and Saudi Arabia). The lack of data regarding these markets, as well as the complex bureaucratic structure, is not encouraging to potential exporters.

Retail market

The fall in oil industry revenues directly reduced consumer demand in this middle- to high-value product market. This has been largely due to the reduced number of foreigners working in the region. Most canned tuna was consumed by foreigners and not by citizens of the region.

Institutional factors

No details are known about Middle Eastern institutional factors.

NETHERLANDS

Wholesale and import markets

The Netherlands' tuna market is young and subject to fluctuation from its present volume of 212,000 cases per year. Imports of raw/frozen and canned tuna supplement the domestic supply in unknown proportions (Table 1). There is a well-developed processing system for salmon. A small part of this system is used to process tuna. Distributors market under their own labels.

Retail market and institutional factors

Specific details regarding the types of tuna products sold, consumer preferences, and institutional factors are not known. However, the Netherlands follows EEC trade policy, and since salmon is favored over tuna, tuna distributors attempt to attract consumer demand through price competition.

SWEDEN

Wholesale and import markets

No tuna harvesting or processing industry exists. Sweden's medium-value tuna market totals 268,000 imported cases, of which 67 percent is imported from Thailand and the Philippines (Table 1). The market is inexpensively accessible to importers through contacts in the retail distribution network. Impod, a Swedish government agency, assists developing countries to introduce their products to the Swedish market.

Retail market

Lightmeat and whitemeat tuna in both water and oil packs are available in the Swedish retail market. Recent changes in consumer health awareness have led to a dramatic shift toward water-packed lightmeat tuna.

Institutional factors

No specific details relating to institutional factors are known; however, Sweden does promote its own version of the EEC's policy of trade-oriented financial assistance for developing countries.

SWITZERLAND

Wholesale and import markets

Switzerland has no domestic tuna harvesting or processing industry and imports 0.5 million cases of tuna to supply its market demand. For the past five years, the import market has remained stable in both volume and structure.

Retail market

The major determinant of Swiss consumer demand for tuna is quality. Swiss quality standards are the highest in Europe, and consumers are willing to pay a premium for high-quality albacore tuna solid packs in oil.

Institutional factors

Government policy and import buyers maintain stringent quality standards for canned imports.

UNITED KINGDOM

Wholesale and import markets

There is no domestic tuna harvesting or canning industry in the United Kingdom. This market is the most dynamic European canned tuna import market, experiencing an average growth of 10 percent annually between 1975 and 1985. A record high rate of 30 percent growth occurred during 1985–86, with the market reaching 3.6 million cases. The major import source is Thailand, accounting for 50 percent of imports in 1986 (Table 1). Imports from Fiji are 10 percent of the market and imports from Solomon Islands account for 3 percent. The buyer/distribution system is concentrated between two market leaders: John West Foods and Princes. The companies prefer to deal through Japanese trading companies. However, some small buyers are deviating from the norm by starting to deal directly with processors and other foreign trading companies.

Retail market

The major determinants of British consumer demand for canned tuna are price and health consciousness. The share of chunk brine packs was 35 percent of the market in 1986 and is still growing. John West Foods and Princes are planning to use innovative packs to promote an even greater market share. This may encompass a new pack size (400 g instead of 200 g) and the introduction of low-salt and sauce packs.

Institutional factors

The United Kingdom follows standard EEC policies.

UNITED STATES

Wholesale and import markets

There is a multi-level tuna industry ranging from harvesting to retail markets in the United States. Recent increased costs of production forced processors out of harvesting and toward greater emphasis on the distribution/marketing level. The U.S. tuna fleet consists of 90 purse seiners and 9 bait boats. The domestic fleet supplies approximately 60 percent of the raw/frozen tuna to U.S. canneries, with 31 other countries supplying the remainder (Tables 2 and 3).

The three largest tuna processing companies, which control 81 percent of the domestic market, are H.J. Heinz Co (a Ralston Purina subsidiary), Van Camp Seafood Co Inc, and Bumble Bee Seafoods Inc. These companies operate eight capital-intensive processing plants: five in Puerto Rico, two in American Samoa, and one in California. The

Table 3. Percentage distribution of U.S. canned tuna imports by exporting country, 1980–85

	1980	1981	1982	1983	1984	1985
Thailand	10.1	14.6	21.3	32.6	55.3	57.3
Philippines	21.7	30.3	31.6	26.2	13.7	14.4
Japan	39.0	30.0	30.2	16.7	16.5	11.1
Taiwan	25.1	22.3	12.2	15.3	11.0	11.0
Ecuador	0	0	0	0	0.5	2.4
Malaysia	< .05	1.0	0.9	2.5	1.0	1.8
Indonesia	0	0.2	0.7	2.2	1.4	0.6
Venezuela	0	0	0	0	< .05	0.4
Singapore	< .05	0.1	0.1	0.3	< .05	0.3
All others	4.1	1.6	3.0	4.3	0.5	0.6

Source: USITC 1986, 188.

offshore plants take advantage of lower labor and capital production costs, as well as more liberal worker health, welfare, and environmental regulations. Even with this domestic production, the United States imports approximately 30 percent of total world canned tuna supply (Table 1).

While some canned tuna is marketed through processor-operated distribution systems, most canned tuna products reach retail and institutional markets through food brokers working on commission. Importers of canned tuna products sell in the housebrand and private-label market and often ship imported products directly to buyers rather than maintain warehouse inventories like domestic canners.

The institutional market (i.e., restaurants, schools, hospitals) grew from approximately 10 percent of the U.S. canned tuna market in 1975 to approximately 14 percent in 1984. However, the success of *surimi*-based imitation crab and shrimp product sales may inhibit further growth of demand for canned tuna in this market.

Retail market

Two major tuna products are sold in retail markets in the United States: (1) whitemeat tuna (albacore and yellowfin) in chunk form, and (2) lower-quality darkmeat, which has soft, flake texture, from skipjack caught in tropical waters. Whitemeat tuna in chunk form is 70 percent of sales in the U.S. market and receives a higher price than the darkmeat tuna.

The major determinants of demand for canned tuna in the United States are changes in disposable income, changes in substitute food prices (poultry and ground beef hamburger), and health consciousness.

Increased health consciousness among consumers has directly affected the packing of tuna products in the United States, resulting in more sales of tuna packed in water than in oil. U.S. consumers exhibit price responsiveness but little brand allegiance. Wholesale and retail promotional discounts, when offered, occur during the slack fall/winter seasons. Overall, canned tuna sales in the United States account for about 52 percent of total world canned tuna consumption. Tuna is the only seafood staple in this country where meat is preferred to fish.

Institutional factors

Importers of raw and frozen tuna are required to have a certificate of approval from the National Marine Fisheries Service (NMFS) and to meet the quality control standards used by the American Tuna Sales Association (ATSA). While no duty is levied on frozen or unprocessed tuna, a tariff of five cents per pound is charged for prepared and preserved tuna such as cooked loins.

The U.S. territories (Puerto Rico, American Samoa, Guam, and the Virgin Islands) ship canned tuna free of duties to the U.S. market. Tuna production in American Samoa affects the tariff paid by other nations because of the way the allowable quota is established. The current tariff schedule for canned tuna packed in water states that "imports not in excess of 20 percent of the preceding year's U.S. pack are dutiable at 6 percent ad valorem" and that those imports "in excess of the quota are dutiable at a rate of 12.5 percent ad valorem." For purposes of establishing the allowable quota on canned tuna in water, tuna packed in American Samoa is not counted as part of the U.S. pack. The quota dutiable at 6 percent, therefore, shrinks automatically with the decline in domestic U.S. canning operations, and more imports will be dutiable at the higher rate of 12.5 percent. Imported canned tuna in oil is subject to a 35 percent ad valorem tariff.

WEST GERMANY

Wholesale and import markets

Fish harvesting and canning in West Germany is primarily devoted to the herring industry, making tuna imports necessary (Table 1). A shift from Japanese and Taiwanese imports in the 1970s to Thailand (54 percent) and the Philippines (19 percent) in the 1980s is largely due to the ability of these newer sources to compete on a price basis—the major determinant of consumer demand in West Germany. Recently there has been an increase in imports from France. Brokers handle the distribution of imports to retail markets.

Retail market

Price is the major determinant of West German consumer demand for canned tuna. Thus competitive market suppliers provide low-cost solid and chunk packs, as well as the dominant low-value flake tuna with added vegetables. The flake/vegetable pack accounts for 40 percent of the market. Neither brand nor quality strongly influences consumer behavior.

Institutional factors

West Germany follows EEC standard policies.

CONCLUSIONS

Although all the markets examined consume and/or import substantial amounts of canned and raw tuna, not all markets are accessible to new entrants. In some countries, entry is precluded by well-established distribution and retail networks. In other cases (France, Italy, and Japan, for example), almost all raw and canned tuna is ordinarily supplied by domestic or colonial fleets or canning industries. Several markets (e.g., EEC countries) are protected by institutional factors. Given these various considerations, the most accessible markets are Australia, Canada, some parts of Europe, Scandinavia, and the United States. Table 4 summarizes import activity by country, using available data. In general, opportunities are greatest at the harvesting and processing levels if the Pacific island countries can meet the quality standards and provide consistent supplies under price competition with other exporting countries. These opportunities may be enhanced by favorable exchange rate conditions.

Relatively inaccessible markets

Of the 13 countries and regions examined in this report, several are considered inaccessible to Pacific island countries.

Table 4. Raw and canned tuna imports for most recent years by importing country

	Imports[a]			
	Raw/frozen		Canned	
	Quantity	% of total domestic supply	Quantity	% of total domestic supply
Australia	small amount	5%	2,495 tonnes	20%
Belgium/ Luxembourg	0	0%	0.5 m cases	100%
Canada	12,340 tonnes	93%	11,000 tonnes	52%
France	25,000 tonnes	unknown	3.9 m cases	56%
Italy	100,000 tonnes	70%	0.5 m cases	15%
Japan	141,119 tonnes	10%	unknown	unknown
Middle East	0	0%	2.0 m cases	100%
Netherlands	unknown	unknown	unknown	unknown
Sweden	0	0%	0.2 m cases	100%
Switzerland	0	0%	0.5 m cases	100%
United Kingdom	0	0%	3–6 m cases	100%
United States	150,000 tonnes	40%	10.9 m cases	30%
West Germany	0	0%	2.0 m cases	100%

Sources: Ashenden and Kitson 1987a; Crough 1987a; Elsy 1987; King 1986; USITC 1986; Waugh 1987b.

[a]Most data are for 1984–86. Cases are standard cases. m = million.

Australia. Australia imports little raw/frozen tuna and relies on domestic processors and established offshore processing operations in Thailand, Indonesia, and Japan to supply canned tuna demand. Some frozen and canned tuna imports are possible depending on relative exchange rates and international prices for bluefin tuna relative to domestic prices. As Japanese bluefin tuna prices increase, Australian fishermen divert catches away from domestic cannery sales.

France and Italy. Markets in France and Italy are heavily regulated and are supplied by domestic, vertically integrated industries; in the case of Italy, they are subject to exchange rate fluctuation. It is important to note that Mexico has recently broken into the Italian market for frozen tuna, even though it is difficult to enter. In addition, some Pacific island yellowfin sales have been made to Italy.

Japan. Japanese markets are at least 80 percent self-sufficient at all industry levels. Revaluation of the yen presents a small opportunity to enter this market with frozen or canned tuna. Quality standards are high at all levels of the market and for all product forms.

Middle East. What little data exist indicates that markets in the Middle East are small, high-value markets and are sensitive to fuel oil production trends.

Relatively accessible markets

United States. The U.S. market, because of its size, openness, and lack of brand allegiance, appears to have a potential for sales. Consideration in this market should be given to avoidance of import tariffs if possible by getting imports in before the annual allowable quota is filled. It is also important to seek integration into the established U.S. production/distribution system.

Europe. While the EEC promotes a policy of financial assistance for developing countries, the EEC also protects the four EEC tuna-producing countries: France, Italy, Portugal, and Spain. As the most diverse, high-unit value markets in the world, European markets offer opportunities for a variety of products, packaging design, and other innovations. Overall, the United Kingdom appears now to have the highest growth potential. Other European countries have relatively small volume but stable market demands.

Canada. The closure of the domestic cannery provides an opportunity for canned tuna imports.

Market strategies

A crucial determinant of demand in many European markets, as well as Canadian, U.S., and more recently, Japanese markets, is the growing reliance on Thai canned tuna imports due to price competitiveness. Thailand is competitive because of its low production costs and adequate quality standards. Therefore, the Pacific island countries must consider their relative strength in competing on the basis of cost and quality standards to gain access to international markets. Because of the high quality of their packs, island countries should target on those markets that are discriminating and whose consumers are prepared to pay a premium for quality. There could also be merit in the island countries coordinating their international tuna sales in an attempt to minimize competition within particular markets.

NOTE

1. This chapter synthesizes the chapters by Ashenden/Kitson, Crough, Elsy, Floyd, King, and Waugh for PIDP's tuna project.

III. IMPORTANT COMPETITORS

10. The Development of the Mexican Tuna Industry, 1976–1986

Linda Lucas Hudgins

ABSTRACT—The Mexican tuna industry has the largest and one of the most productive fleets in the world as of late 1986. This development has surmounted a domestic financial crisis and the imposition of a six-year embargo by the United States against Mexican tuna. In the future the industry is likely to seek new fishing grounds, and the central and western Pacific are possible alternatives.

INTRODUCTION

The Mexican purse seine fleet by 1987 will be the largest in the world, surpassing in capacity the fleet of the United States. The Mexican fleet in 1986 had the largest catches in the eastern Pacific of the 11 countries reporting to the Inter-American Tropical Tuna Commission (IATTC). These countries, ordered by size of catch in 1985, are the United States (89,900 tonnes), Mexico (78,083 tonnes), Ecuador (32,451 tonnes), Venezuela (27,088 tonnes), and Costa Rica (3,363 tonnes), as well as Colombia, Panama, Peru, Cayman Islands, USSR, Spain, and Vanuatu (131,259 tonnes combined) (Joseph 1986).

Mexico and the countries of the south Pacific share common conditions with respect to tuna industry development. These conditions are (1) the availability of abundant tuna resources and (2) a commitment to exploit these resources as a source of employment, foreign exchange, food protein, and economic development revenue. The strong commitment of the Mexican government to develop its tuna resources played a crucial role in the survival of the country's industry.

Some caveats can be made regarding the applicability of the Mexican model of tuna development to the economies of the Pacific island countries. The Mexican economy is massive in relation to those in the region. Mexico's real gross domestic product in 1983 was $97.3 billion (United Nations 1984a). Therefore, because of size and relative degree of economic development, Mexico has access to those domestic and international credit markets foreclosed to smaller economies. In addition, Mexico's population of over 81 million offers the potential for a large domestic market not available to smaller economies.

The first sections of this chapter provide an overview of the role of government policy in the development of the Mexican tuna industry, which is followed by a review of industry conditions and an examination of domestic and international markets. Both the imposition of the U.S. embargo against Mexican tuna imports (1980–86) and the financial crisis in Mexico (1982) severely affected the country's developing tuna industry. The effect of these events, as well as internal industry problems, is described and analyzed in the last sections of this chapter.

GOVERNMENT POLICY

Administrative background

Two government agencies are involved in the direct implementation of fisheries policy in Mexico: the Subsecretario de Pesca ([SEPESCA] Subsecretariat of Fisheries) and Banco Nacional Pesquero y Portuario, S.N.C. ([BANPESCA] National Fishery and Ports Development Bank). National industrial policy is developed by SEPESCA, with BANPESCA providing funds to the projects identified by SEPESCA. BANPESCA was specifically established in 1979 to provide credit to the country's fishing industry and to facilitate the financing of new fleets and processing plants.

Other legislation relevant to fisheries concerns the ownership of Mexican business and nationality of crew. Foreign ownership of business in Mexico cannot exceed 49 percent, although recently this restriction has been ignored in an attempt to attract foreign investment into the Mexican economy. In addition, the registered captain of a fishing vessel must be a Mexican citizen. There are no legal provisions that permit foreign crew members, although each vessel may employ up to five foreign instructors for periods of up to two years.

Private sector incentives

The Mexican government was instrumental in encouraging private investors, many of whom had formerly owned shrimp vessels, to move into the tuna industry when shrimp vessels were nationalized and sold to cooperatives in 1979. The major incentives for these private investors were (1) a potentially lucrative U.S. tuna market for exports and (2) the abundant tuna resources in the eastern tropical Pacific Ocean. The long-term objectives of tuna fleet development were the exploitation of tuna resources in the eastern Pacific Ocean with a fleet of 115 superseiners (over 1,000 gross registered tonnes [GRT]) and the sale of round and processed (canned) tuna in international markets, particularly in the United States.

Private sector investors put up 15 percent ($60 million) of the tuna vessel purchase price and between 1978 and 1982 purchased 54 new purse seine vessels for $400 million to be built abroad. BANPESCA financed 31 of these vessels—about 40 percent of the total fleet debt—and further guaranteed the debt on the remaining 23 vessels (about 60 percent of the fleet debt). Of the 54 new vessels, 14 were less than 900 GRT in size while the rest were 1,000 GRT or larger. Initially, the government gave income tax concessions to those willing to invest in the tuna industry by taxing industry profits at a rate of 7 percent. By 1986 these concessions were removed. Industry profits are currently taxed at regular industry rates (about 42 percent).

Government operations in the tuna industry

Direct participation by the government in the development of Mexico's tuna industry includes direct purchase of vessels, construction of vessels, and operation of canneries. In 1986 the government owned 25 vessels outright, representing a total operating capacity of about 10,745 GRT. Five of these vessels were small bait boats. The government has plans to build five 75-GRT bait boats (DELFIN series) and fourteen 750-GRT purse seine vessels (ATUN series). Construction plans for ten of the DELFIN and two of the ATUN series along with two 1,200-GRT vessels were, however, uncertain in 1986 because of the domestic financial crisis.

Mexico has active state participation in the fleet, but that participation (state and social sector combined) is a relatively small percentage (22 percent) of the total. Although state-owned canneries have 55 percent of the installed processing capacity, Mexico did not increase its financial and risk exposure with respect to the tuna industry by building additional plants. The Mexican government has an almost nonexistent role in sales, distribution, and marketing, which are predominantly handled by large food conglomerates.

Government political activity

The development of the tuna industry in Mexico was affected by government initiatives other than those associated with SEPESCA and BANPESCA. In particular, the declaration of the Mexican exclusive economic zone (EEZ) in 1976 indirectly led to Mexico's withdrawal from the IATTC in 1978 and the subsequent imposition of the U.S. embargo on all tuna and tuna products from Mexico in 1980.

In January 1980 President Lopez-Portillo issued a decree requiring that a license fee be paid for tuna fishing in the Mexican EEZ. In July 1980 the Mexican navy arrested U.S. tuna seiners for fishing in the Mexican EEZ without permits. The United States imposed an embargo against importation of all Mexican tuna and tuna products. Negotiations between the United States and Mexico over access rights for the U.S. tuna fleet to the Mexican EEZ had been unsuccessful for three years prior to imposition of the embargo. Mexico terminated all fishing treaties with the United States in December 1980.

In Mexico City the "tuna war" with the United States was seen as an issue of national integrity, while enforcement of the U.S. embargo was primarily motivated by domestic protectionism. In April 1986 Mexico proposed voluntary annual export restraints of 20,000 tonnes to the U.S. tuna market. Government sources in both countries believe that this proposal provided the impetus for resolution of the embargo in August 1986.

Impact of the 1982 financial crisis

The debt for the 54 new tuna vessels was, except for one case of Spanish pesetas, denominated in U.S. dollars, and these vessels were built in various shipyards in Europe and the United States, with delivery dates scheduled for the early 1980s. In the time interval between vessel orders and vessel delivery, the Mexican peso was allowed to float relative to other world currencies. By 1982 a combination of falling oil prices and revenues, the currency devaluation, and a large external debt exceeding $80 billion drove the Mexican economy into a fiscal crisis and severe recession. The tuna industry began to feel the crisis immediately. Several vessels were out of service during 1982–83, in part because dollars were not available to purchase replacement parts or make repairs. The debt burden in pesos on the new vessels increased by over 700 percent between 1980 and 1984. During the same period the world price of tuna declined by over 30 percent.

In late 1983 BANPESCA proposed to the vessel owners a vessel-refinancing scheme that was formally implemented in May 1984. The plan involved BANPESCA's assuming the external debt with foreign shipyards and, in turn, refinancing the vessels with the Mexican owners at an average of 175.5 pesos to the dollar. The exchange rate at the time of the vessel purchases in 1980 was 22 to 25 pesos to the U.S. dollar but had climbed to over 300 pesos to the dollar by 1984.

A requirement of the refinancing plan was that vessels would have to meet production goals to qualify for continuing participation in the program. Debt amortization was extended in some cases. Once the refinancing plan was operational, all funds were channeled through BANPESCA, including proceeds from tuna sales to the state-run canneries. A recent government evaluation of the success of the refinancing through April 1986 gives mixed results. Few vessels were able to meet production goals and hence have not made regular debt payments. Repayment of the vessel debt is a continuing source of friction between government and industry.

Three direct effects on the Mexican tuna industry related to the country's financial crisis can be considered outside the controversy of vessel debt payment: (1) the devaluation of the currency would have made Mexican exports relatively cheaper, but the U.S. embargo prevented the Mexican tuna industry from taking advantage of its currency position; (2) the inability on a national level to acquire hard currency for debt service and operating expenses had negative impacts on industry development and efficiency; and (3) the general economic recession in Mexico reduced incomes and national consumption of all goods, thus reducing the domestic demand for canned tuna.

INDUSTRY OVERVIEW

Resource availability

Almost 100 percent of Mexican tuna fleet operations occur within the Mexican 200-mile EEZ, although some vessels have begun to fish farther south along the central and south American coast. The fleet is expected to continue to follow this pattern in the immediate future as long as the tuna resources remain plentiful in the eastern tropical Pacific. According to industry sources, all the yellowfin tuna they need can be caught within 150 miles of the Mexican coastline.

Mexico's estimates of annual yield of all tuna in its own EEZ is 170,000 tonnes per year. Seventy-five percent of the yellowfin and skipjack (129,000 tonnes) is concentrated in the northern area of the zone off the Baja California peninsula and the central southern region off the tip of the Baja peninsula. The remaining 25 percent of yellowfin and skipjack is located on the Atlantic coast in the area of the Yucatan peninsula in the Gulf of Mexico.

By the end of 1987 the Mexican tuna fleet will have an annual catching capacity of nearly 140,000 tonnes. This amount is about 30,000 tonnes less than the estimated resources in the Mexican EEZ. The Mexican government recently announced that it will issue more permits in the future to foreign vessels to fish the EEZ (Fishing News International 1986:56). Thus the resource availability to the Mexican fleet will increasingly depend on the potential competition from foreign vessels.

Fleet capacity and production

Mexico's original tuna development plan in 1977 called for a total fleet of 120 vessels. Although the Mexican fleet is now expected to be rationalized at 89 vessels, the following discussion is based on the still-current fleet projection of 104 operational vessels by the end of 1987.

The projected Mexican tuna fleet has 31 vessels of less than 400 GRT in size, 27 between 401 and 750 GRT, and 46 between 751 and 1,200 GRT. Since 1980, 75 vessels have been built, making the Mexican fleet the newest in the world. As of July 1986, 16 of the 104 vessels were still under construction, while 39 were inactive awaiting re-outfitting, repairs, or parts. Six vessels had sunk. Of the total projected tuna fleet of 104 vessels, 70 percent will be purse seiners and 30 percent will be bait boats. In 1985 there were 61 vessels, representing 18,900 tonnes of carrying capacity, actively fishing in the Mexican fleet. As of July 1986, 59 vessels were fishing.

The current fleet capacity in use totals about 68,500 GRT, catching over 90,000 tonnes a year. There are 59 vessels in operation in the Mexican tuna fleet, with 16 others under construction or tied up awaiting repairs, parts, or leasors. Sixty percent of the fleet is owned by the private sector, 16 percent by cooperatives, and about 24 percent by the government.

On average the Mexican catch was composed of about 70 percent yellowfin tuna and about 30 percent skipjack tuna. This proportion varies by year, depending on yellowfin availability. For example, in 1986—a good fishing year—the catch was composed of about 90 percent yellowfin, and it appears that the fleet could catch over 110,000 tonnes. The years 1982 and 1983 essentially represented lost years to the fleet because of the dollar crisis, the effects of El Niño on resource availability, and the closure of the U.S. market.

Crew costs for a Mexican vessel are less than those of a pro forma 1,200-GRT vessel operating at full capacity and represent about 26 percent of total costs. Fuel consumption and insurance costs are greater (about 32 percent of total costs) than expected for a typical vessel of the 1,200-GRT size. Actual short-run production costs for a 750-GRT vessel in the Mexican fleet are about $469 per tonne caught, while costs for a 1,200-GRT vessel in the Mexican fleet are approximately $670 per tonne caught.[1] These figures do not include an account for debt service or depreciation. Production costs per tonne as compared with world tuna prices provide a measure of the economic efficiency of the Mexican fleet. At $750 a short ton, the Mexican vessels are competitive, exclusive of debt service.

Ports and processing capacity

Tuna is unloaded at eight ports on the Pacific coast. Only Ensenada and Mazatlan have enough skilled labor and suitable handling equipment. As a result, these two ports are subject to overcrowding. The handling capacity at other ports could be almost doubled with an increase in equipment and labor.

The unloading capacity appears to be slightly less than that needed to utilize total catches of the projected fleet of 104 vessels. Available unloading capacity is about 528 tonnes per eight-hour shift, or 126,700 tonnes annually. Several ports have inadequate handling equipment, skilled labor, and other complements for efficient unloading. These problems have been identified by the Mexican government, and plans are under way to alleviate them.

The total refrigeration capacity across the eight ports (149,000 tonnes annually) appears to be adequate. The cost of refrigeration in Ensenada is about $15 per tonne per month. The refrigeration capacity is 30 percent owned by the public sector and 70 percent by private and public sector canners. Port congestion increases the cost of processed tuna in Mexico because catches have to be held in refrigeration while waiting for processing capacity to become available.

Currently 18 canneries are packing tuna in Mexico. These general food-product canneries also process fruit, vegetables, and other fishery products. This nonspecialized processing sector has been identified as a bottleneck and a source of inefficiency. These general product plants operating at full capacity could can approximately 132,500 tonnes of tuna per year. If the fleet operated at full capacity and if markets were available, additional seafood-canning capacity would be required to accommodate landings. An alternative would be to operate the existing canneries on a 24-hour basis at the expense of canning other food products.

On average the Mexican canneries produce about 0.48 tonnes of canned tuna for each tonne of input. The estimated costs of canning a standard case of tuna with existing canneries are about $23.30. These costs represent average costs across a variety of different size and quality plants. The estimated cost is within the range of wholesale list prices for comparable cases produced by various countries. Production costs per standard case reported in Infopesca (1986) by country are Ecuador $24.00, Fiji $31.30, Japan $24.40, Taiwan $23.75, Thailand $20.00, and Venezuela $31.00. While the wholesale price in Mexico may be competitive with other international canners, the domestic retail price of a can of tuna in Mexico (203 pesos [$0.30] in July 1986) is expensive relative to incomes and other protein sources.

Alleged inefficiency in the state canning operations is attributed to poor management, lack of specialization in fishery products, and relatively high administrative costs. To avoid these sources of inefficiency, two new seafood plants are under construction. These plants will process only seafood and have a larger-scale operation (18,000 tonnes each) than the current multiproduct plants. The plants are being designed in cooperation with investors and government agencies in France. French canning technology will be used, and the plants will be in operation by the end of 1987. The plants will be managed as quasi-public Mexican corporations under the name Pesca Industrial Corporación, S.A. ([PICOSA] Industrial Fish Corporation) and will be located in the states of Colima (Manzanillo) and Chiapas (Puerto Madero). Mexican industry sources report that in return for financing and technological advice, the French investors will receive 3 percent equity from the plant operations.

INTERNATIONAL AND DOMESTIC MARKETS AND PRODUCT DISTRIBUTION

International tuna trade

Mexico exported about 11,000 tonnes of raw tuna in 1977 to the United States. The embargo stopped all exports to the United States from 1980 onward. As of 1985, Mexico had found alternative markets in which to sell about 36,000 tonnes a year. The industry took about five years to establish new marketing channels for exports. Recently the industry, using private brokers and the offices of SEPESCA, completed sales to Thailand, Canada, Italy, and Costa Rica at American Tuna Sales Association (ATSA) prices FOB (free on board) Ensenada. Other marketing channels are currently being pursued.

A continuing complaint of the tuna industry has been the divergence between the price set by the Mexican government (official price) and the world market price. Under conditions of the government vessel refinancing, the landings must first be offered to domestic canners before being offered on export markets. The domestic price increased by 73 percent between 1984 and 1986, but it is still only 90 percent of the quoted world price. A domestic price increase for purse seine vessels will provide an incentive for them to operate at full capacity. At the same time, any domestic price increase will increase costs of the final canned product, which would exacerbate the already low level of domestic sales.

Domestic market conditions

There are 14 domestic canned tuna labels being marketed by seven firms in Mexico. Herdez is a private-sector food distributor with the largest market share (36 percent) and has the most aggressive marketing approach. In 1984 the state-run canneries of Productos Pesqueros Mexicanos ([PPM] Mexican Fishery Products) had 35 percent of the market, and the private sector had about 65 percent, which was an increase of 22 percent for PPM over its share in 1980.

Based on information from a BANPESCA report, the relative shares accruing to the various levels of the industry are as follows: of the final price, 33 percent is received at the ex-vessel level, 45 percent by packers and wholesalers-distributors, and 22 percent by retail outlets. In the U.S. industry about 55 percent of the retail price is received at the ex-vessel level, about 28 percent at the wholesale level, and 17 percent at the retail level.

Canned tuna distribution bottlenecks occur in Mexico, with inventories accumulating in the larger cities and no product being available in the rural areas. The total consumption is the highest in Mexico City and the larger cities, but this may be a function of product availability. Storage of inventories for lengthy periods of time by large food wholesalers also distorts any estimates of current domestic consumption. Mexican government sources admit that reliable information on domestic sales is not available and that canned or raw product not exported usually is attributed in official calculations to domestic consumption. The Mexican media report that domestic tuna consumption is nearly 30,000 tonnes (round weight) per year, but this is probably an overstatement reflecting both consumption and inventories. The minimum annual domestic consumption is estimated to be about 9,000 tonnes.

The two identifiable sources of low domestic consumption, aside from general recessionary conditions, are (1) the high price of tuna relative to other protein sources and incomes and (2) the poorly developed internal marketing and distribution systems. As a result, the potential sales to the domestic market are difficult to predict. A population of over 81 million certainly offers the potential for tremendous sales. To stimulate domestic sales, industry sources recommend that the product form be differentiated in the future to provide consumers with a greater range of choices.

DEVELOPMENT POTENTIALS AND PROBLEMS

The success of development enterprises can be evaluated in terms of how well the enterprise fulfilled its goals, the condition of its financial health, or its ability to respond to changing economic and political conditions. Industry development over the long run also critically depends on the environment created by government. The interaction between government policy and industry success is an important element when the Mexican tuna case is evaluated. For example, if the development of the tuna industry had been less closely linked to national economic conditions, the financial situation of the industry might be healthier than it is now. Conversely, without the strong commitment of the Mexican government, the industry might have collapsed under the pressure of the U.S. import embargo. This section assesses the successes of the industry and identifies continuing short- and long-run problems.

Government and industry strategy

The national fishery plan stated several goals for the fishery sector, including the tuna industry. Catches have increased, and production of fishery products in general has increased. The goals of increased employment in the fishery sector and development of new domestic and international markets have been only partly met. A new purse seine fleet has been established, and the country's fisheries catch has been diversified to include tuna.

Several government and industry strategies clearly contributed to these successes and can be identified as follows:

• *Declaration of the Mexican EEZ (1978) and the presidential decree requiring license fees for foreign fishing vessels in the EEZ (1980).* These events signaled to the Mexican private sector that the government would support investment in fisheries. The response of the private sector was strong and immediate. The development of a tuna industry was envisioned as a profitable business opportunity, as well as a potential source of export earnings, employment, domestic tax revenues, and food protein.

• *Tax concessions.* The Mexican government granted income tax concessions to the industry to further encourage its development. This policy is consistent with protecting domestic industry during the start-up phase of development. As these concessions are removed, the industry will be increasingly exposed to world tuna market competition.

• *Free-trade zone status.* The tuna industry development has primarily taken place in Ensenada, Baja California. Ensenada is a free-trade area exempted from import and export tariffs levied on Mexican industries located elsewhere in the country. This policy is consistent with the plans that targeted the tuna catch for the international market. The Mexican tuna industry could export or import freely. There is little evidence that the tuna industry was ever intended to generate large export or import tax revenues. In fact, the free-trade status of the tuna industry occurred by accident as the Baja peninsula has enjoyed free-trade status for several years.

The government and industry currently recognize that free trade gives the Mexican industry an advantage over other countries' industries (for example, the Philippines) where tariffs are levied on exported raw tuna. An additional advantage of the free-trade status is the ability to import tax-free parts for vessels. It is expected that a free-trade zone status will be requested for the new canneries under construction in Colima and Chiapas states, which will give the canned products from these sites an advantage in world and domestic markets.

• *Technology acquisition*. The industry, with the support of the Mexican government, has sought out and acquired technological expertise worldwide. The Mexican industry has capitalized on its long-standing business contacts in the U.S. tuna industry and developed new contacts in the European markets. The proximity of the industry to San Diego, California, also allowed it to employ displaced crews from the declining U.S. tuna industry. Although several former U.S. vessel captains have been employed as instructors in the Mexican fleet, this situation is considered temporary. These instructors often want to be paid in dollars, which imposes a financial burden that the industry wishes to avoid. The highest-producing vessel in the industry in 1985 had an all-Mexican crew, including the helicopter pilot.

• *Industry structure*. The Mexican tuna industry is structured much like the U.S. tuna industry. The purse seine owners are sophisticated entrepreneurs, often engaged in a wide range of business activity besides tuna production. General managers located at the port cities oversee the daily operations of vessels and crews. These managers are versed in business and usually have extensive backgrounds in the tuna industry. The SEPESCA and BANPESCA offices in the port cities also oversee daily operational matters related to finance, state cannery purchases, and policy. In addition, the fishermen's cooperatives have both local and national offices.

Policy issues are generally transacted in Mexico City among vessel owners, the national association of cooperatives, SEPESCA, BANPESCA, and other government agencies. In some respects, this structure has significantly enhanced vessel productivity in Ensenada. Except for financial matters, there is relatively little intervention in the day-to-day operations of the vessels in the harvesting sector.

Development problems

Despite careful planning, the industry and the government now face a crucial turning point in tuna industry development. The major problems are primarily the result of the national financial crisis and the imposition of the U.S. embargo.

The most important short-term problem facing the Mexican tuna industry in 1986 is cash flow. There are difficulties in providing the fleet with operational monies and resolving the external debt, both of which emanate directly from the domestic financial crisis.

The finances involved in support of the tuna industry have become complex and reflect the direct linkages between the industry and the national economy. The short-term problem is the delay in payments to vessels from government-operated cannery sales. These payments come through BANPESCA, which is also trying to collect vessel debt payments. Under the terms of the vessel-refinancing and loan guarantees, BANPESCA required the vessels to meet specified performance standards. Several vessels have not met these standards. As a result, BANPESCA has had to coordinate with the national treasury to ensure foreign vessel debt payments. At the same time, BANPESCA must decide whether or not to foreclose on those vessels that do not make debt payments. If foreclosures occur, BANPESCA will become the de facto owner of tuna vessels, a situation that most banks wish to avoid.

SEPESCA and BANPESCA are attempting to resolve the debt servicing problems of the vessel owners. It is politically difficult for either SEPESCA or BANPESCA to seek special concessions for the fishing industry when so many other export-oriented industries in Mexico face similar financial problems. The decision to refinance tuna vessels in 1984 and assume the external debt no doubt subsidized some inefficient vessel operations that otherwise would have gone bankrupt. Without the U.S. embargo, operations will be less political, enabling the Mexican government to enforce production requirements and discourage fishing by the marginal vessels. Both government agencies fully recognize the challenge and remain fully committed to establishment of a domestic tuna industry.

As the fleet is rationalized, longer-term problems related to world markets and resource availability will become apparent. Mexican tuna exports, either round or canned, could depress world tuna prices in the future, forcing some Mexican vessels out of operation. These vessels would have to be converted or sold at a probable loss. In addition, as Mexico allows more foreign fishing in its EEZ, resource problems may become relevant.

IMPLICATIONS FOR THE PACIFIC ISLANDS REGION

The significance of the development of the Mexican tuna industry for the Pacific islands region is twofold. First, the tuna industry in Mexico has dealt with economic and political problems relevant to many developing countries. The lessons applicable to other countries that manage or develop tuna resources include credit limitations, strength of the national economy, trade-offs between public and private sector well-being, and government/industry collaboration in investment. Second, the Mexican industry now has the potential to become a major agent in the world tuna market. Decisions made by the industry in Mexico could affect investment and marketing opportunities for industries in the Pacific islands region.

Lessons from the Mexican case

Economic development, even of an export-oriented industry, takes place within a national context. The Mexican tuna industry survived the domestic financial crisis and the U.S. embargo because the Mexican government was able to provide heavy subsidies. These subsidies took several forms: credit guarantees, tax-free imports and exports in Ensenada, purchase and inventory of tuna catches, biological and economic research, and domestic raw tuna price stability.

In the early planning stages, the government chose to let the private sector operate where it was the most efficient—in the harvesting sector. This decision minimized the political favoritism that is often observed in state-run enterprises, leading to charges of inefficiency and corruption.

Both government and industry had sufficient institutional flexibility to decrease the size of the planned fleet, organize and implement the vessel-refinancing program, and maintain political integrity in responding to the U.S. embargo. It is yet to be seen whether this flexibility will succeed in overcoming the remaining problems of bottlenecks in ports and processing and the development of a domestic demand for tuna.

In at least three instances, however, planning and cooperation between government and industry appear to have failed.

First, any industries that export primary commodities are sensitive to fluctuations in the world prices of those commodities. The Mexican planners apparently did not foresee either the collapse of tuna prices or the international restructuring of the tuna industry. Over the long term, however, both these events possibly could work in favor of the Mexican industry; in other words, the industry, now in its intermediate stage of development, may be equipped better than a younger industry to withstand further price decreases.

A second instance of failure to collaborate occurred with the declaration of the Mexican EEZ and the foreign licensing requirement, which led to imposition of the U.S. embargo. The Mexican government, to fulfill its commitment to the domestic tuna industry, then had to subsidize the industry for survival. It is still debatable why the embargo lasted as long as it did, but clearly the Mexican government policy was in conflict with industry development goals.

In the third instance, the national debt crisis generated complex internal problems for the Mexican tuna industry; however, in 1986 the government appeared willing to participate in the resolution of these problems to the extent possible, given its fiscal constraints.

Mexico and the Pacific islands region

Little debate exists over the fact that sometime in the future the Mexican tuna fleet, rationalized at an efficient size, will become a distant-water fleet seeking alternative fishing grounds. The central and western Pacific offers one of these alternatives. Several conditions will determine the level of interaction between the Mexican industry and the tuna-rich nations in the Pacific islands.

• *Size of rationalized fleet*. As previously discussed, the lifting of the U.S. embargo will help to normalize operations of the Mexican industry. In addition, as Mexican government subsidies and concessions are removed, the operating fleet will be increasingly exposed to the competitive pressures of world tuna markets. This pressure will probably drive some Mexican tuna vessels out of operation. The ultimate size of the Mexican fleet relative to the eastern Pacific resources will determine whether or not the fleet seeks alternative fishing grounds.

• *Domestic market development*. If the domestic market in Mexico is sufficiently developed, the industry may seek alternative fishing grounds to supply this market. Market expansion is likely to require greater amounts of the raw product as well as transshipment facilities and additional canning capacity. The most efficient location for this processing capacity would be near the distant-water fishing grounds. Therefore, another condition, which may generate interaction between the Mexican tuna industry and the Pacific island nations, would be the need to supply an increased domestic demand for tuna.

• *Location choice*. Either scarcity of domestic tuna resources or increased domestic demand could induce the Mexican fleet to fish outside the Mexican EEZ. The choice of location will depend on political considerations as well as economic costs. Several Latin American countries, including Mexico, have been negotiating treaties for fishing rights under the organizational auspices of Organización Latinoamericana de Desarrollo Pesquero ([OLDEPESCA] Latin American Organization for Fishery Development). The negotiating countries include several that have neither fleets nor effective means of enforcing EEZ claims. The Mexican fleet could in theory become a "pirate fleet" along the central and south American coasts. The probability of this will ultimately depend on the Mexican government's policy with respect to regulation of domestic fleet activities outside the Mexican EEZ.

Even if the Mexican fleet were to fish farther south along the Pacific coast of central and south America, the fleet will face increasing fuel and transshipment costs as distances increase. It is highly improbable that joint ventures in fuel depots, transshipment facilities, or processing will develop between Mexico and those Latin American countries that have domestic capabilities for these inputs in tuna production. More likely, the Mexican industry (and government) will look to joint ventures with countries that have little or no domestic industry. These opportunities are available predominantly in central America (Honduras, Nicaragua) and in the Pacific islands region.

Consequently, a third condition, which could encourage interaction between a Mexican distant-water fleet and the Pacific island nations, would exist either if the OLDEPESCA negotiations break down or if the costs of fishing the central and western Pacific relative to the Latin American coast are less.

CONCLUSIONS

Joint-venture opportunities between Mexico and Pacific island countries could take any number of forms, ranging from direct purchase of fishing access rights to intercountry industrial development projects. Whatever the actual outcome, two points must be considered: credit availability and fuel requirements. The status of Mexico's national economy will continue to affect the operations of the tuna industry even in a distant-water capacity. Foreign exchange and credit availability will be continuing problems.

Barter among developing countries is increasingly being used in trade packages. It would be realistic to suppose that fishing deals with Mexico could include a barter component. Barter has obvious drawbacks for the Pacific island countries in that Mexico may not produce the goods and services that are needed by these countries. However, the advantage is that the barter arrangement can circumvent credit and foreign-exchange problems, which can be severe.

The ability of the Mexican tuna industry to become a major distant-water fishing power in the central and western Pacific will depend on continued Mexican government support. One example of potential problems with government support is related to fuel requirements. It has been suggested that Petroleos Mexicanos ([PEMEX] Mexican National Petroleum Company) underwrite the placement of fuel depots around the Pacific for use by a distant-water fleet. Although theoretically attractive, this plan effectively requires special treatment for the tuna industry. Therefore, although economically feasible, such a plan may not be politically acceptable.

Conditions in the future probably will be conducive to collaborative efforts in fishing between Mexico and the Pacific island nations. The success of any interaction will be affected by the fact that the parties involved will be those developing countries that need to conserve foreign exchange and credit. Creativity in negotiations will be called for in a way not required in negotiations with the industrialized countries (for example, the United States). This creativity in negotiations has the potential of enhancing the economic development of the countries involved, as well as freeing them from the credit constraints imposed by domestic and international debt problems.

NOTE

1. Calculations used 1984 costs and an exchange rate of 300 pesos to the U.S. dollar.

11. The Development of the Philippine Tuna Industry

Jesse M. Floyd and David J. Doulman

ABSTRACT—This chapter traces the evolution, expansion, and decline of the tuna industry in the Philippines. The industry in the Philippines has implications for Pacific island countries for two reasons. First, it is a competing industry for these countries, and changes in the Philippine industry could affect investment and marketing opportunities for industries in the Pacific islands. Second, the reasons for the decline of the Philippine industry may provide guidance for island countries in developing and managing their own industries.

INTRODUCTION

The Philippine tuna industry's success was due to improved catching methods, which were accompanied by improved preservation and storage facilities and by modern processing techniques. The overall result was a dramatic increase in the export trade of frozen and canned tuna commodities from the Philippines. In terms of foreign exchange earnings, the Philippine tuna industry appears to have peaked in 1980–81. Since then the annual value of exports has declined, and the total volume of frozen tuna exports has decreased. Deteriorating conditions in overseas markets have contributed to the decline of the Philippine tuna export industry, but internal factors have had a more significant impact on the industry's long-term development potential.

INDUSTRY STRUCTURE

The tuna fishery emerged as the largest fishery sector in the Philippines during the five-year period from 1976 to 1980. Total recorded production of tuna and tuna-like species increased from less than 25,000 tonnes in the early 1970s to over 200,000 tonnes in 1980. Since 1980 total tuna production has consistently exceeded 200,000 tonnes per year.

Production reached its highest level of 242,000 tonnes in 1983 and declined to 226,000 tonnes in 1984.

Almost one-half of the tuna catch in the Philippines consists of skipjack (*Katsuwonus pelamis*) and yellowfin (*Thunnus albacares*). Frigate tuna (*Auxis thazard*) and eastern little tuna (*Euthynnus affinis*) are equally abundant, but they are lower-value species and generally are sold as fresh fish on the local market. Several other tuna and tuna-like species are occasionally caught in non-commercial quantities in Philippine waters.

Tuna fishing operations are undertaken by both the municipal and commercial fishing sectors. The distinction between the two sectors is made solely on the basis of vessel size. Municipal operations involve small-scale artisanal fishermen operating boats less than 3 gross tonnes (GT), most of which are wooden double-outrigger canoes known as *bancas*, 4 to 11 m in length. It is estimated that there are about 110,000 motorized *bancas*, typically equipped with 10–16 hp inboard gasoline engines, and 90,000 non-motorized *bancas*. All fishing boats over 3 GT are classified as commercial vessels and require an operating license from the Bureau of Fisheries and Aquatic Resources (BFAR). In 1981 there were 2,349 registered commerical fishing vessels and 675 accessory vessels. The majority of the commercial fishing vessels were small purse seiners, ringnetters, bagnetters, and trawlers less than 20 GT. Among the larger fishing boats, 131 were over 100 gross registered tonnes (GRT), 89 were purse seiners operating primarily in the offshore tuna fishery, and 25 were trawlers. Most of the accessory vessels were fish carriers.

Municipal operations

Tuna fishing operations in the municipal sector occur year-round, but the fishing effort and production in different regions of the country is seasonal. The major factor governing operations is the prevailing monsoon. Although the BFAR compiles some data on the total number of municipal fishermen and *bancas* per region and on tuna production by region, the data do not indicate the number of fishermen, boats, and units of gear involved in catching tuna on a full-time or part-time basis, and it is not possible to estimate the catch-per-unit-of-effort (CPUE).

Despite the nature of their gear and the limited size and range of their fishing vessels, municipal fishermen have consistently accounted for the majority of tuna production in the Philippines. In 1976 the municipal sector landed 85 percent of total tuna production and 65 percent of skipjack, yellowfin, and bigeye production. Since 1976 the relative proportion of tuna landings by municipal fishermen has declined as a result of the rapid development of the commercial tuna fishing industry. Municipal fishermen landed 51 percent of total tuna production in 1982 and approximately 50 percent of the skipjack and yellowfin production from 1982 to 1984.

The most common fishing gears used by municipal tuna fishermen are handlines, gill nets, and ringnets. The most productive of these gears is the handline. Catches by handline gear accounted for 80 percent of skipjack and yellowfin production by municipal fishermen and 60 percent of total municipal tuna production in 1981. Gill nets accounted for 10 percent of skipjack and yellowfin production and 20 percent of total municipal tuna production. Ringnets accounted for the majority of the remaining municipal tuna catch, but various other gears such as the longline, troll line, and fish corral were also used.

Municipal handline operators frequently catch large, deep-sea yellowfin tuna as well as some bigeye tuna and billfish. These operations occur throughout the year with a peak season during the summer months. They are particularly common in the Moro Gulf and the Sulu Sea, the most productive tuna fishing grounds in the Philippines.

The fish are purchased by local tuna exporting companies and are either frozen for further processing or packed in ice for immediate air transportation to the Japanese fresh-fish market. These procedures are duplicated in Zamboanga and Davao and, to a lesser extent, in Cagayan d'Oro on the northern side of Mindanao and in Puerto Princessa on Palawan, where tuna exporting companies have representatives and where freezing facilities are available. Some tuna exporting companies also operate a fleet of ships—up to 300 GRT each—that purchase tuna from municipal fishermen on the fishing grounds to ensure adequate supply, proper handling, and product quality. In many instances, however, municipal landing sites are isolated and lack suitable preservation and storage facilities. As a result, much of the tuna landed by the municipal sector is consumed by communities in the vicinity or transported to regional marketing centers for local consumption.

Commercial operations

The commercial tuna fishing industry grew rapidly in the 1970s as Philippine producers developed the capacity for export operations in response to the increasing demand for tuna on the international market. Most of the growth was the result of increased production by small purse seine and ringnet operators using bamboo fish-aggregating devices (FADs). The success of these operations prompted the Food and Agriculture Organization (FAO) to conduct fishing surveys in late 1974. These surveys confirmed that tuna resources were substantial in Philippine waters and that FADs could be modified and used commercially. The FAO team estimated that the potential catch of representative seiners (420 GRT and 283 GRT) would be at least 1,500 tonnes if they fished commercially in Philippine waters for 25 days per month, ten months a year.

Commercial production of skipjack and yellowfin tuna increased from 13,000 tonnes in 1976 to its highest level of 63,000 tonnes in 1983. As a result of this success, the Philippines became one of the top five producers of albacore (*Thunnus alalunga*), yellowfin, skipjack, bigeye (*Thunnus obesus*), and bluefin (*Thunnus thynnus*) tuna in the world in the second half of the 1970s.

GOVERNMENT POLICIES

The introduction of purse seining in combination with FADs was accompanied by two important changes in Philippine government policy. The most important change in its development policy regarding fisheries occurred as a result of Presidential Decree No. 704 (1975). A second and related policy change is reflected in the Philippine investment incentives legislation. Both policies played a critical role in the rapid expansion of the Philippine tuna industry.

Presidential Decree No. 704 provides the basis for all government policies and programs in the Philippine fishing industry. It authorizes foreign investment in the fishing industry in contrast to earlier legislation that limited the exploitation of the country's marine resources to Filipino citizens and corporations.

The Investment Incentives Act (1968), amended in 1977, is known as the Agricultural Investment Incentives Act. The purpose of the amending decree was to foster foreign participation and investment in the Philippine fishing industry by making those joint venture corporations and foreign-owned firms incorporated in the Philippines eligible for the investment incentives. Under the Agricultural Investment Incentives Act, those individuals, partnerships, and domestic corporations—whose voting capital stock is 60 percent owned by Philippine nationals, cooperatives, and other entities organized and existing under Philippine laws—qualify for investment incentives.

Investment incentives available to qualifying enterprises are listed in the appendix to this chapter. The major incentives applicable to fishery enterprises include (1) duty-free entry of boats and gear; (2) tax deductions for pre-operating expenses and approved training costs; (3) net operating loss carryover; (4) accelerated depreciation on boats; and (5) for exported products, tax credit for taxes and duties paid on supplies, such as fuel, and tax reduction for five years for the cost of local labor and materials.

INDUSTRY OPERATIONS

Frozen tuna exporters

The potential of tuna as a major export commodity in the Philippines prompted representatives and officials of several tuna exporting companies to form the Philippine Tuna Producers and Exporters Association (PTPEA) in 1973. The principal objectives of the PTPEA were to (1) establish confidence in the quality of Philippine tuna products in foreign markets, (2) to foster fair and friendly competition among members, and (3) to promote cooperative studies with government and international agencies for the effective development and exploitation of the country's fisheries resources. The membership of the PTPEA included 22 companies in 1982–83. Twelve of these companies accounted for 95 percent of the volume and value of frozen tuna exports in 1980.

Five of the major frozen tuna exporters in 1981 were officially recognized as joint ventures. They were Philippine Tuna Venture, Peninsula Fishing Corporation, Fortuna Mariculture Corporation, Eastship Fishing Corporation, and World Marine & Development Corporation. Each of these companies was classified as a joint venture and registered with the Fishery Industry Development Council because they involved charter agreements to lease foreign fishing vessels. Dole Philippines Inc, a wholly owned U.S. company, was the second largest exporter of frozen tuna in 1981, but it is not classified as a joint venture by Philippine fishery authorities. As a group, these six companies exported 12,943 tonnes of frozen tuna worth $15.8 million in 1981, approximately 36 percent of the total volume and value of the Philippine frozen tuna exports.

Frozen tuna exports declined in 1981 despite an increase both in total tuna production and in the number of companies exporting tuna. In 1981, 35 companies were exporting fresh and frozen tuna and were operating 80 purse seiners (14 of which were over 500 GRT) and 56 longliners between 20 and 79 GRT.

Distant-water fishing operations

Several purse seine operators, which are still active in the Philippine tuna industry, have dealt with the problems of excess fishing capacity, resource depletion, and financial losses by extending the range of their fishing operations through access arrangements with neighboring countries. These access arrangements date back to 1981, and Philippine companies have operated in Palau, Sabah, Borneo, Papua New Guinea, and the Federated States of Micronesia.

Canned tuna exporters

The Philippines has processed and exported canned tuna since the early 1970s, but the volume and value of exports were quite small until 1978. That year Judric Canning Corporation became fully operational and began processing tuna exclusively for export markets. Judric's entry into the Philippine canned tuna industry motivated other established canners to intensify their tuna processing operations, and by 1980 export production amounted to over 11,000 tonnes and almost $30 million.

In 1980 five canners were active in the Philippine canned tuna export industry besides Judric: Century Canning Corporation, Premier Industrial & Development Corporation, Pure Foods Corporation, South Pacific Export Corporation, and Santa Monica Canning Corporation. Judric produced almost one-half of the estimated total export volume in 1980. Pure Foods was the next largest producer and accounted for 19 percent of the estimated export volume. Of its capital stock, 20 percent is owned by the American Hormel International Corporation; most of the remainder is held by the Ayala Corporation, a Filipino-Japanese joint venture. Century, the largest locally owned corporation in the industry, accounted for 13 percent of the estimated canned tuna export production in 1980, Santa Monica 9 percent, Premier Industrial 6 percent, and South Pacific 5 percent.

Deteriorating conditions in overseas markets in the early 1980s had a serious impact on Philippine canned tuna operations. All of the 17 domestic fish canners that had intended to diversify their production and enter the canned tuna export industry in 1982–83 canceled their plans. Some companies such as Standard Foods and Majescan stopped processing tuna for overseas markets in 1982. Others reduced their Manila-based operations because of tuna supply shortages and higher fish prices and then decentralized their activities either by building new canning facilities or by expanding their existing facilities in the southern Philippines.

The adjustments made by processors in the canned tuna industry and the expansion of facilities in the southern Philippines led to further increases in Philippine canned tuna exports in 1983. In 1983 canned tuna exports increased to 2.1 million cases, of which 72 percent consisted of retail cases and 28 percent institutional cases.

One reason behind the rapid expansion of Philippine canned tuna exports and the number of exporters in the industry was the ban on canned mackerel and sardine imports in August 1983. The ban was imposed by government as a means of conserving foreign exchange. But it also had the beneficial effect of encouraging processors, which previously had canned tuna only for export markets, to diversify their production of canned mackerel and sardines for the local market without competition from overseas processors, primarily Japan. Mar Fishing was the only company that did not take advantage of the import ban, while most of the other canners began processing more mackerel and sardines than tuna.

Although the import ban has had a positive effect, conditions have not improved for the Philippine canned tuna industry. Exports declined in 1984 to 1.9 million cases, approximately the same level as in 1982, and the industry utilized only 40 to 45 percent of its estimated 110,000 tonnes of processing capacity.

INTERNATIONAL TRADE

The proliferation of companies entering the Philippine tuna industry since the introduction of industrial tuna purse seining in 1976 resulted in the rapid expansion of international trade in tuna commodities. Increasing exports of frozen tuna in the late 1970s were accompanied by exports of dried and smoked tuna (*katsuobushi*). Bonito and undersized skipjack weighing less than 1.5 kg are generally used for *katsuobushi* because of their lower oil content, which facilitates drying during the manufacturing process.

Exports of *katsuobushi* reached their highest level in 1980, totaling 608 tonnes and valued at $2.6 million. At that time 18 companies were active in the *katsuobushi* industry. The largest company, accounting for approximately one-third of the volume and value of exports, was Orient Marine and Fishing Resources. In 1981 production declined by almost 50 percent to 341 tonnes and $1.2 million, and the number of major companies in the industry fell to six. The following year the industry declined further, and exports amounted to only 193 tonnes and $0.5 million.

Frozen tuna exports also declined dramatically in 1981–82 from their highest level in 1980 to only 17,731 tonnes and $15.9 million. Although exports increased by a small margin in 1983 to 18,559 tonnes and $21.2 million, they declined to their lowest level since 1976 to 13,759 tonnes in 1984. Of the export volume in 1983 (1,742 tonnes), 9 percent was sashimi-grade yellowfin destined for fresh fish markets in Japan, 20 percent (3,774 tonnes) was frozen skipjack, and 70 percent (13,043 tonnes) was yellowfin suitable for canning. The composition of frozen tuna exports in 1983 reflects both a steady increase in the volume and proportion of sashimi-grade exports since 1980 and a general decline in the proportion of frozen skipjack in total exports.

The United States has consistently been the largest importer of Philippine frozen tuna in terms of volume. In 1979 the United States imported 23,536 tonnes of frozen tuna worth $20.4 million from the Philippines. This amounted to 72 percent of the total volume and 64 percent of the total value of Philippine exports that year. In 1980 and 1981 the volume of frozen tuna exports to the United States remained fairly stable, but the proportion of Philippine exports destined for the United States decreased due to the overall increase in the volume of exports from the Philippines to the rest of the world. In 1983 Philippine frozen tuna exports to the United States declined to their lowest level since the mid-1970s and amounted to only 4,796 tonnes and $2.6 million, or 26 percent of the export volume and 12 percent of the export value.

As the proportion of exports to the United States declined, the role of Italy and Japan as trading partners increased. In 1983, the last year for which data are available, Italy imported 45 percent of the volume and 49 percent of the value of Philippine chilled/frozen tuna exports. Most of the exports to Italy were frozen yellowfin tuna for canning. Japan imported 26 percent of the volume and 36 percent of the value, approximately one-third of which was sashimi-grade tuna.

The agreement among government officials and industry representatives to allow annual frozen tuna imports of 14,000 tonnes has been in effect since 1982. In April 1985, however, the Tuna Canning Association of the Philippines (TCAP) requested that the government increase the level of imports to 30,000 tonnes annually to reflect more recent increases in the industry's capacity. According to the TCAP processors, 30,000 tonnes represents the difference between the industry's operating capacity (approximately 110,000 tonnes) and the volume of fish available for processing (70,000 to 80,000 tonnes). BFAR preliminary data show that there were no frozen tuna imports in 1983, and industry sources indicate that there were only 2,000 tonnes of imports in 1984. No tuna was imported in 1985 for canning.

Several factors account for the industry's failure to take advantage of the tuna import authority. The most important factor is related to the country's lack of foreign exchange and its high interest rates. According to some industry representatives, another factor is the bureaucratic time delay involved in actually issuing and implementing the import permit. Increased production of distant-water fishing operations may have also been a factor, but low prices for canned tuna in the international tuna market and a protected domestic market for canned mackerel and sardines have probably been among the most important factors influencing the industry's decision not to import tuna and instead to use its excess processing capacity to can mackerel and sardines for the local market.

INDUSTRY DECLINE

Internal problems

The Federation of Fishing Associations of the Philippines (FFAP) first expressed concern about signs of overfishing and resource depletion in 1981. In a paper commenting on the rapid expansion of the country's tuna fishing fleets, an FFAP representative noted that dozens of purse seiners were operating in close proximity with one another in the limited expanse of the Sulu Sea and the Moro Gulf and stated that many boats were reporting low yields and signs of resource depletion. Similar concerns were also expressed by two of the country's largest tuna operators: Philippine Tuna Venture and RJL Martinez Fishing.

In response to the overfishing and resource depletion problems, the FFAP recommended (1) adopting a policy that would impose a limit on the number of tuna seiners operating in the country at any given time and (2) requiring an operator to phase out an equivalent older vessel if a more modern vessel is built or imported as a means of maintaining the established tonnage and number limit. No limits were proposed for longline fishing vessels.

Several factors have contributed to the resource depletion problem. One factor has been the proliferation of companies entering the Philippine tuna industry, as well as the increasing number of large purse seine vessels over 500 GRT operating in the fishery. This has been accompanied by numerous reports (and apprehensions) of foreign fishing vessels poaching in Philippine territorial waters. The Philippine Coast Guard apprehended 144 Japanese and Taiwanese fishing vessels for illegal fishing from 1972 to 1977, an average of 2.3 vessels a month. Taiwanese fishery authorities, in turn, acknowledge that five fishermen have been killed and that 53 fishing boats and 167 fishermen have been detained by Philippine authorities. Given the country's long coastline and the Philippine Coast Guard's limited surveillance capability, the actual number of foreign boats engaged in illegal fishing in Philippine waters is presumably quite large. The FFAP has estimated that the illegal catch may be as high as 50,000 to 100,000 tonnes a year. The FFAP has also called on the government to take a stronger position against vessels caught fishing illegally in Philippine waters.

Another factor affecting the resource depletion problem has been the introduction and widespread adoption of purse seine/FAD fishing and its apparent effect on the catch of juvenile fish. In General Santos City, for example, about 100 tonnes of small skipjack, weighing 70 to 100 g each, are unloaded daily during the height of the tuna fishing season from March to June. Available catch reports from commercial export operators also suggest that the Philippine purse seine tuna fishery is catching a significant quantity of small tuna. Data from four exporters indicate that almost two-thirds of their production in 1981 consisted of undersized tuna.

Moreover, FAO researchers claim that more than 95 percent of all skipjack landed in the Philippines is less than 30 cm in length and that fish are routinely landed at a length of 14 cm. The problem is further aggravated by the fact that large numbers of small tuna are not landed but rather are used as bait for longline and handline fishing around FADs.

Overfishing, resource depletion, and shortages in supply to the local tuna processing industry were not the only problems confronting the Philippine tuna industry in the early 1980s. Continued growth in the tuna industry was also constrained by increases in the costs of operation, particularly increases in the cost of fuel, which accounts for between 50 and 60 percent of the total operating costs of distant-water fishing vessels. Both the cost of imported fuel to the fishing industry and the industry's right to a subsidy on its fuel expenditures are controversial issues in the Philippines. The controversy arises over the interpretation of Section 106(a) of the Tariff and Customs Code that provides for a refund (drawback) of up to 99 percent of the duty paid on all imported fuels used for the propulsion of vessels engaged in coastal trade. Industry representatives argue that fishing vessels are entitled to the fuel duty drawback and that the failure to grant this drawback to industry has prevented it from operating at full capacity.

Another setback for Philippine tuna producers occurred when the Philippine government proposed the imposition of a 4 percent tax on frozen tuna exports. The proposed tax was later reduced to 2 percent and implemented in November 1983. The principal purpose of the export tax was to capture windfall profits gained by exporters from the devaluation of the peso in October 1983 and not necessarily to capture the resource rent from the exploitation of the country's tuna resources. The tax was a controversial issue between government and industry, but it also led to a split between the tuna producers and processors, who were both represented in the PTPEA at that time. The break occurred because the processors in the newly formed TCAP endorsed the 2 percent tax on exports, while producers felt that the export tax would place them at a disadvantage in negotiating prices with local tuna processors.

The export tax was the first of three taxes recently levied on exporters by the government to bolster the troubled Philippine economy. An excise tax of 10 percent on the purchase of foreign exchange and payment for freight, interest, and other charges was imposed in 1984, followed by a stabilization tax of 30 percent on the exchange rate differential resulting from the June 1984 devaluation of the peso. In addition, tuna exporters have had to pay a customary inspection fee to the BFAR equivalent to 0.5 percent of the free on board (FOB) value of the products.

The export, excise, and stabilization taxes were intended to promote economic stability in the Philippine economy by compensating for adjustments in the peso's exchange rate. In the industry's view, the taxes added to operating costs and undermined the competitiveness of the Philippine tuna industry compared with the processing industries of other developing countries. Moreover, the taxes were imposed at a time when production, manufacturing, and packing costs were increasing and tuna commodity prices were declining. Philippine canned tuna processors claim that they lost an average of $2.49 per case of canned tuna on sales made from November 1983 to August 1984. The new taxes and the increased cost of imported goods (i.e., fuel, tinplate, and spare parts) caused by the successive devaluations of the peso have had a serious effect on the profitability of tuna operations in the Philippines.

Another factor affecting the operating costs of the Philippine tuna industry is the high cost of freight charges on exports of canned tuna to the west coast of the United States, the major destination of Philippine tuna products. According to PTPEA representatives, members of the Philippine-North American Conference (PNAC) charged $198 per tonne for canned tuna freight to the U.S. west coast in 1982, approximately 30 percent of the FOB Manila export sales price of $650 per tonne of tuna. This rate compared with only $160 per tonne from Singapore and Guam to the U.S. west coast and $95 from Japan. The disparity in freight rates has prompted PTPEA members and officials to claim that commercial freight rates from the Philippines are disproportionately high and that high freight rates place the Philippine tuna industry in a non-competitive position on the U.S. market.

External problems

Tuna exporters faced several problems in overseas markets that affected the profitability of the Philippine tuna industry. The problem receiving the most attention was the sharp decline in tuna prices on the international market in general and in the United States in particular. For example, the value of frozen yellowfin exports from the Philippines to Japan and Europe fell from $1,487 per tonne in 1980 to $950 in 1981, and the value of frozen skipjack to the United States fell from $1,180 per short ton in 1981 to $450 in 1982. Despite recovery from the 1981–82 worldwide economic recession, international market conditions for frozen tuna have not significantly improved. Oversupply in relation to demand has been the major reason behind the continuing slump of the world tuna market since 1982.

The glut of frozen tuna on the international market has been accompanied by higher import duties on canned tuna to the United States. Imports of canned tuna not in oil arriving in the United States before the 20 percent quota is filled are dutied at 6 percent ad valorem and thereafter at 12.5 percent. Given that imports of canned tuna not in oil to the United States have consistently exceeded the quota by progressively larger amounts since 1980, Philippine exporters of canned tuna have had to pay higher duty on an increasing proportion of their canned tuna exports to the United States. This has added to the operating costs of Philippine processors and exporters when they are already disadvantaged by declining prices and weakening consumer demand for the Philippine product in the U.S. market.

Besides the U.S. import duty, canned tuna from the Philippines is subject to a countervailing duty. The countervailing duty was imposed in August 1983 by the U.S. government on Philippine canned tuna imports in response to a petition submitted to the U.S. Department of Commerce by the U.S. Tuna Foundation (USTF) on behalf of the major American tuna processors. The petition successfully argued that the Philippine government subsidized the local tuna processing industry by (1) granting investment incentives to export-oriented enterprises, (2) providing preferential financing benefits for exports, (3) holding equity participation in at least one Philippine tuna cannery, Diamond Seafoods, and (4) issuing policies of insurance and certificates of guarantee against credit risks arising out of or in connection with export transactions.

The petition also called for a 10 percent ad valorem countervailing duty, but the U.S. International Trade Administration (USITA) imposed a 0.72 percent duty. After recently reviewing the case, the USITA instructed the U.S. Customs Service to forgo collection of the countervailing duty effective 8 March 1985. A deposit is still being collected pending the USITA's final report.

The U.S. countervailing duty was imposed at a time when Philippine canned tuna imports were losing their dominance in the U.S. market. After successfully competing with Japan and Taiwan and becoming the largest canned tuna exporter to the United States in 1981 and 1982, the Philippines fell behind Thailand as the largest exporter to the United States in 1983. That year Thailand exported 33 percent of the total volume of U.S. canned tuna imports, and the Philippines exported 26 percent. The following year Thailand expanded its exports to the United States by 125 percent, from 18,108 tonnes in 1983 to 40,763 tonnes in 1984, and increased its share of the U.S. canned tuna market to

55 percent. The market share of the Philippines declined to 14 percent in 1984, and its exports to the United States dropped from 14,502 tonnes in 1983 to 10,079 in 1984.

Several other factors have contributed to Thailand's success and enabled it to improve its competitive position as compared with the Philippines in the canned tuna industry. Thai processors are not subject to any import fees, export taxes, or foreign duties. Labor, electricity, and freight charges, as well as can, carton, and label costs, are less. Interests rates are lower, and foreign exchange to import fish, tinplate, and spare parts is more accessible. The sole advantage that the Philippines enjoys is lower fish prices. However, Philippine tuna producers are more inclined to export their fish to Thailand, where prices are higher. Philippine tuna producers must also export some of their production to earn foreign exchange so that they can purchase essential import items not available in the Philippines (e.g., nylon nets, electronic equipment, and spare parts for their vessels).

IMPLICATIONS FOR PACIFIC ISLAND COUNTRIES

Despite its decline, the Philippine tuna industry remains a relatively important supplier of canned tuna to the world market. The industry competes with Pacific island countries in the marketing of their canned tuna and related products. Island countries wanting to establish tuna processing industries will also face competition from Philippine producers in the future, even if these countries concentrate on the more discriminating and high-value segments of the market.

Economic conditions, both within the Philippines and within the tuna industry, are significantly different from those conditions found in the Pacific islands region. The reasons underlying the development and subsequent decline of the Philippine tuna industry do not seem to be directly relevant to policymakers in the islands region.

Central to the problem of the Philippine tuna industry is the inability of canners to secure adequate supplies of domestic imported frozen tuna to meet their processing requirements. The domestic fishery is being overexploited, and despite extremely large tuna harvests, much of the tuna landed is unsuitable for canning because it is too small. Thus this tuna is sold on local fresh-fish markets; however, it would probably not be harvested if these markets and the canning industry were not competing for fish.

The difficulties associated with obtaining foreign exchange have meant that Philippine processors are unable to import frozen tuna to meet the shortfall in the domestic catch. This situation is ironic, given the world's oversupply of frozen tuna in recent years. Philippine exports of frozen tuna to Europe and Thailand have also exacerbated conditions in the industry.

Tuna industries in the Pacific islands region are not generally faced with raw material shortages or foreign exchange restrictions that inhibit industry development and growth. Domestic fresh-fish markets are not competing with canneries, and industries have "first call" on domestically caught fish before exports are made.

Policies designed to promote the growth of the Philippine tuna industry were initially successful, but other macroeconomic policies, which were implemented to grapple with a deteriorating economy, restricted their effectiveness. Incentives are provided to the tuna industry, but they now have little meaning because other policies (e.g., exchange rate restrictions) prevent the incentives from taking effect and having the intended impact. If some sectors of the economy are to be actively promoted, policymakers should ensure that other policies adopted reinforce, or at least not constrain, those policies intended to promote particular sectors.

CONCLUSION

The Philippine tuna industry is plagued by numerous problems and is struggling for survival. Dwindling tuna resources, escalating costs, increasing taxes, and declining export prices are some of the problems that Philippine tuna producers and processors faced in 1985. These problems were complicated by (1) the national economic crisis, (2) the high interest rates, (3) the lack of foreign exchange, (4) the continuing slump in the international tuna market, and (5) Thailand's effectiveness in the international canned tuna industry.

The measures taken to assist the tuna industry in 1985 are a positive indication of the Philippine government's intentions. Nonetheless, operators in the tuna industry have reason to question government policy. The taxes imposed on Philippine tuna exporters and those still in effect eroded the industry's competitiveness, weakened its morale, and destroyed the cooperation that existed between industry and government. Moreover, the industry's confidence in government has been shaken by the government's failure to grant the duty drawback for fuel oil for all fishing vessels. Added to this is the industry's contention that the Ministry of Agriculture and Food (MAF) has purposely delayed issuing tuna import permits.

By far the biggest and most contentious issue looming over the future of the Philippine tuna industry is the government's intention to liberalize trade for several commodities, including canned and frozen fish. If the government decides to implement the liberalization plan—whether on its own or as a result of pressure from international financial institutions—and if the plan authorizes the importation of canned mackerel and sardines, Philippine tuna processors will once again have to face competition from cheaper Japanese canned fish imports in domestic markets. This situation will undermine the success that tuna processors have had on the local market since the ban on imported fish products in 1983, and it may spell the end of the Philippine tuna canning industry because processors now depend on local sales of their canned mackerel and sardine production to sustain tuna operations.

Government policies and inaction, however, are not the only factors behind the demise of the Philippine tuna industry. The canning industry is overcapitalized, the resource is showing signs of depletion, and supply shortages are a common complaint of the processing sector. There are also reports that the tuna industry is inefficient and particularly that the yield from processing canned tuna in the Philippines is lower than it is in many other countries. This inefficiency is due to several factors such as the smaller size of fish processed in the Philippines, the lack of motivation among the work force, and the poor quality control both at sea and during the manufacturing process. The overall result is a product that many importers feel is inferior to other products on the international tuna market.

In view of this situation and the other problems affecting the Philippine tuna industry, a significant resurgence of the industry is unlikely in the near future.

APPENDIX: Incentives Available to Fishery Enterprises in the Philippines

Incentives Act (1968)

1. Deduction of organizational and pre-operation expenses from taxable income over a period of not more than ten years from start of operation.
2. Deduction of labor-training expenses from taxable income equivalent to 0.5 percent of expenses but not more than 10 percent of direct labor wage.
3. Accelerated depreciation.
4. Carryover deduction from taxable income of net operating losses incurred in any of the first ten years of operation deductible for the six years immediately following the year of such loss.

5. Exemption/reduction and/or deferment of tariff duties and compensating tax on importation of machinery, equipment, and spare parts.
6. Tax credit equivalent to 100 percent of the value of compensating tax and customs duties that would have been paid on machinery, equipment, and spare parts (purchased from a domestic manufacturer), had these items been imported.
7. Tax credit for tax withheld on interest payments on foreign loans, provided that such credit is not enjoyed by lender-remittee in its own country and that the registered enterprise has assumed liability for tax payment.
8. Deduction from taxable income in the year that reinvestment was made of a certain percentage of the amount of undistributed profits or surplus transferred to capital stock for procurement of machinery and equipment and other expansion.
9. Protection from government competition.

Export Incentives Act (1971)

1. Additional deduction from taxable income of direct labor costs and local raw materials utililized in the manufacture of export products but not exceeding 24 percent of total export revenues for producers, 10 percent of total export sales for traders, and 50 percent of total export fees for service exporters.
2. Preference in the grant of government laws.
3. Exemption from export and stabilization taxes.
4. Additional deduction from taxable income of 1 percent of incremental export sales.

Agricultural Investment Incentives Act (1977)

1. Accelerated depreciation of breeding stock. Under this decree, breeding stock is considered to be a fixed asset or capital equipment subject to depreciation.
2. Additional deduction from taxable income of 25 percent of research and development expenses and 25 percent management training expenses of Philippine nationals, provided the deduction shall not exceed 10 percent of taxable income within seven years from date of registration.
3. Tax exemption on breeding stocks, fish, plants, and genetic materials imported within seven years from date of registration.

4. Additional deduction from taxable income of 30 percent of freight and transportation expenses within seven years from date of registration of enterprises established in a preferred geographical area for fishery/agricultural development, where transportation facilities are deficient and such freight and transportation expenses are incurred in the course of transporting registered products from the enterprise's project area to the nearest economic marketing center.

12.
The Development of the Tuna Industry in Thailand

Greg J. Crough

ABSTRACT—This chapter analyzes the development of Thailand's tuna industry. It reviews the structure of the country's fishing and tuna industry, examines government policies toward the industry, discusses tuna processing and trade, evaluates prospects for Thailand's tuna industry, and describes implications for the Pacific islands region.

FISHING INDUSTRY OVERVIEW

Agriculture and fisheries dominate the economy of Thailand and account for more than 70 percent of employment, 70 percent of foreign exchange earnings, and 24 percent of the gross domestic product (GDP). In addition, agricultural and fisheries production has led to the development of a wide range of manufacturing activities based on chilling, freezing, drying, pickling, and canning. Food processing represents about one-fourth of all manufacturing activity. Thailand is one of the five net food exporters in the world, and it is the only one in Asia.

Since the mid-1960s the fishing and processing sectors have grown rapidly, and Thailand is now one of the ten largest fishing countries in the world, with a total catch of 2.1 million tonnes in 1984 that was valued at $675 million. Marine fishery production accounted for more than 90 percent of the country's total catch. Thailand is also one of the ten largest exporters of fishery products, with exports valued at $633 million in 1984.

Thailand's coastline of about 2,600 km borders on the Gulf of Thailand and the Andaman Sea; of the country's 73 provinces, 23 are coastal. The fishing industry provides the primary source of animal protein for most of Thailand's population, and this industry has long been the main source of income of the people living in Thailand's coastal regions. With about 20,000 fishing vessels, the Thai fleet ranks seventh in the world.

Although fisheries production increased significantly during the past decade, recent production has rarely exceeded the peak level achieved in 1977. The declaration of exclusive economic zones (EEZs) by neighboring countries is estimated to have caused annual losses of up to 800,000 tonnes of potential catch by Thai fishermen.

With about 70 percent of Thailand's total marine landings coming from outside the country's national waters, the other countries are likely to increase surveillance and enforcement efforts in their own waters, and the catches by Thai vessels are likely to decrease. Many Thai fishing vessels have strayed, often inadvertently, into the EEZs of other countries, and hundreds of fishermen and their vessels have been seized. Because Thai joint-venture agreements have been concluded with a number of countries, including Indonesia, Bangladesh, China, Saudi Arabia, and Australia, a significant proportion of the country's marine catch now comes from these joint ventures.

Due to the loss of about 40 percent of their traditional fishing grounds, Thai fishermen intensified their fishing efforts in domestic waters (particularly in the Gulf of Thailand). Consequently, Thai fish stocks are now being overfished. In both the Gulf of Thailand and the Andaman Sea, fishing in excess of the estimated maximum sustainable yield has been occurring for several years. Moreover, the catch composition has been changing toward smaller and less valuable species.

The rapid development of the commercial trawling and purse seining fleet and the loss of large areas of traditional fishing grounds have combined to produce increased conflicts between Thailand's offshore fishing fleets and the coastal fishermen who have been suffering extreme economic hardships as a result. Some fishermen have had to resort to illegal fishing—such as nearshore push netting—which has caused additional resource depletion due to the higher proportion of juvenile shrimp and fish species caught with this gear.

In addition, the deforestation of mangrove swamps and inland forests has resulted in heavy siltation of coastal regions and the degradation of important nursery grounds for marine fish and shrimp species. The Thai government has also been encouraging major heavy industrial development at various coastal locations, including petrochemical industries on the eastern seaboard in Chon Buri province. The resulting high levels of urban and industrial pollution have caused serious problems for the marine environment of the Gulf of Thailand.

A final problem is common to many developing countries. Many of the facilities at the country's 19 major landing ports and private jetties (including ice production facilities) are unsanitary, and on-board handling of catches is often inadequate. These problems have contributed to a decrease in the proportion of food fish in total production and an increase in trash fish. It has also encouraged the canning industry to utilize increasing quantities of relatively higher quality imported marine products.

TUNA FISHING INDUSTRY

Until the late 1970s, tuna was still a relatively minor fishery in Thailand. Tuna was not a major component of the diet of the Thai people, and with only a small number of canneries, fishermen had few outlets for any tuna they caught. The total catch of all tuna between 1974 and 1980 averaged about 12,000 tonnes a year, but after 1980 the catch increased dramatically, to a peak of over 85,000 tonnes in 1983. The dramatic increase in the tuna catch was in the Food and Agriculture Organization (FAO)-defined fishing area of the western central Pacific (Area 71), which includes the Gulf of Thailand, while the eastern Indian Ocean (Area 57) catch recorded a more modest increase. Despite the record catch in 1983, recent unpublished estimates suggest that the catch of longtail tuna (*Thunnus tonggol*) declined to less than 25,000 tonnes in 1986.

Although Thailand's fishing fleet has made record catches in recent years, it still accounts for only a relatively small proportion of the total tuna catch in these FAO-defined fishing areas. During the period 1974–84 the tuna catch of Thailand averaged only about 3 percent of the total catch by all countries in both the western central Pacific and eastern Indian Oceans.

The high proportion of small fish and the lack of quality control on many fishing vessels have meant that despite the large catches in recent years, significant quantities of tuna caught are small and unsuitable for canning. The small fish are generally used for fishmeal.

The fishing season in Thailand covers the whole year, with only small annual variations due to the changes in the two monsoon seasons. There are five important landing ports for tuna fishing vessels (Songkhla, Fhuket, Pattani, Satun, and Rayong). These five ports accounted for more than one-third of Thailand's marine landings in 1984.

Numerous types of fishing gear are used in the tuna fisheries in Thailand. The principal methods of catching tuna are purse seines and gill nets.

Comprehensive catch-and-effort statistics for the Thai tuna fleet for the period 1971–83 have been published by the Indo-Pacific Tuna Development and Management Program in Colombo, Sri Lanka.

For purse seine gear between 1971 and 1981, the catch-per-unit-of-effort (CPUE) (average catch per day) averaged only about 53 kg per day in the Gulf of Thailand. In the following two years there was a dramatic increase to 567 kg per day. In the Andaman Sea the purse seine catch rate was higher between 1971 and 1981, averaging about 150 kg per day, although catch rates were more variable. In 1982 the catch rate increased to 465 kg per day, declining to 177 kg per day in 1983. Of the record catch of 85,000 tonnes in 1983, almost 80 percent was taken by the 691 fishing vessels registered as using luring purse seine gear.

GOVERNMENT POLICIES FOR THE FISHING INDUSTRY

The fisheries sector has always had an important place in Thai government policy, although the relative contribution of this sector to the overall GDP of Thailand has not increased significantly due to the rapid growth of the manufacturing and service sectors in the last decade. The fisheries sector makes an important contribution to export earnings and employment, as well as provides the principal source of animal protein in the diet of the Thai people.

An important emphasis in the Thai government's policies toward the fishing industry has been the promotion of commercial aquaculture. Part of the reason for this emphasis has been the recognition of the difficulties facing the marine fisheries sector and the increasing inability of that sector to meet the demands of domestic consumers and the export-oriented processing industry. Thailand has a well-developed smallholder aquaculture tradition, and freshwater production in 1985 accounted for over one-third of fisheries production. Foreign investment in aquaculture projects is encouraged, particularly in brackish water shrimp and prawn farming.

The fishing industry is one of 37 business activities classified as Category B businesses according to the Alien Business Decree of November 1972. The government does not allow the formation of companies that are majority-owned by foreigners to engage in business activities listed under this category unless they are companies promoted by the Board of Investment (BOI). Companies that were operating prior to 26 November 1972 are allowed to continue their operations if an appropriate certificate is obtained from the Thai Business Registrar.

Most foreign direct investment in Thailand now takes place through joint ventures. There are relatively few instances of fully owned subsidiaries or branches of multinational corporations (MNCs). Equity joint ventures are the preferred mode where foreign ownership is involved, although, as in many other countries, there is a proliferation of nonequity or contractual arrangements with MNCs.

The BOI is under the Office of the Prime Minister, and its role is to encourage foreign and domestic investment. The board may promote particular investments or companies when the output of the project

(1) is not available locally in sufficient quantity, (2) is currently being produced by outdated processes, or (3) is regarded as being nationally beneficial from economic, social, or security viewpoints. Particular emphasis is placed on export-oriented production and resource development, as well as processing, employment generation, decentralization, energy conservation, and development of basic industries.

The deep-sea fishing and offshore fishing industries are among those industries eligible for promotion. The vessels used for deep-sea fishing must be a minimum of 150 gross registered tonnes (GRT) and incorporate sophisticated radar and other equipment, while offshore fishing vessels must be at least 50 GRT. By the end of 1984, a total of seven companies had received promotional status, including one of the largest seafood processing and cold-storage companies in Thailand (Thai Seri Universal Food Co Ltd).

The tuna fishing industry is now receiving more attention from the government, and several policy initiatives are being formulated to increase the quantity and quality of domestic production. To some extent, these efforts seem to have been successful with the dramatic increases in the domestic tuna catch, but since 1984 the catch has declined significantly. In addition, due to the tuna canning industry's rapid rate of expansion, increasing quantities of imported tuna are required to maintain output. Clearly, the reliability of supplies of tuna is important for the canning industry, and a more stable domestic tuna catch is desirable. The raw material costs of the canning industry must be kept as low as possible because it exports virtually all of its production to some of the most competitive tuna markets in the world. Although Thailand is a major buyer of tuna in the world market, it cannot control the price of its imports to any significant degree, and the prices of domestically produced marine products have increased considerably (although they are still below the cost of imported tuna).

The government's objective is to increase the domestic catch to about one-third of the requirements of the country's canneries. To do so, a major effort will be necessary to upgrade the skills and technology of the domestic fishermen and the fishing fleet. The government is already strongly encouraging local fishermen to merge together and form companies so that they can negotiate supply agreements with the domestic canneries. The canneries could also finance either the upgrading of existing vessels or the acquisition of new vessels. Some 20 local boats are estimated to have contracts with the canneries; between 50 and 80 vessels sometimes provide tuna. The government hopes that the canneries will give more consideration to the interests of local fishermen and will assist them in improving the quality and quantity of the domestic catch.

The Thai fleet already contains a significant number of medium-sized purse seiners that use sophisticated fishing equipment. The government's intention is that this fleet will be encouraged to fish for tuna in certain confined coastal areas of neighboring countries, including Indonesia (a fishing agreement was signed in October 1986) and Papua New Guinea, and perhaps further into the western Pacific and Indian Oceans. Joint-venture agreements with several countries have been negotiated for these purposes. Given the size of the tuna canning industry in Thailand, the supply of frozen tuna to such a market on a more regular basis would be an attractive proposition for some small countries of the western Pacific that may be having difficulties in marketing their catch.

The government is also examining the possibility of assisting fishermen and fishing companies to acquire larger deep-water purse seiners and carrier vessels. However, due to the lack of experience of Thai fishermen in this type of tuna fishing, a more fruitful approach might be to enter into joint ventures with some of the developed countries that fish in the Pacific and Indian Oceans and that have excess fleet capacity. An important aspect of these joint ventures would presumably be the transfer of technology.

TUNA PROCESSING INDUSTRY

Although agriculture still plays an important role in Thailand's economy, food processing has assumed the dominant role in the country's rapidly expanding manufacturing industries and now accounts for about 30 percent of the manufacturing sector.

The canned seafood industries developed along with large-scale fruit and vegetable processing. Following the establishment of the BOI in 1972, canned seafood was recognized as an important potential export earner for Thailand. In the same year the industry was started with the establishment of an Australian-Thai joint venture company, Safcol (Thailand) Ltd, which began to process and export a range of products including tuna. This company has been particularly successful and is Thailand's largest exporter of processed seafood products.

The major growth period for the industry, however, was in the late 1970s and early 1980s, and Thailand is now the world's largest exporter of canned tuna. The tuna canning industry in Thailand is the third-largest in the world, after the United States (American Samoa and Puerto Rico) and Japan.

Thailand is also the world's largest exporter of canned crabmeat, and other important products include canned squid, baby clams, and shrimp. In the period 1977–84 the seafood canning industry as a whole recorded an average annual rate of growth of 47 percent.

Several reasons caused the growth of the tuna canning industry in Thailand. The first was the existence of a significant industrial and commercial infrastructure, which generated the rapid growth of the manufacturing and service industries in the last decade. The government placed strong emphasis in its national plans on the development of a wide range of infrastructure facilities, particularly communications and transportation.

The second reason is that Thailand's geographical location is central to three main fishing areas in the Indian and Pacific Oceans, and this aspect has become important in recent years as the canning industry has had to rely on large and increasing quantities of imported tuna. Although the initial development of the industry in the early 1970s was based on an apparently large tuna resource in the Gulf of Thailand, problems with the quality and reliability of the domestic catch have forced Thailand to become one of the world's largest importers of tuna. The fishing areas of particular importance include the waters near the Philippines and Indonesia, the western Pacific region, and the Indian Ocean (the Seychelles and Maldives).

The third reason has been the low wage structure, which is important in a labor-intensive activity such as seafood processing and canning.

An additional factor, which partly explains why the Thai industry has grown rapidly only in the last few years, was the economic and political problems of the Philippines. In 1982 the Philippines was the largest canned tuna exporter to the United States, accounting for 32 percent of U.S. imports. Thailand accounted for 21 percent of imports in 1982, but by 1985 Thailand's share had increased to 57 percent and the Philippines had declined to 14 percent.

The diversified nature of many canneries in Thailand is important for their viability, particularly given the competitive nature of the tuna processing and marketing industry. Many companies produce a wide variety of canned products, depending on market conditions and availability of raw materials. For some companies, the canning of pet food, particularly using domestically caught sardines and pilchards, is the most rapidly expanding and profitable segment of their business.

One of the main constraints facing the industry in Thailand, as in many other countries, is the extent of processing overcapacity. Clearly, the success of the industry in developing major export markets has attracted a considerable number of new entrants, which has resulted in reduced margins for many companies.

The government is concerned about the problems resulting from the overcapacity in the tuna canning industry, particularly those associated with the quality of the product and the extent of the price competition that affects overall export returns. It has been suggested that the government seek to limit the expansion of the industry's production capacity to about 15 million cartons (the present capacity is about 12 or 13 million cartons).

The processing and preservation of food, including seafood products, is an activity that is eligible for promotion by the BOI, and the promoted companies are entitled to various investment incentives. For many companies, which have invested in the industry in recent years, these incentives have considerably improved the profitability of their operations.

By 1983, 19 of the 30 to 35 canned seafood producers that catered largely to exports were promoted by the BOI, and with a combined capacity of 60,000 tonnes these companies accounted for about 80 percent of the country's canning capacity. The BOI's 1985–86 directory listed 17 promoted companies, including most of the largest seafood processors and tuna canners in Thailand. A small number of these companies are joint ventures with foreign companies or individuals, including the company that established the first tuna cannery in Thailand, Safcol (Thailand) Ltd. However, most of the largest companies are apparently locally owned and involve some of Thailand's prominent family groups.

Large tuna companies from other countries (e.g., the United States) have not directly invested in the industry in Thailand but have instead preferred to purchase the canned product packed under their own labels from existing canneries.

MAJOR TUNA OPERATORS

Although many companies operate tuna canneries in Thailand, the industry is dominated by a few large companies. Despite the importance of tuna canning and the tremendous growth in production and exports in recent years, most of the companies process a wide range of seafood products, and the diversified nature of their business is an important source of stability for the industry. The large companies operating in the industry are reviewed below.

Unicord Co Ltd

Unicord Co Ltd is the largest tuna canning company in Thailand, with the second largest tuna cannery in the world (capacity of 350 tonnes per day). The company claims to account for about 8 percent of world production. Unicord, one of the largest 100 companies in Thailand, is a joint venture between two prominent families involved in the domestic whiskey distillery business. It began manufacturing various seafood products in 1978. Canning tuna for human consumption was estimated to account for about 90 percent of gross revenue. Other canned products include pet food (made from tuna and sardine by-products) and canned sardines for human consumption.

Unicord is a major importer of skipjack tuna (*Katsuwonus pelamis*), and of the 120,000 tonnes imported into Thailand in 1985, about 80,000 tonnes were purchased by the company. Unicord claims that it accounts for 60 percent of Thailand's tuna exports to the United States, although as with the other companies, strenuous efforts are being made to diversify market outlets for the canned product, particularly in Europe.

Thai Union Manufacturing Ltd

The second-largest tuna canning company is Thai Union Manufacturing Ltd. Predominantly owned by Thai shareholders, the company began production in 1980 as a BOI-promoted company.

Safcol (Thailand) Ltd

Safcol was the first company to be established in Thailand to undertake tuna canning and seafood processing, and it became a BOI-promoted company in 1972. The company is a joint venture between Safcol Holdings Ltd (Australia's largest seafood processing and exporting company), local Thai family interests, and several Chinese investors. Safcol processes more than 60,000 tonnes of marine products per year.

The company operates six tuna canneries and seafood processing plants at various locations in Thailand, and tuna canning accounts for about 25 percent of the company's business. At the end of 1986 the company announced plans to invest $3 million in a new processing plant near Bangkok, designed to be the company's largest facility, employing more than 1,000 workers.

Safcol's gross revenue has increased rapidly during the last decade, primarily due to diversified operations. A wide range of seafood products is processed and exported, including clams, squid, lobster, pilchards, and sardines, as well as tuna. However, pet food canning, which is the largest segment of the company, is continuing to grow, and probably has the best long-term growth potential. The company is concerned about future expansion of the tuna canning industry, primarily because of protectionist moves in some of the main markets (particularly the United States) and because of Thailand's processing overcapacity.

Thai Seri Group

Another important seafood fishing and processing company is the Thai Seri Group, which began operations in 1932 as an agency for fish buying and selling. The group operates about 20 fishing boats and employs some 3,000 workers in its diversified seafood processing, cold-storage, and fishing operations. The company has entered into joint fishing ventures in Bangladesh and India and has been involved in negotiations for fishing rights in the South China Sea, the Persian Gulf, and the Pacific Ocean.

Apart from the companies described above, the tuna canning and seafood processing industry has stimulated a considerable amount of associated industrial development. Of particular importance is can manufacturing, printing, and packaging. Thailand has more than 60 can-manufacturing factories and 70 manufacturers of paper boxes and corrugated cartons.

INTERNATIONAL TRADE IN TUNA PRODUCTS

Thailand is the world's largest exporter of canned tuna products, a position achieved in only the past few years, and it is the third-largest importer of fresh and frozen tuna, after the United States and Japan.

The rapid expansion of the tuna canning industry has necessitated increasing volumes of imported tuna because the domestic tuna fishing industry is unable to supply sufficient tuna for processing. However, data relating to imports are not generally reliable.

The growth of imports from the United States accounts for the largest component of the growth in total tuna imports in recent years. Other important suppliers include Japan, the Maldives, and France. In times of tight supply, some Thai canneries have purchased tuna from as far away as Mexico.

Total imports of fresh and frozen fish were valued at $28.5 million in 1983 and $63.7 million in 1984. Despite the dramatic increase in imports, they still accounted for less than 1 percent of the country's total imports of all commodities.

In 1985 tuna exports totaled more than 87,000 tonnes and were valued at $175 million, although these exports still accounted for only about 3 percent of Thailand's total exports.

The companies exporting tuna from Thailand are overwhelmingly contract packers, and production is generally sold under brand names of major producers, distributors, and retailers in individual markets. The western European markets have a greater tendency to use the brand names of the Thai canneries. Although some of the major U.S. tuna canners do buy the canned product from Thailand, they are primarily reliant on their own production in Puerto Rico and American Samoa to supply the continental U.S. market. The Thai canneries supply a significant proportion of the production of many known brands in the U.S. market, including Chicken of the Sea, Geisha, and 3 Diamonds; they also supply large quantities of tuna to institutional users. John West Foods sources at least 60 percent of its tuna from Thailand. While Safcol is largely a contract packer, it also markets many seafood products under its own brand name.

Thailand's exports to several developed countries have been particularly impressive. In the United States prior to the 1980s, imports of canned tuna accounted for a comparatively small share of the domestic market. However, as demand among U.S. consumers changed from tuna packed in oil to tuna packed in water, imports of the latter product increased rapidly.

Thailand's share of U.S. imports of canned tuna increased from 10 percent in 1980 to 57 percent in 1985 and to 64 percent for the first eight months of 1986. Canned tuna exports accounted for about 8 percent of Thailand's total exports of $1.54 billion to the United States in 1985.

Although the United States represents a major proportion of the total business of Thai canneries, many have been trying to diversify their markets. For Unicord, the largest producer, U.S. exports have fallen from 70 percent to about 55 percent of the company's total exports, with European sales constituting about 45 percent. For Safcol, U.S. exports accounted for about 25 percent of the company's tuna business in 1986, with the remainder directed to Europe and the Middle East. The price sensitivity of the tuna market means that the share of particular markets can vary considerably, depending on changes in currency exchange rates.

The European market is becoming important to the Thai industry, and several companies indicated that the product shipped to Europe was of a somewhat higher quality than that sent to the United States because many European consumers were willing to pay more money for a better-quality product.

In a potentially significant development, Safcol (Thailand) Ltd made its first sales of canned tuna to the Japanese market in mid-1986. The opportunity to achieve such sales was provided by both the strong appreciation in the value of the Japanese yen and an ability to produce a canned product that would meet the stringent quality standards demanded by the Japanese. These sales were arranged directly with a Japanese supermarket chain and were packed under that chain's own label, effectively bypassing the Japanese wholesaling system. Undoubtedly, the revaluation of the yen is beginning to cause dramatic structural changes in the Japanese economy, including its tuna canning industry, which is suffering from overcapacity. It has even been suggested that the Japanese canning companies seek to relocate part of their production facilities in lower-cost offshore locations, probably on a joint-venture basis. Thailand would be a strong contender to attract some of this investment.

Concerns have recently been expressed about the quality of canned tuna exports from Thailand, but the quality of the product from most companies is generally believed to be satisfactory.

DEVELOPMENT PROBLEMS AND THE FUTURE OF THE INDUSTRY

Despite the extraordinary growth and development of the tuna industry in Thailand during the 1980s, some continuing problems may affect the industry's future growth prospects. The first is the possibility of increased protectionism in the United States. This protectionism may be aimed specifically at canned tuna imports or more generally may constitute part of a trade package for dealing with the country's trade deficit. Considerable concern already exists in Thailand and, of course, in other countries about the implications of some recent agricultural legislation passed, or introduced, in the United States.

The rapidly increasing volume of imports of canned tuna in water from Thailand and other countries has led to rising calls for protection for the domestic industry, even though only one tuna cannery remains in the U.S. mainland.

Despite calls for more protection, the U.S. industry can no longer present a unified position on many issues because of the changes that have occurred in the fishing and processing sectors in the last decade. On the one hand, some producers, including the most aggressively protectionist company, Star-Kist, have expanded their production facilities in lower-cost locations, such as Puerto Rico and American Samoa, and are already "exporting" tuna to the United States. Other companies have imported the canned product from Thailand.

Against these protectionist moves, Thailand has become more assertive and effective in presenting its case in Washington. The Thai government has also been quick to point out that Thailand is a "frontline state" in Southeast Asia and that the "security" of U.S. economic, political, and military interests in this part of the world require an economically strong Thailand.

The imports of frozen tuna pose an additional problem for the future development of the Thai industry. Although exports of canned tuna are growing strongly, a corresponding growth has occurred in imports of tuna because the domestic catch is insufficient to meet the canning industry's requirements. The cost of imported tuna accounts for about one-third of revenue from exports. This problem is compounded by the undervaluation of exports by some companies in order to reduce their payments of withholding tax, which is set at 25 percent.

Export revenue is further depressed due to intense price competition from other countries, as well as to competition among the large number of producers in Thailand itself.

The increasing technological sophistication of the industry in Thailand will undoubtedly assist the country in maintaining its premier position in the world export industry. Many companies are utilizing some of the latest and most expensive canning technology available, although from the country's overall point of view, the installation of such equipment is at the expense of taxation revenue, and the equipment generally has to be imported.

One important advantage of the industry in Thailand is that most of the major companies are diversified and thus are not totally reliant on the export of canned tuna. To some extent, resource availability problems also affect other parts of the seafood processing industry, but the range of products of the companies can insulate them from market disturbances in individual countries or products. The industry is also well integrated and provides an important economic stimulus to other sectors of the economy.

IMPLICATIONS FOR THE PACIFIC ISLANDS REGION

The competitiveness and success of the tuna canning industry in Thailand has implications for any country in the Pacific islands region that either operates tuna canneries or contemplates the development of canneries. Because such canneries would have difficulty in competing against Thailand in the lower-priced segment of the tuna market, they would probably need to concentrate on the higher-quality and higher-value canned product.

The growing level of imports, however, does provide a market outlet for tuna caught in the Pacific Ocean. Now that the United States and the Pacific island countries have agreed on a tuna treaty, market access for the U.S. fleet in Thailand should provide opportunities for additional exports of tuna from the western Pacific Ocean. If Thailand either significantly increases its domestic catch or relies more on its own joint ventures for future supplies, then the market opportunities for some Pacific island countries will become more restricted. The Thai government has already indicated that it will seek to achieve a greater measure of self-reliance for the tuna industry, but the success of this strategy remains to be seen.

Recent catch statistics indicate that the domestic tuna catch of Thailand has fallen sharply since 1984, and despite the Thai government's policy measures, the western Pacific remains the most productive resource area for skipjack tuna. At least three countries, the Federated States of Micronesia, Solomon Islands, and Papua New Guinea, will probably exploit these market opportunities.

13.
A Summary of Tuna Industries in Mexico, the Philippines, and Thailand: A Comparative Analysis

Linda Lucas Hudgins

ABSTRACT—This summary chapter compares the tuna country studies of Mexico, the Philippines, and Thailand. This synthesis describes the importance of these three countries in the global tuna market and analyzes the development patterns chosen by the governments of Mexico, the Philippines, and Thailand for their respective tuna industries.

INTRODUCTION

Mexico has the largest and newest purse seine tuna fleet in the world, and since 1980 it has become a major harvester competing with the fleets of the United States, Japan, the Philippines, and Spain in the international tuna market.[1] The Philippine tuna industry, formerly an international leader in tuna markets, has contracted due to resource depletion problems and inappropriate government policy. Thailand, as of 1983, has become the major exporter of canned tuna to the United States competing with processing industries in the Philippines, the United States, Japan, and Taiwan.

The tuna industries in Mexico, the Philippines, and Thailand represent three diverse cases of industry development at both the harvesting and the processing levels. In each case, government policy affected industry performance. Furthermore, because these industries have become major actors in the world tuna market, the actions taken by industries in Mexico, the Philippines, and Thailand will have repercussions for tuna industries in the Pacific islands region.

The tuna harvests of Mexico, the Philippines, and Thailand represent about 18 percent of the world's tuna catches and 26 percent of all tuna caught in the Pacific Ocean. These countries together have over 323,000 tonnes of tuna canning capacity, which is about 15 percent of the world's total.

The four issue areas included in this synthesis are (1) an industry overview with discussion of fleet capacity, tuna resource availability, and processing activity for each country; (2) government policy as related to tuna industry development with attention to both national and industry-specific promotional policies within each country; (3) position of each country relative to international tuna markets with discussion of tuna exports, imports, and international relations with the United States; and (4) concluding remarks that identify lessons from the tuna industry development in the three country studies.

INDUSTRY OVERVIEW

Fleets

The tuna fleets of Mexico, the Philippines, and Thailand reflect those differences between fleets that have developed from artisanal-based fisheries and those that have been developed for the commercial export market. Mexico concentrated on building a modern purse seine fleet with distant-water capabilities, while Thailand and the Philippines concentrated on development of domestic processing. The Mexican fleet sells raw tuna primarily on the international market for processing elsewhere, while the fleets in the Philippines and Thailand sell to their domestic canning industries, which then export tuna, primarily to the United States. Direct comparisons of fleet productivity in the three countries are not useful because the compositions of the fleets by vessel size and number of vessels engaged in fishing vary widely in each country. A critical difference, however, is that the Mexican fleet has distant-water capabilities, which enable it to fish well out of its own territorial waters. For this reason, domestic tuna resource availability is not the same immediate problem for the Mexican fleet that it is for the fleets of the Philippines and Thailand.

The differences in vessel productivity between the commercial fleet of Mexico and the fleets of the Philippines and Thailand are given in Table 1. For example, in 1984 the Mexican fleet of 59 purse seiners caught about 72,800 tonnes of tuna, the Philippine fleet of over 112,000 artisanal and commercial vessels caught about 225,700 tonnes, and the Thai fleet of 20,000 multipurpose and multispecies vessels caught about 76,800 tonnes.

The Mexican fleet consists of both bait boats and purse seiners. About 30 percent are bait boats of less than 400 gross registered tonnes (GRT). These vessels fish close to the Mexican coastline. The majority of the fleet's vessels are purse seiners. About 26 percent are between 400 and 750 GRT, and 44 percent are 750 GRT or larger. The seiners are engaged in a full-time commercial fishery enterprise in Mexico's exclusive economic zone (EEZ) and along the Pacific coastline of Central America. The Mexican fleet has the capacity to catch over 110,000 tonnes of tuna annually. The catches have been between 70 and 90 percent yellowfin tuna (*Thunnus albacares*), with the remainder being mostly skipjack tuna (*Katsuwonus pelamis*). The fleet is 60 percent owned by private investors and is based at Ensenada, Baja California, Mexico, about 40 km south of San Diego, California (United States).

The tuna fleet in the Philippines has over 110,000 artisanal vessels, which are mostly less than 3 GRT in size. The fleet also has 2,349 identified commercial vessels, mostly less than 100 GRT. Artisanal fishermen harvest about 51 percent of the country's total tuna catch consisting of several species, although 50 percent are yellowfin, skipjack, and bigeye tuna (*Thunnus obesus*). The fleet fishes year-round in various parts of the country with a variety of gears including handlines, gill nets, ringnets, and purse seines. The introduction of fish-aggregating devices (FADs) in the late 1970s contributed both to dramatically increased tuna catches and to eventual resource depletion problems.

The Thai fishing fleet ranks seventh worldwide in size with about 20,000 vessels. All vessels are multipurpose and multispecies, and almost all are less than 100 GRT. Tuna accounts for between 3 and 6 percent of total fleet catches. The major species, eastern little tuna (*Euthynnus affinis*) and longtail tuna (*Thunnus tonggol*), are caught year-round with purse seines, gill nets, and troll lines. Most of the tuna caught by fleets is by vessels of less than 25 m in length fishing in Thailand's territorial waters. In recent years, the fleet has lost a significant portion of its fishing area due to maritime disputes with neighboring countries. The size of tuna caught has also declined, and because the small fish is unsuitable for canning, it is used in the domestic

Table 1. Tuna catches in Mexico, the Philippines, and Thailand, 1976–84

	000 tonnes			
Year	Mexico	Philippines	Thailand	Total
1976	—	124.9	9.7	134.6
1977	19.5	215.9	12.9	248.3
1978	25.4	183.9	10.3	219.6
1979	31.9	197.3	16.8	246.0
1980	33.1	200.8	13.6	247.5
1981	70.5	203.7	22.2	296.4
1982	41.4	216.6	49.3	307.3
1983	27.7	242.2	85.3	355.2
1984	72.8	225.7	76.8	375.3

Sources: Crough 1987b; Floyd 1986b; Hudgins 1986d.

fishmeal industry. The fleet provides about 30 percent of the tuna utilized by the domestic canning industry, with the remainder being imported.

Resource availability

No immediate resource availability problems appear to be related to the tuna stocks in the eastern tropical Pacific (ETP), the area that includes the Mexican EEZ. This is in contrast to the resource depletion problems in the Philippines and Thailand. The source of this depletion is largely due to territorial restrictions in Thailand (i.e., several overlapping EEZs) and to illegal fishing and catches of juvenile tuna in the Philippines.

Tuna stocks in the ETP are primarily yellowfin and skipjack tuna. Stocks in the ETP have been managed since 1950 by the Inter-American Tropical Tuna Commission (IATTC), although political conflicts beginning in the early 1970s have weakened the organization's strength as a regulatory body. In particular, several Latin American countries have withdrawn from the IATTC in protest over quota allocations that they believe to favor countries with large historical catches (e.g., the United States) over those with newly developing fleets (e.g., Mexico). Some Latin American countries, led by Mexico, are forming an alternative management regime, the Eastern Pacific Tuna Organization, under the auspices of the Latin American Organization for Fishery Development ([OLDEPESCA] Organizacion Latinoamericana de Desarrollo Pesquero).

The IATTC estimates that the maximum sustainable yield (MSY) of all tuna in the ETP is 545,400 tonnes, which can support the current levels of fishing effort. The Mexican government, however, intends to license more foreign vessels than it has done previously, and this will place an added pressure on stocks. The potential catch of the Mexican fleet, when fully developed, is about 140,000 tonnes, some 30,000 tonnes less than the estimated annual sustainable yield of tuna in the Mexican EEZ. The Mexican fleet is prepared to fish farther south along the Pacific coastline of the Americas or in the central and western Pacific if faced with resource problems.

The tuna resource availability around the Philippines is in question, and overfishing has been cited as a cause for the contraction of the Philippine tuna fishery since 1980. Between 1981 and 1985 at least 3,000 FADs were placed in Philippine waters. In addition, several companies entered the industry with large purse seine vessels, leading to overcapitalization. Reports also indicate that there is a serious illegal fishing problem caused by foreign vessels. These vessels could be catching up to 100,000 tonnes of tuna annually. Depletion of the domestic resource has led to increased imports of frozen tuna to supply the Philippine canning industry.

In Thailand the declaration of 200-mile EEZs by neighboring countries resulted in the loss to Thai fishermen of 40 percent of their traditional fishing areas. Nowhere does the Thai EEZ extend the full 200 miles offshore because of competing claims from neighboring countries. As a result of the decreased fishing area available to Thai fishermen, the estimated MSY of several marine species has been exceeded for years. Thai fishermen also are reportedly catching younger fish, as is happening in the Philippines. Thailand's fishing fleet accounts for only a small proportion (from 3 to 5 percent) of the total tuna catches from the western and central Pacific and from the eastern Indian Ocean.

Processing

The tuna processing sectors in the Philippines and Thailand developed rapidly and have been successful in exporting to the U.S. market and in gaining a large share of the market over the past ten years. The exports are the most important market because there is no significant domestic demand for canned tuna in the Philippines or Thailand.

The Philippines was a major exporter of raw frozen tuna to the United States until the early 1980s, when resource depletion became a serious problem. The canning sector expanded rapidly in 1980 and absorbed much of the frozen tuna (which previously had been exported) but still required additional imported tuna to fully supply input needs. The major exporters of canned tuna from the Philippines in 1980 were (1) Judric (which by 1982 was fully owned by Safcol, an Australian based company); (2) Pure Foods, a joint venture between American Hormel International and Filipino-Japanese investors with an estimated 19 percent of the export volume; and (3) Century, a locally owned corporation. Between 1980 and 1983 the Philippines gradually displaced Japan as the most important exporter of canned tuna to the United States. However, as the general economic conditions in the Philippines began to deteriorate, currency problems developed. It became increasingly difficult for the Philippine domestic canning sector to acquire hard currency for tuna and equipment imports. Today the industry in the Philippines has about an 11 percent share of the U.S. market, down from 32 percent in 1982, and it is expected to contract even further (Table 2).

The decline of the tuna industry in the Philippines is attributable to resource depletion and currency problems, as well as to several institutional policies implemented by the government in response to unfavorable economic conditions in the country. For example, the import ban placed on canned mackerel and sardines in 1983 was intended to conserve foreign exchange. The ban induced domestic canners to process more mackerel and sardines for domestic sales. However, this resulted in declines in canned tuna production and exports because capacity utilization was shifted toward mackerel and away from tuna.

The canning industry in Thailand now holds about 57 percent of the U.S. market for canned tuna (Table 2). Three major companies process and export tuna from Thailand: (1) Unicord Co Ltd with sales of about $46 million annually (of which about 90 percent is tuna); (2) Thai Union Manufacturing Ltd with sales of $50 million annually from various seafood products and pet foods; and (3) Safcol (Thailand) Ltd with the largest exports of processed seafood products from Thailand and annual sales of $70 million. Safcol is a joint venture with Australian interests, which until 1984 had substantial holdings in the Philippine canning industry. A fourth company, Thai Seri Group, has sales of $18 million annually and is the largest of the vertically integrated fishing operations with vessels, cold storage, and canning capacity.

More than 50 seafood canneries are estimated to be in Thailand as compared with 18 multipurpose plants in Mexico that process tuna. The Thai canning sector grew by over 47 percent per year in 1977–84, has diversified from fruit and vegetable canning, and employs at least 10,000 persons in direct canning operations, with hundreds of others employed in producing cans, printing and packaging, and making boxes and cartons. It has been suggested that the government seek to limit

Table 2. Tuna canning production in the Philippines and Thailand relative to global canned tuna production and U.S. imports, 1979–86

Year	Philippine production (canned tuna) tonnes	Thailand production (canned tuna) tonnes	U.S. imports (canned tuna) tonnes	Global production (canned tuna) tonnes
1979	4,079	—	23,634	611,000
1980	11,151	—	29,088	648,000
1981	18,033	8,181	31,815	747,000
1982	19,411	15,453	39,996	702,000
1983	23,537	28,179	55,449	761,000
1984	22,725	39,862	74,538	856,000
1985	21,816	87,134	97,263	—
1986	—	92,591[a]	—	—

Sources: Crough 1987b; Floyd 1986b; USITC 1986, 72 and 201.

[a]January–August only.

the expansion of the industry to about 15 million cartons of tuna annually, or about 2 million more than the present industry capacity.

Mexico has adequate processing capacity to can the domestic catch (100,000 to 140,000 tonnes annually), but there is little domestic demand for canned tuna. These canneries are used for fruits, vegetables, and fishery products. The Mexican national fishery development plan targeted the harvesting sector for the development and sales of frozen (unprocessed) tuna in export markets and therefore did not plan to expand the cannery sector. Two new seafood-only canneries, however, are under construction. These canneries utilize French technology and capital and are expected to alleviate the inefficiencies that exist in the multiproduct canneries. There are no plans for Mexico to export canned tuna.

Mexico, the Philippines, and Thailand are competitive in world markets for canned tuna partly because of their low wage structures. Wages for workers in the canning industry in all three countries are about $3 per day, about one-eighth of those in the United States and U.S. territories (Puerto Rico and American Samoa), exclusive of benefits. Wages in Thailand represent about 4.5 percent of total production costs, while wages in Mexico are about 6 percent of total canning costs. In both cases, fresh fish represents the largest cost component, 58 percent in Thailand and 64 percent in Mexico.

GOVERNMENT ACTIVITIES RELATED TO TUNA INDUSTRY DEVELOPMENT

The governments of Mexico, the Philippines, and Thailand have pursued different strategies in promoting their respective tuna industries. Mexico concentrated on the harvesting sector of the industry, a strategy designed to develop export markets for frozen tuna rather than processed tuna. Thailand concentrated on the processing sector of the industry with the intention of developing export markets in canned tuna. The Philippine government at different times directed policy at both the harvesting and the processing sectors. Despite industry planning, the industries in both Mexico and the Philippines were severely affected by national economic crises that led to currency devaluations and shortages of foreign exchange needed to purchase equipment and unprocessed tuna.

The difference between official industry promotion in Mexico, the Philippines, and Thailand is that the Mexican policies were more directly tailored to the fishing industry, whereas the policies of the Philippines and Thailand were generally directed at the exporting industries. However, there are some exceptions. The Thai government, for example, wanted to encourage local fishermen to increase the fishing effort both in Thai waters and in neighboring waters to supply tuna for the canning industry. To do so, the government negotiated access treaties with neighboring countries permitting Thai fishermen to legally fish in these areas (and presumably permitting them to make higher tuna catches). The Philippine government also allowed the canning sector special import concessions to import tuna for processing because domestic catches were insufficient to meet domestic demand.

The lesson from these different approaches seems to be that consistency in policy application predicts success more than any particular policy orientation. The Philippine government policies were uneven with respect to the fishing industry because the government was attempting to deal with larger macroeconomic problems. Mexican policies, on the other hand, were sufficiently flexible to support the development of the industry within the context of a national financial crisis.

Although different in orientation, the three countries share some common policies. In general, the tuna industries in each country are private sector operations, with government support being provided through legislated preferences. For example, to enhance employment, all three countries imposed restrictions on vessel crewing, with preference being given to their respective nationals.

Mexico has been the most aggressive of the three countries in directly promoting the development of the tuna industry (including vessel financing and vessel debt guarantees with foreign shipyards). Only Mexico has significant government ownership (22 percent) of vessels and canning capacity (55 percent). After declaring its EEZ and targeting the fishery sector for development, the Mexican government provided strong support for the industry. In 1980, for example, Mexico seized U.S. vessels fishing without licenses in the Mexican EEZ. The seizure led to the imposition of a U.S. embargo on Mexican tuna imports that lasted from 1980 to 1986. The embargo cost the Mexican industry at least $200 million in lost sales. During this period the Mexican economy went into a deep recession. The government refinanced the tuna fleet and essentially assumed a $400 million debt with foreign shipyards. Moreover, for the duration of the embargo, the Mexican government canneries bought, canned, and inventoried any catches that were not sold on the international market.

The Philippines and Thailand have across-the-board legislation that promotes export-oriented industries by giving preferential treatment over a wide range of tax, tariff, employment training, and capital depreciation issues. Some major legislative initiatives for Mexico, the Philippines, and Thailand are summarized in Table 3.

Table 3. Selected legislation related to tuna industry development in Mexico, the Philippines, and Thailand

Country	Legislation/promotional activity
Mexico	National Fishery Development Plan (1977, 1986) Establishment of Fishery Development Bank (1979) Fiscal incentives • Five-year reduced income tax rate • Exemption from import/export taxes on vessels or equipment in free-trade zone • Vessel debt guarantee • Vessel debt financing
Philippines	Investment Incentives Act of 1968 • Generous deduction of start-up and labor training expenses • Accelerated depreciation • Exemption/reduction or deferment of duties and taxes on machinery and equipment • Generous tax credits on domestic equipment purchases • Certain income tax deductions and exclusions • Protection from government competition Export Incentives Act of 1971 • Tax credits, exemptions, and deductions related to export activities Agricultural Investment Incentives Act of 1977 • Accelerated depreciation on fixed capital stock • Tax deduction for transportation expenses from targeted areas to encourage agricultural development • Tax deduction for training of Philippine nationals
Thailand	Investment Promotion Act (1977) • Guarantees against nationalization and competition from the state • Protection from imports • Permission to own land, remit foreign currencies, bring in foreign technicians • Reduction in or exemption from import duty on machinery, raw materials • Corporate income tax exemption for three to eight years • Withholding tax exemptions Negotiation of access for Thai fishermen with neighboring countries Support for fishermen and training for cannery workers

Sources: Crough 1987b; Floyd 1986b; Hudgins 1986d.

Political problems or currency depreciation or appreciation, particularly in relation to other tuna-exporting country currencies, has usually affected the competitiveness of tuna exports from these three countries. In the case of Mexico, for example, the peso was devalued, but the U.S. embargo against Mexican tuna imports prohibited Mexico from taking advantage of this situation.

POSITIONS IN WORLD TUNA MARKETS

Although international markets for frozen and canned tuna remain in flux, some trends are apparent. The world's tuna resources are overwhelmingly located in the territorial waters of the developing countries of the central and western Pacific and Central and South America. Access to these resources will continue to be an important issue for any harvesting country with few tuna resources.

Japan still leads as the major tuna harvesting country, followed by the United States, Spain, Indonesia, the Philippines, France, Taiwan, and Mexico (Table 4). When fully developed, the Mexican fleet could easily become one of the world's top four tuna producers. About 65 percent of all tuna caught worldwide is canned in the United States and its territories, in addition to Japan, Thailand, Italy, and Ghana (Table 5). The market for frozen tuna is extremely competitive, and the strength of the market depends on the overall supply conditions for tuna as well as the final demand for canned tuna. All harvesting countries, including Mexico, will be competing to supply the processors. Future expansion in the Philippines is constrained by resource availability problems.

Although a domestic demand for raw tuna exists in the Philippines and Thailand, domestic demand for canned tuna is small in Mexico, the Philippines, and Thailand. Thus these countries will continue to be subject to fluctuations inherent in export markets for primary products. Processors in these countries will be competing to supply the world demand for canned tuna. Thailand has clearly become a leader in this market in recent years, displacing the Philippines in the U.S. market in particular. Trends are given in Table 6. The Philippines and Thailand, however, must continue to seek out low-cost sources of raw tuna to remain competitive.

Table 4. Major tuna harvesting countries by percentage share of world catches, 1980–84

	Percentage of global tuna catches (all oceans)				
	1980	1981	1982	1983	1984
Japan	40	36	37	36	38
United States	13	12	11	14	13
Spain	6	7	7	6	6
Indonesia	4	5	5	5	5
Philippines	4	5	6	6	5
France	4	4	4	4	5
Taiwan	6	5	6	5	5
Mexico	2	4	2	2	4[a]
Other	21	22	22	22	19
Total	100	100	100	100	100

Sources: United Nations. Various years. King 1986.

[a]By 1986 this percentage had increased to 5 percent.

Table 5. Major importing countries of fresh and frozen tuna by percentage of global imports, 1980–84

	Percentage of global imports of fresh and frozen tuna				
	1980	1981	1982	1983	1984
United States	48	46	38	35	30
Japan	16	17	22	25	18
Italy	14	12	13	13	13
Thailand	—	—	—	5	16[a]
Ghana	5	5	3	3	2
Other	17	20	24	19	21
Total	100	100	100	100	100

Sources: United Nations. Various years. Crough 1987b.

[a]By 1986 this percentage had increased to 18 percent.

Table 6. Percentage distribution quantity of U.S. canned tuna imports by exporting country, 1980–85

	1980	1981	1982	1983	1984	1985
Thailand	10.1	14.6	21.3	32.6	55.3	57.3
Philippines	21.7	30.3	31.6	26.2	13.7	14.4
Japan	39.0	30.0	30.2	16.7	16.5	11.1
Taiwan	25.1	22.3	12.2	15.3	11.0	11.0
Ecuador	.0	.0	.0	.0	.5	2.4
Malaysia	a	1.0	.9	2.5	1.0	1.8
Indonesia	.0	.2	.7	2.2	1.4	.6
Venezuela	.0	.0	.0	.0	a	.4
Singapore	a	.1	.1	.3	a	.3
Other	4.1	1.6	3.0	4.3	.5	.6
Total	100.0	100.0	100.0	100.0	100.0	100.0

Source: Calculations based on official statistics of USITC 1986, 188.
[a]Less than 0.05 percent.

CONCLUSIONS

The market for tuna is truly an international one, involving imports and exports of both processed and unprocessed tuna. In the future, more countries like Mexico, with little domestic demand but with large tuna resources, are likely to enter this international market. For this reason, the future production of Mexico, the Philippines, and Thailand will affect any tuna activities undertaken by countries in the Pacific islands region. The impact will be especially felt on the supply side of the market if the supplies of tuna increase with new entrants.

The experiences of Mexico, the Philippines, and Thailand highlight some issues that are important to the development of industrial tuna operations. Each country was able to specialize at a particular level (harvesting or processing). Mexico has abundant resources and therefore chose to specialize in harvesting. The Mexican fleet is not constrained by resource availability in its own EEZ. Because the fleet has distant-water capabilities, it can operate in other areas of the Pacific. Resource availability, however, is a problem for the Philippines and Thailand. Consequently, although their processors are competitive in wages and other inputs, their operations depend on imports of frozen fish.

Each case of potential tuna industry development should be examined individually with respect to domestic conditions. However, three general observations can be made. First, self-sufficiency in the production process depends on resource availability, fleet and processing efficiency, and marketing capabilities. Second, a fleet needs to be capitalized relative to the country's available resource and the market that has been targeted for sales, either domestic or international; a domestic market would normally support a smaller fleet. Third, the domestic economic conditions have predictable effects on the tuna industry's productivity. In Mexico and the Philippines, for example, the domestic financial crisis produced a severe currency shortage that affected the ability of vessel owners and processors to import parts and equipment for their operations.

NOTE

1. This chapter synthesizes the chapters by Crough, Floyd/Doulman, and Hudgins for PIDP's tuna project.

IV. INTERNATIONAL BUSINESS

14. The Development of the Australian Tuna Industry

Greg J. Crough

ABSTRACT—This chapter provides a historical account of both the development of the Australian tuna industry and the evolution of Australian government policies toward the industry. It also analyzes the involvement of multinational corporations in the industry.

INTRODUCTION

The Australian tuna industry, although not a particularly large tuna industry by world standards, is an important component of the $350 million-per-year seafood industry in Australia. The industry is predominantly based on domestic catches of southern bluefin tuna (*Thunnus maccoyii*) (SBT), a migratory species that travels between Java, Australia, New Zealand, and South Africa. The SBT fishery is a major industry for parts of the Japanese distant-water fishing fleet, and small catches are also taken by New Zealand vessels. In addition to SBT, increasing quantities of other tuna species are now being exported, particularly to Japan for its sashimi market. Because of shortfalls in the domestic availability of tuna for human consumption, some of the Australian tuna canneries are reliant on imports of skipjack tuna (*Katsuwonus pelamis*) from the western Pacific region.

The tuna fishing and processing industry in Australia is geographically dispersed. The SBT fishing fleet is concentrated in Port Lincoln, South Australia. The east coast tuna fishing fleet, based on the airfreighting of sashimi-grade tuna to Japan and the United States, is primarily based in several southern New South Wales ports, while a smaller diversified fleet operates off the coast of Western Australia. Four states have small tuna canneries.

There is a significant degree of government involvement in the regulation and management of the tuna industry. In the case of the SBT fishery, the present Australian government has played a particularly important role in the development of national and international management arrangements.

The industry is also characterized by the operations of several large international companies in the processing and marketing sectors of the industry. These companies include one of Australia's largest seafood processing and marketing companies, Safcol Holdings Ltd, which has extensive operations in Australia and parts of Southeast Asia. Several foreign companies, which also have widespread international interests in the fishing and food processing and marketing industries, operate in the Australian tuna industry, including the H.J. Heinz Company, John West Foods, and Marubeni.

INDUSTRY DEVELOPMENT AND EVOLUTION OF GOVERNMENT POLICIES

Early years

The establishment of two small tuna canneries in New South Wales and South Australia during the 1930s represented the beginnings of the commercial tuna industry in Australia. But the introduction of pole-and-line fishing from the United States in 1950–51 led to the major expansion of the industry. This method proved superior to existing methods and was easily adapted to Australian conditions.

The early pole-and-line vessels were predominantly multipurpose in nature. Most vessels were engaged in the east coast trawl fishery during the off-season. In South Australia the lack of alternatives for boats of the size required to fish SBT, as well as the natural complementarity of the New South Wales and South Australian seasons, encouraged several South Australia-based operators to specialize in SBT fishing. This involved operating off New South Wales from about October to December and then returning to South Australia for a season that extended from December to May or June. Regular seasonal activities of this nature had been established by the end of the 1950s.

Small-scale SBT fishing activities gradually expanded during the 1960s using trolling gear off the southwestern part of Western Australia. No government management restraints were imposed in the Australian SBT fishery during this period.

With experience and the introduction of new fishing techniques, large quantities of SBT began to be harvested. However, given the potential impact of unrestrained purse seine activity on pole-and-line fishermen and on the stocks of SBT, a freeze was announced by the Australian government in May 1975, until the effect of such fishing methods on the stocks of tuna in Australian waters could be fully evaluated.

Tuna fishing activities off Western Australia had also gradually increased in the late 1960s and early 1970s, and thus a total ban on purse seining was imposed in that state. This ban was partly because of the concentration of small-size SBT in those waters and partly because of the potential impact of a purse seine fleet on the emerging small-scale local fishery.

By the mid-1970s Australian SBT production was reasonably stable at about 10,000 to 12,000 tonnes per year.

The canning sector of the industry was firmly established by this time and was based on the H.J. Heinz Company Australia Ltd cannery in Eden (New South Wales), the South Australian Fishermen's Cooperative (Safcol) canneries in Melbourne and Port Lincoln, the Hunts Foods Pty Ltd cannery in Albany, and the West Ocean Canning Pty Ltd cannery in Perth. These canneries process about 7,500 tonnes of tuna annually, and all of their production is consumed in Australia.

Limited-entry management

During the 1970s the size of tuna vessels gradually increased, and the fishery's success varied from year to year. Single-purpose vessels depended on high profits in good years to carry them over periods when prices were poor or catches were low. However, multipurpose vessels turned to other fisheries, which led to pressure for the introduction of a limited-entry management policy in the SBT fishery.

In March 1976 the Australian government introduced a freeze on further entry to the southeastern fishery. The freeze was intended as a temporary measure to allow an assessment of the biological situation of the fishery and to stabilize investment following two relatively poor seasons. Fishermen felt recovery would be helped by controlling the number of vessels entering the fishery. In the meantime both prices and catches improved.

Many fishermen took advantage of the security offered by participating in a limited-entry fishery and the fairly buoyant climate, which are associated with better prices and catches, to upgrade or replace their vessels with larger-purpose vessels. As a further incentive to build larger vessels, any vessel of over 21 m in length earned a shipbuilding bounty from the Australian government. As a consequence, the fishing power of the SBT fleet based in southeastern Australia increased dramatically during the late 1970s, tripling in size. Simultaneously, activities off Western Australia expanded with fishermen devoting more time to SBT activities. The range of their operations expanded steadily, and as catches increased, many fishermen replaced their small trolling vessels with larger, farther-ranging vessels.

This expansion of fishing effort and efficiency culminated in 1982–83 with a total catch of over 21,000 tonnes. These large catches were far in excess of the requirements of the domestic canneries. One result of these large catches in the late 1970s and early 1980s was a substantial increase in exports of frozen SBT, particularly to Italian canneries. Since the mid-1970s, SBT export prices have been in excess of the prices paid by Australian canneries. This situation has created a shortage of domestically caught tuna, and canneries have been forced to import skipjack tuna for processing.

Exports of sashimi-grade tuna to Japan have grown since 1975, initially based on SBT from Port Lincoln, but more recently on yellowfin (*Thunnus albacares*), albacore (*Thunnus alalunga*), and bigeye (*Thunnus obesus*) tuna caught in the waters off New South Wales and northern Queensland. It is estimated that about 30 percent of Australian tuna exports now consists of the higher-priced sashimi tuna.

Termination of limited-entry management

Based on the advice of scientists working for the Commonwealth Scientific and Industrial Research Organization (CSIRO), the Australian government decided that the previously imposed freeze on the entry of vessels to the SBT fishery would be lifted in 1981. The decision met with varying responses from different sections of the industry. In general, the New South Wales, Victoria, and Western Australia sections of the industry wanted the freeze lifted, while the South Australian industry did not. The New South Wales and South Australian pole-and-line boat owners had reservations about more purse seiners being allowed into the SBT fishery, although they supported the incentives being given for the larger vessels to increase their efforts in catching skipjack tuna. Indeed, at the same time that the freeze was lifted, the government announced that applications were being sought from Australians interested in conducting feasibility studies using purse seine vessels to fish for skipjack and other tuna in the Australian fishing zone (AFZ) and more distant waters.

Trilateral international management

Concerned that a continuation of the declining trend in biomass could put the SBT fishery and the SBT stock at risk, Australia, Japan, and New Zealand initiated discussions about the biological condition of the stock and the need for international management. In 1982 scientists of the three countries recommended that urgent steps be taken to ensure that the spawning stock not fall below the 1980 level (about 220,000 tonnes). The three countries agreed as an interim measure to restrict further growth in their SBT fisheries and to pursue discussions concerning the implementation of an international management regime.

Prior to these meetings the Australian SBT catch had increased significantly. Moreover, the CSIRO warned that if catches continued at the 1982 level, the biomass would be reduced to only 20 percent of its virgin levels. In addition, the government was committed to restrict further growth as a result of the tripartite meetings. Thus the Australian government announced in 1983 that interim management measures would be introduced.

In addition to these immediate pressures, the Australian government had a clear responsibility to impose management measures pursuant to the Fisheries Act (1952), which stated that the objectives of fisheries management were to ensure that the living resources of the AFZ are optimally utilized and not endangered by overexploitation. Furthermore, since the 1970s, recognition had been growing within the international community that where a stock of a highly migratory species such as SBT is shared between a coastal state and international waters, coastal countries like Australia and New Zealand should be obliged to cooperate with distant-water fishing nations (DWFNs) to ensure effective conservation.

Interim management plan

The introduction by the Australian government of the interim management plan for SBT in 1983 eventually led to the introduction of the present system of individually transferable quotas. It was the first time that the Australian government had taken a leading role in the introduction of new national management arrangements for a major Australian fishery, thereby integrating the biological and economic aspects of the fishery. To improve the effectiveness of these management arrangements, the policy required the development of national consultative arrangements between the industry and all governments concerned.

Under these interim measures, the Australian government introduced a national quota of 21,000 tonnes of SBT: 15,000 tonnes for the eastern sector of the fishery and 4,000 tonnes for the western sector, with the remaining 2,000 tonnes held as a reserve quota for special use in both sectors. Purse seine operators were restricted to 5,000 tonnes and were to be prohibited from transshipment operations, except to licensed carrier vessels. To address the 1982 recommendations of the scientists for a reduction in exploitation of small fish, size limits were established.

National management plan

The introduction of a management plan for SBT in 1984 involved extensive discussions with authorities. The issues to be resolved were complex, including the most appropriate type of management regime, the state of the SBT stocks, and the distribution of entitlements between Australian states and different sectors of the fleet within states. The Australian government's role in managing fisheries resources was strengthened by the Fishing Legislation Amendment Act (1985), which gave the government the power to implement management plans, and by the companion Fisheries Levy Act (1984), which authorized the imposition of a levy on quotas created under the management plan to partly offset management costs.

Negotiations with Japan

The Australian government was not prepared to impose stringent management measures on its own SBT fleet while permitting Japanese fishermen to continue to take SBT in the AFZ. The problem for Japan was that it did not want to create a precedent that could be followed by other countries in voluntarily restricting the operations and catches of the Japanese fleet. In the absence of an acceptance by Japan of international catch quotas, Australia was not willing to risk the benefits of its catch restraint being reflected simply in improved catch rates by Japanese fishermen. Thus Japanese longliners were excluded from access to the AFZ south of 34°S from 1 November 1984.

Since 1979 the operations of the Japanese fishing fleet have been covered by annual access arrangements. These access agreements include restrictions on the number of vessels and areas of operation, payment of access fees, details relating to port access, and catch reporting requirements.

International review of catch levels

As a consequence of the continuing international scientific concern over the state of the SBT stocks, Australia, Japan, and New Zealand agreed to limit the international catches of SBT to 31,000 tonnes per year for three years beginning in 1986. Australia's catch levels will be reduced from 14,500 to 11,500 tonnes per year, Japan's from 23,150 to 19,500 tonnes, and New Zealand's will remain at 1,000 tonnes.

Japan's willingness to provide several million dollars in financial assistance to the Australian SBT tuna fishermen permitted the Australian industry to reduce its catch by 3,000 tonnes.

Although the package of financial assistance is quite large from the Australian industry's point of view, it is a relatively low-cost means of ensuring the future of Japan's SBT industry, which is valued at about $175 million per year. The reduction in the Australian catch should significantly reduce the catch of the smaller fish, and the negotiations between Japanese industry organizations and the Australian Tuna Boat Owners Association will undoubtedly improve their bilateral relationship.

Transferable catch quotas

By the early 1980s the Australian SBT tuna fishing fleet was dominated by large, technologically sophisticated vessels based predominantly in South Australia. The implementation of the interim management plan in 1983 and the introduction of the system of individually transferable quotas in the SBT fishery by the Australian government in 1984 forced a major restructuring and rationalization of the Australian fleet. The fleet has been reduced by over 50 percent, and there has been a significant transfer of quotas between fishermen. This has occurred because fishermen are entitled to sell their individual allocations if they so desire.

The major rationalization has taken place in Western Australia, where the number of boats with quotas has fallen by more than 55 percent. However, South Australia has experienced a 39 percent increase in the total quota owned by fishermen in the state. Such an increase was expected because of the large investments in the tuna fleet in South Australia, coupled with the fact that the initial quota allocations for some of the state's large vessels were insufficient to maintain their financial viability. It has been estimated that up to $4.8 million has been paid for the purchase of SBT quotas by the South Australian industry since the management plan was introduced.

East coast tuna management

Since 1982 a small-vessel longline fishery for sashimi-grade tuna has developed off New South Wales. The fishery concentrated on yellowfin tuna, but fishermen are also catching some bigeye and SBT. Since 1979 the export of fresh chilled yellowfin and bigeye tuna to Japan has prompted growth in the east coast tuna fishery. It is estimated that in 1986 at least 150 boats were fishing for tuna for Japan's sashimi market, although the level of involvement in this fishery fluctuates because many fishermen have diversified into other fisheries.

The biological status of this developing fishery has not been researched in the same detail as that of the SBT fishery. However, it is believed to be a small but integral component (both geographically and biologically) of the tuna fisheries of the broader southwestern Pacific region. The species of interest to commercial fishermen operating off eastern Australia are highly migratory and are generally thought to range well beyond the AFZ into the western Pacific Ocean.

The Australian government is involved in discussions with the industry and state governments for the introduction of a management plan for the fishery. Although largely an inshore fishery at present, there is potential for development of offshore tuna resources.

U.S.–South Pacific tuna negotiations

At the South Pacific Forum meeting of heads of government in Tuvalu in 1984, it was agreed that the member countries should attempt to negotiate a multilateral tuna treaty with the United States. In 1986, 16 Pacific island countries and the United States reached agreement on a multimillion-dollar fishing rights package.

From the Australian government's point of view, unless Australia becomes a signatory to the treaty, it will not receive protection from the trade sanctions prescribed by the U.S. Fishery Conservation and Management Act (1976). It is now likely that the operation of the treaty will bring U.S. fishing vessel activity adjacent to the AFZ. The possibility of Australian interception and/or seizure of a U.S. vessel fishing for tuna in the AFZ will increase, as will the risk of U.S. prohibitions on the import of Australian seafood products. In 1984–85 Australia's seafood exports to the United States totaled $100 million.

U.S. industry representatives have indicated an interest in gaining access to tuna resources in the Coral Sea off Australia's northeast coast. U.S. purse seine vessels have been operating in waters adjacent to the AFZ, and the vessels have been using Cairns (Queensland) for repairs and refits.

CORPORATE INVOLVEMENT IN THE FISHING AND PROCESSING SECTORS

Considerable changes have occurred recently in the relative positions of some Australian fishing and processing companies in the tuna industry. Changes have accelerated following the introduction of the SBT management plan, the resulting concentration of SBT quota ownership in South Australia, and the increased ability of the canneries to import tuna for processing following the Industry Assistance Commission (IAC) inquiry. Australia's major tuna fishing companies are reviewed below.

Safcol Holdings Ltd

Safcol Holdings Ltd is Australia's largest seafood processing and trading group (six plants in Australia, one in the Philippines, and seven in Thailand, with sales offices in the United States, New Zealand, and

Asia). The group produces a wide range of processed seafood products and exports to a large number of countries. Its turnover in Australia in 1985–86 was approximately $90 million with Australian employment of more than 1,000. The company's Thai minority-owned operations recorded a turnover of about $80 million in 1986. The holding company is a wholly owned subsidiary of the Adelaide-based food processing conglomerate called Southern Farmers Group Ltd, which is itself a 62 percent-owned subsidiary of Australia's fourth largest listed company, Industrial Equity Ltd.

The Safcol group of companies has undergone important changes over the last 20 years, including a major restructuring of its ownership in the early 1980s. From its origins as a locally based fishermen's cooperative in South Australia, the group is now a major multinational corporation.

Australia. Safcol Seafoods Pty Ltd was incorporated in 1964 to operate a newly acquired Melbourne cannery. Prior to this, the cooperative had been producing its own brands of tuna at its Port Lincoln cannery. But after the 1979 closure of the Port Lincoln cannery, the Safcol group's tuna canning capacity in Australia was concentrated solely in Melbourne. Safcol Seafoods Pty Ltd took over the administration of the group's overall tuna operations and was run essentially as an autonomous unit within the group.

Safcol was also involved in the initial introduction of Japanese carrier vessels to Australia to improve the performance of the Australian fishermen in supplying sashimi to the Japanese market. Safcol continues to have a direct involvement in this aspect of the industry through its chartering of a New Zealand carrier vessel.

Despite the changing role of the overall Safcol group and some previous difficulties, Safcol Seafoods Pty Ltd has been successful in obtaining satisfactory quantities of the Australian tuna catch for its own canning purposes. One important reason for this has been that the company has facilities in Port Lincoln, which are capable of storing large quantities of tuna even at peak season. The company is thus able to buy the tuna at competitive prices as it is landed. Apart from benefiting Safcol, this procedure also provides security of sales for the fishermen.

Safcol Seafoods Pty Ltd has been steadily but moderately profitable, and during the period 1980–85 profits declined substantially (a small loss occurred in 1983–84). To some extent, this is to be expected because the company operates as a contract packer, and there have been increasing difficulties in obtaining access to sufficient quantities of tuna caught in Australia. The devaluation of the Australian dollar has drastically increased the cost of importing tuna, and it is likely that the canneries will pay higher prices for domestically landed fish.

Thailand. In 1972 Safcol Holdings Ltd entered into a joint venture with a Thai prawn processing company (Kim Tong Exports) to establish what is now a major tuna/diversified seafood processing, canning, and exporting operation. The Thai government's foreign investment policy required Safcol to hold a minority shareholding in the new company, Safcol (Thailand) Ltd. The company was classified as a "promoted company" by Thailand's Board of Investment. There have been some changes in the joint venture partners since the company was registered, and Safcol Holdings Ltd now has a 33.3 percent interest. Thai family interests hold 52.1 percent of the shares, while the remaining 14.6 percent is held by Chinese interests.

The attraction of Thailand was the very low wage rates and an apparently large supply of tuna and other fish species available for processing. Considerable quantities of tuna were being caught in Thailand's coastal waters, but the fish were generally dried and exported to Japan. In addition, canned seafood was recognized as an important potential export industry for Thailand and was actively promoted by the government. The first exports of canned seafood took place in 1972, and the industry has expanded substantially since then.

By 1981 Safcol's Thai operations were spread over three locations (Bangkok, Phuket, and Songkhla). They included five canneries, two frozen products plants, and a major can-making plant (producing 110 million cans a year). Total sales had risen to $36 million, and employment was more than 2,000. In addition, the group operations were supplied by more than 3,000 Thai fishermen.

Of the 17 seafood companies granted promotion certificates by the Board of Investment, the Safcol operation was the second largest in 1980, accounting for 15 percent of total production capacity. It was by far the largest importer of raw materials, accounting for almost 50 percent of the Thai industry's imports. The range of seafood products being processed included tuna, clams, squid, lobster, pilchards, and sardines. One reason for the success of Safcol's Thai operations is its diversification in seafood processing.

Safcol has indicated that the tuna canning component of its operations is marginally profitable and is almost totally reliant on imports of skipjack tuna sources from the distant-water fishing fleets in the western Pacific and Indian Oceans. The company's operations now have a turnover of more than $80 million, or about 15 percent of Thailand's total seafood exports.

The primary emphasis in the tuna operations is on trading large volumes. Sales to the United States have been increasing, as have exports to the United Kingdom. The dramatic fall in the value of the Australian dollar, as well as the relatively small size and competitiveness of the Australian market for tuna for human consumption, has made exporting to Australia a far less attractive proposition.

Philippines. In 1977 Safcol Holdings Pty Ltd decided to enter into a tuna canning joint venture in Manila with the Judric Seafoods Company of Hong Kong. The Philippines had processed and exported canned tuna since the early 1970s, but the value of exports was quite small until 1978, when the Judric Canning Corporation in the Philippines became fully operational. Safcol Holdings Pty Ltd initially held a 20 percent shareholding in the Judric Canning Corporation (through a subsidiary, Safcol Hong Kong Ltd) and was contracted to provide the technical expertise in canning and marketing. The Judric Canning Corporation was registered as a Board of Investment-preferred non-pioneer enterprise and was thus entitled to certain tax concessions.

By 1980 the Judric Canning Corporation was by far the largest producer of canned tuna in the Philippines, accounting for about one-half of the country's exports. Its net sales had risen to approximately $5.2 million by the end of 1979.

Safcol expanded its operations in the Philippines in 1981 with the establishment of another joint-venture company with Judric Seafoods Company called the Philippines Tuna Canning Corporation, which operated a cannery in Zamboanga. By 1982 these two canneries accounted for about 18 percent of the total canned tuna exports of the Philippines. It has been estimated that by 1982 these two canneries together exported more canned tuna to the United States than any other single cannery in the world. As a condition of their Board of Investment approval, neither company was permitted to can products other than tuna. Nor could either country sell more than nominal quantities to the domestic market. Unlike the company's operations in Thailand, which were based on a wide range of seafood processing activities, the operations in the Philippines were entirely dependent on the world tuna market, particularly the U.S. market.

In June 1982 Safcol Holdings Ltd acquired all the shares in both the Judric Canning Corporation and the Philippines Tuna Canning Corporation for approximately $4.9 million. Two years later the company valued the net tangible assets of the two companies at only $1.89 million.

For much of the tuna canning industry in the Philippines, the prospects appeared to be bright. During the period 1979–82 the country's canned tuna exports to the United States grew from 13 percent of the U.S. market to 32 percent. Despite the apparent success of the tuna canning company in the Philippines, major problems began to develop, some of which were specific to Safcol's operations, but others affected the entire tuna industry in the Philippines. One year later Thailand had the largest share of the U.S. tuna market, 33 percent, while the Philippines' share had fallen to 26 percent.

Many of the tuna canners in the Philippines sought to diversify their production. Safcol, however, decided in July 1983 to merge its Philippine operations with the LIG Marine Group of companies, thereby reducing its shareholding in the two companies to 50 percent. However, Safcol canneries suffered as a result of increased competition from the many new canneries that had been established in the Manila area. Its share of export production fell, and it had difficulties in obtaining fish supplies for processing and tin for can making.

Safcol believed that by broadening the base of its operations in Zamboanga to include canned fish for the domestic market and that by installing new management, the company could improve its operations. But this did not happen, and Safcol has been negotiating to sell its 50 percent share of the Philippines Tuna Canning Corporation for at least a year.

The failure of Safcol's operations in the Philippines had a serious impact on the financial position of the Safcol group as a whole. In 1982–83 Judric and the Philippines Tuna Canning Corporation reported combined losses of $400,000, and the group as a whole plunged to a loss of $600,000. In 1983–84 the group reported a loss of $4.9 million. The total investment of $5.8 million in the Philippines had been written off. By 1984–85 the group had at least financially recovered, and a profit of $1.4 million was reported.

Indonesia. After 18 months of negotiations, the Indonesian government approved the establishment of a tuna and pet food cannery at Bitung in 1981. Safcol held a 40 percent shareholding in the joint-venture company called P.T. Safcol Indonesia. The new $1.4 million cannery was designed to process 5,000 tonnes of tuna (the same capacity as the Zamboanga cannery) and to employ 150 workers. The facility commenced production in April 1982.

The new cannery, however, was in a difficult situation from the start. Its location was remote, and Safcol had major problems in obtaining sufficient supplies to keep the cannery operating near capacity. Substantial losses were incurred, and in 1984 the cannery closed. This resulted in extraordinary trading losses of $750,000 in Safcol's 1983–84 accounts.

H.J. Heinz Company Australia Ltd

H.J. Heinz Company Australia Ltd is a subsidiary of one of the world's largest diversified food processing and marketing corporations, the H.J. Heinz Company. Heinz was established in 1935.

In 1973 Heinz tried to take over Safcol. Heinz entered the tuna industry in Australia a year later when it paid $6.7 million to the U.S.-controlled food corporation, Kraft Holdings Ltd, for its Greenseas division, which consisted of a processing plant at Eden (New South Wales). The plant produced canned tuna, salmon, and cat food. The Greenseas brand had about 40 percent of Australia's $18 million tuna market (Safcol held a 50 percent market share).

Between 1977 and 1979 Heinz made further attempts to extend its involvement in the tuna industry. Since 1974 the company has invested more than $2.5 million in improving the basic fish-handling facilities, new freezers, aerial fish spotting, a new pet-food-canning plant, and the Greenseas factory. In 1979 Heinz also purchased a 25 percent shareholding in Australian Fisheries Development Pty Ltd, which owned one of the five tuna purse seiners then in operation.

Heinz provided a wide range of services to the Australian tuna fishing fleet, including aircraft spotter services. The service was terminated in mid-1984. The reasons were that the company was having difficulties in obtaining local supplies of tuna and that it was able to import tuna for processing.

Of the 4,000 to 5,000 tonnes of tuna annually processed at the Eden cannery, about 90 percent is now imported, although the devaluation of the Australian dollar is making imports more costly. Imported tuna is predominantly skipjack and is obtained from the Japanese fishing fleet operating in the western Pacific Ocean.

The Heinz facilities at Eden are valued at more than $8 million. They are part of the fisheries division of Heinz, which is not reported separately in the accounts of the Australian subsidiary. Approximately 75 percent of the cannery's tuna production is for human consumption. The other products include canned cat food and by-products such as fishmeal and fish oil. Heinz has stated that the tuna canning component of its Australian operations is marginally profitable.

Port Lincoln fishing companies

One finding of the tuna industry surveys recently conducted in Australia was the importance of the corporate form of ownership of the tuna fishing fleet in South Australia, particularly in comparison with other sections of the fishing industry. Such an ownership structure has largely been necessitated by the financial requirements of larger fishing vessels.

The corporate form of ownership adopted by the fishing companies in the SBT fishery is largely based on private family owned organizations. Some of these families are closely involved in processing and marketing through their joint ownership of several other companies.

Port Lincoln Tuna Processors Pty Ltd. Port Lincoln Tuna Processors Pty Ltd was formed in 1973 by ten Port Lincoln fishermen who had previously been fishing for Safcol. Port Lincoln Tuna Processors is probably the single most important company in the Australian tuna industry. The shareholders of the company own fishing vessels that have accounted for a significant proportion of the SBT catch for many years. These boats supply canneries in Port Lincoln and Melbourne and the Japanese sashimi market.

In 1980 Port Lincoln Tuna Processors constructed a new cannery for tuna storage and prawn processing at Port Lincoln. The cannery, valued at $1.4 million, began production early in 1981. Almost the entire output of the cannery is marketed by Seakist Foods Ltd, which has a significant share of the Australian tuna retail market. The cannery also produces the canned product for John West Foods.

Also of interest is the ownership of prawn licenses by tuna fishermen. The prawn fishery has been particularly profitable, and the profits made from prawning were crucial for the development of South Australia's tuna industry.

Australian Southern Bluefin Exporters Pty Ltd. This company owns approximately 2,000 tonnes of the national SBT quota, although in the most recent season it was supplied with approximately 4,500 tonnes, or more than one-third of Australia's total catch.

The company was incorporated in early 1982, but its origins date back to 1969 when the Australian Bight Fishermen's Society was formed by 13 fishermen. The history of the society (as with Safcol) reflects the apparently inevitable transition from a cooperative to a corporate form of organization. The company owns several purse seiners and spotter planes. Its turnover in 1985 was approximately $4.4 million.

Lukin family group of companies. The family companies, Lukin and Sons and Karina Fisheries Pty Ltd, were among the first groups in Australia to recognize the potential for sashimi exports to Japan. In November 1983 the family expanded its involvement in the sashimi export industry with a freezer storage and fish-processing plant (storage capacity of 1,000 tonnes of tuna at –65°C). The intention was to export about 150 tonnes of sashimi-grade tuna per year to Japan.

Sashimi development. One of the most important developments in the tuna industry in Port Lincoln was the introduction of the Japanese carrier vessels to export frozen sashimi tuna to Japan. The Japanese sashimi market was not easily penetrated by the Australian fishermen, mainly because of the intricacies of the Japanese marketing system.

The most important Japanese company to be involved in the carrier vessel arrangements is the trading company, Marubeni Corporation.

Marubeni's initial contact was made with Port Lincoln Tuna Processors Pty Ltd, which wanted to export greater quantities of sashimi-grade tuna.

In addition to the tuna exported by boat, an increasing quantity of tuna is being air-freighted to Japan, with the supplies coming from three states. Market acceptance of the tuna is good, and prices have been rising. Air-freighted exports of fresh and chilled tuna, mainly from Australia's east coast, totaled 530 tonnes in 1985–86 and were valued at $3,361 per tonne.

Fishing/processing companies in Western Australia

The major rationalization in terms of the number of boats fishing for SBT occurred in Western Australia after the introduction of Australia's management plan. In addition, Western Australia is the only state to have one of its tuna canneries close (in 1985), despite the purchase of a portion of quota by the Western Australian government. Only one company is now involved in the canning industry.

Kailis and France group. The Kailis and France group is one of Australia's largest integrated fishing industry organizations. The companies in the group include

- Australian Seafood Producers Pty Ltd, which operates in four states in processing, packaging, and distributing various seafood products to both domestic and export markets;
- KFV Fisheries (Qld) Pty Ltd, which operates several large refrigerated prawn trawlers in northern Australia;
- Kailis and France Pty Ltd (formerly West Ocean Canning Pty Ltd), which processes tuna, salmon, and other fish products.

Prior to the introduction of SBT quotas, the canneries in Western Australia were generally able to produce the canned product at a somewhat lower cost than both Safcol (South Australia and Victoria) and Heinz (New South Wales). One advantage was that the Western Australian canneries were not required to provide aerial spotting services for the fishing fleet. With the vastly increased tuna catch in Western Australia in the early 1980s, these canneries started producing large quantities of generic and housebrand canned tuna for supermarkets, and the general effect was to depress prices throughout Australia.

With the introduction of tuna quotas, however, the Western Australian catch has fallen dramatically, and Kailis and France Pty Ltd is required to import tuna. The company previously purchased some tuna in Port Lincoln, but transportation costs generally restricted the size of these purchases. The declining tuna catch has to some extent been offset by improved catches of salmon in recent years, probably as a result of the new management measures introduced in both Western Australia and South Australia. The Kailis and France cannery has been profitable since 1980, and the company recorded a sales revenue of $5.6 million in 1984–85.

The other cannery in Western Australia, which closed in 1985, was operated by Hunts Foods Pty Ltd. The cannery produced various tuna and salmon products.

15.
Japanese Tuna Fishing and Processing Companies

Geoffrey P. Ashenden and Graham W. Kitson

ABSTRACT—This chapter reviews Japan's tuna fishing and processing companies. The chapter examines the industry structure, its interaction with the Japanese government, and the market situation for tuna products. In conclusion, the chapter evaluates the implications for the Pacific islands region.

INDUSTRY STRUCTURE AND OVERVIEW

Structure of the tuna fishing industry

Japan has fewer than 200 large-scale fisheries enterprises. Medium and small enterprises number about 10,000, and coastal fisheries enterprises total about 190,000. The heavy bias toward small fisheries in the structure of the Japanese fishing industry as a whole is also characteristic of the tuna and skipjack catching sectors.

The majority of the "far seas" tuna and skipjack operators fall into the medium or small class with an average of two or three boats per enterprise. These companies rely largely on the Federation of Japan Tuna Fisheries Co-operative Associations (Nikkatsuren) to act on their behalf.

Major fishing company activities

The most notable change in the Japanese fishing industry over recent years has been the deliberate policy of gradual diversification by large fishing companies away from engaging solely in fishing. This shift occurred after the creation of exclusive economic zones (EEZs) during the 1970s.

The major Japanese fisheries companies have chosen to develop their efforts in distribution (including importing) rather than fishing alone; they have built on their strength in seafood and other frozen food distribution through the control of refrigeration facilities and distribution companies. The activities that have been maintained or expanded are those operations in which the major companies have special strengths—for example, in large factory-boat *surimi* production—and in which they have had preferential access to resources, particularly those in foreign waters. While joint-venture projects do keep the major fisheries companies actively involved in fishing, they prefer to have access to foreign resources without necessarily having to operate the fisheries themselves.

The emphasis in new ventures tends to be concentrated outside Japan. Often the projects involve fish farming. Most of the major fishing companies also have moved into processed foodstuffs and are planning new construction or expansion. One example of an expansion operation is crab stick processing, which is based in the United States. The diversification into processed food frequently involves non-seafood products. Diversification is expected to continue, with the fisheries companies becoming highly integrated food manufacturers.

In sales, Taiyo Gyogyo is the largest Japanese fisheries company, with annual sales (including non-fishing activities) totaling $3,605 million in the period to January 1986. Nippon Suisan (also known as Nissui) is the second largest company, with total sales of $2,962 million for the period to March 1986. The annual sales of Taiyo and Nippon Suisan are each more than double those of the next largest companies, Nichiro ($1,366 million to November 1985) and Kyokuyo ($1,258 million to October 1985). Hosui (sometimes Hohsui), the sixth-largest company, is 63 percent owned by Nippon Suisan.

In fishing capacity, Nippon Suisan is larger than Taiyo, with vessels grossing 204,000 gross registered tonnes (GRT) against 99,400 GRT for Taiyo. Taiyo, however, has a larger number of vessels. The gross tonnages of the other major companies are all below 36,000 GRT. The major capacity for all companies is in trawling, with limited involvement in tuna and skipjack fishing.

The participation by the major companies in tuna and skipjack fisheries is greatest within the purse seine sector; 8 of Japan's 35 licensed medium- to large-scale purse seine vessels are owned by two of the fisheries majors (Taiyo has 5 vessels and Kyokuyo owns 3 vessels). Only one of the 6 majors operates longline vessels (Taiyo with 6), and only one company has skipjack pole-and-line operations (Hosui, the Nippon Suisan subsidiary, owns 2 vessels). Thus Taiyo has the greatest commitment among the major fishing companies to tuna and skipjack fishing, with both purse seine vessels and longliners. However, within the total Taiyo operation, the tuna and skipjack operations are minor components.

In the processing sector, no major fishing company operates its own tuna canneries within Japan. Nippon Suisan has one cannery that produces meat products, and two Hoko canneries process mackerel and squid.

Of the major fishing companies, three (Nippon Suisan, Kyokuyo, and Taiyo) have become active in tuna distribution within Japan, working either directly or through affiliated specialist tuna trading companies (*maguro shosha*) or through closely related wholesalers or subwholesalers from the fish markets.

Catch. Of tuna species that have a high value in the Japanese market (bluefin and bigeye), Japanese vessels take approximately two-thirds of the total world catch. Japanese fishermen take approximately one-fourth of the world catch of other tuna species and nearly one-half of the total world catch of skipjack.

Catching methods. Japanese tuna fishing technology is advanced. Three basic techniques are used: pole-and-line, longlining, and purse seining. Historically, pole-and-line fishing has been the main skipjack catch method both by far-seas and offshore fleets, but vessel numbers and volumes have declined over time as economies of the method have suffered and as availability of baitfish has declined. The decline in far-seas pole-and-line fishing has been countered by a relatively recent increase in purse seining for skipjack and other tuna species, primarily small yellowfin.

The purse seine skipjack catch in 1985 exceeded the far-seas pole-and-line skipjack catch. Growth in the purse seine sector has coincided with use of fish-aggregating devices (FADs) and has paralleled similar developments by other purse seining fleets.

For sashimi tuna, Japanese fishing and processing techniques are unique. Because of the value of individual fish and the demanding nature of the Japanese market, the catching and processing methods have become specialized. Longlining is the fishing technique used, and supercold storage facilities are necessary to preserve quality.

Handling of frozen sashimi tuna. Techniques for handling frozen tuna for sashimi are highly refined and have an important influence on price. The principal objective is to preserve freshness and color of the flesh. To preserve freshness, it is important to preserve glycogen levels and prevent undesirable by-product accumulation by killing the fish quickly and then bleeding and gutting it.

Tuna species, as distinct from skipjack and billfish, have myoglobin in muscle and blood. Exposure to air results in oxidation to oxymyoglobin, which has an attractive fresh color, and then to metamyoglobin, which takes on a dark, dry appearance. Metamyoglobin accumulation is diminished considerably when temperatures are lowered, so that storage at -55°C or -60°C is possible for long periods without color deterioration.

Normal storage periods for frozen sashimi tuna in Japan are 2 or 3 months, with storage on the vessel up to 14 months for the bluefin tuna longliners. At higher temperatures, storage capability is diminished. As a practical guideline, it is essential for frozen sashimi tuna that the temperature at the body center be reduced to -30°C to -40°C within four hours. Increased body temperature resulting from high levels of metabolic activity produces a discoloration of the meat known as burning. For tuna destined for processing, temperatures of -20°C are regarded as satisfactory for storage.

Handling of chilled tuna. The handling of fresh chilled tuna is a specialized activity; the temperature must be lowered as quickly as possible. The fish must be in 0°C water within five or six hours and can remain at this temperature for up to 45 days. A fishing company in Japan claims that chilled tuna can be held for more than one month with ice at -0.5°C to -1.0°C, making possible surface shipment in ordinary chilled containers or in chiller lockers in dry cargo vessels.

This claim is validated by the North American jumbo bluefin experience. Commercial preparation of jumbo *maguro* consists of making cuts through the fifth fin from the tail, at the anus, and at the head, and then removing the gut through this cavity. Next, the fish is held in a tank of crushed ice for two days and then packed in a vinyl-lined box, with ice in the box and within the stomach cavity. Temperature variation outside the -0.5°C to -1.0°C range needs to be avoided—although wider temperature ranges have been noted in other instances but for shorter periods.

Tuna sashimi loins. Recently, sashimi tuna has been shipped in the form of loins rather than whole tuna.

There are eight options for processing loined tuna. In the initial stages the options include the location of the loining activity (at sea or onshore) and whether the tuna or skipjack is processed in a fresh/chilled or frozen form. The options in subsequent stages are whether the resultant loin is transported chilled or frozen.

Currently, there is actual experience in four of these options for tuna and five of the options for skipjack. For tuna, there is no known experience of loining frozen fish at sea for commercial purposes. Nor is there any experience of loining chilled tuna onshore with loins subsequently frozen, nor of loining frozen tuna onshore and transporting it chilled, except for those times when it will be used immediately.

Recent supplies of sashimi loins from outside Japan have come from several sources, including two successful Japanese operations and one

Korean operation that failed. The Korean operation (a company called Kyodo Suisan) caught primarily yellowfin in the Indian Ocean by longliners. The loins were processed in Pusan and shipped in refrigerated trucks, via ferry, to the Osaka wholesale market.

One reason for the failure of the Pusan operation was the problem of color change (*henshoku*) in processing; another was the selection or grading of loins for the market. Although precise details are not available about the processing methods used, it appears that yellowfin tuna held at -55°C in Korean longliners was partially thawed onshore, cut into loins, packed in polystyrene boxes, and refrozen for shipment at -55°C. Color changes occurred as a result of temperature increases during processing and subsequent refreezing. The *henshoku* resulted in low prices.

Another company (Japan Orchid Fisheries) was unsuccessful in the 1970s in importing fresh yellowfin loins from the Philippines because of the country's inadequate grading capability at that time. A view, widely supported in Japan, is that grading for the Japanese market must be done by skilled Japanese personnel or, as a minimum, that training in grading techniques must be by Japanese nationals.

Successful loining operations are conducted by several companies from the Mediterranean. These enterprises process large bluefin tuna (over 200 kg) caught by purse seiners or set nets belonging to Taiyo, Toshoku, or Sumiyoshi based in Gibraltar or Palermo, Sicily, between May and July. Loining occurs on board for the Gibraltar-based operation and on land for the Palermo operation after normal post-catch killing and bleeding. Where loining is done onshore, the fish is chilled in brine on board. The packaging of loins for shipment is in ice in special cartons, and the temperature is held at -0.5°C to -1.0°C during storage and shipment by air to Tokyo. Bluefin loins are also shipped by sea from the Mediterranean to Japan. The fish is caught by Japanese longline vessels, loined on board, and frozen to -55°C. Transportation to Japan is by supercold carrier vessels.

Skipjack sashimi loins. Attempts by Taiyo to ship frozen skipjack loins for sashimi use from Solomon Islands to Japan do not appear to have been commercially successful. A trial shipment of about two tonnes made in 1983 fetched a price of only $1.25 per kg against a ruling price at the time of $0.94 per kg for skipjack in the round. Difficulties occurred in transit with twisting and warping of loins because they were packed loosely, and trimming was necessary because of surface discoloration. Taiyo envisaged a much better result with the fish wrapped in vinyl (either loose or in shrink wrap) and packed in a container freezer-tray. This operation was discontinued when low prices of about

$3.75 per kg failed to meet the higher costs associated with the extra labor requirements. (The upper end of the skipjack loining market is indicated by prices of about $5 per kg for chilled loins recently cut at sea by Japanese operators.)

There is no record of skipjack for fresh consumption, in loins or whole, being processed outside of Japan for shipment to the Japanese market. Industry sources believe this is because of inadequate processing and handling technologies rather than difficulties during transportation. All skipjack for fresh consumption is caught by the pole-and-line method and frozen in brine at between –25°C to –30°C. An important aspect of freezing is the adjustment of fish density in the freezer to avoid rubbing of the skins and thus to maintain a better appearance in frozen form for buyers.

Canning. The canning techniques used in Japan for tuna are similar to techniques used elsewhere in the world.

Skipjack is loined for canning purposes. This procedure is sometimes carried out in Japan at locations other than the cannery itself. Often the plant where loining occurs is located close to the canning plant, and loins are chilled rather than frozen during shipment to the cannery factory. This procedure is similar to that formerly employed between Ensenada (Mexico), where loining occurred, and the Star-Kist canning plant in San Diego. Production of frozen skipjack loins for subsequent transportation to American canneries is carried out in Korea.

Other processed products. Skipjack processed products include the traditional dried/smoked product known as *katsuobushi* and its variants: *kezuribushi*, *arabushi*, and *namaribushi* (described below). Other traditional products include *tataki* (heat-seared) and *shiokara* (fermented/salted), and there are some relatively new products such as *tsukudani* (tuna pickle).

Katsuobushi is traditionally sold in the form of skipjack loins or fillets; now it is marketed more frequently in the form of thin shavings (*kezuribushi*). The product is used directly by consumers or processors as a flavoring additive to various dishes. *Katsuobushi* is important in Japanese cuisine.

In the *katsuobushi* process, the skipjack loins are boiled, then broiled over charcoal and dried. Next, the loins are inoculated with mold several times in humidity- and temperature-controlled rooms to develop flavor. Dried *katsuobushi* fillet shaved into small pieces is known as *kezuribushi*.

Arabushi is a semi-processed form of *katsuobushi*; it is produced by drying and smoking skipjack loins. Various techniques are used for this, but all involve initial loining and trimming of skipjack and laying loins out on trays for subsequent insertion into a smoking chamber. The period that the trays are in the chamber depends on the moisture level required.

A variant—*namaribushi*—is similar to *arabushi*, but the basic softness of the product is retained by reduction of smoking periods.

If *arabushi* is produced outside Japan, it is shipped in this form; its subsequent processing into *katsuobushi* or *kezuribushi* requires using aspergillus fungus, and to date this has been done only in Japan. After grinding off the burned outside layer, the *arabushi* is inoculated with the mold.

Imports to date of *fushi* products (mainly dried, smoked skipjack and mackerel) have been of *arabushi*, with the final processing into *katsuobushi* being done in Japan. The principal reason that imports are solely *arabushi* is because the mold inoculation procedures are technically more demanding, and while the presence of aspergillus on the imported product is not prohibited by the Ministry of Health, the presence of the often-associated microtoxins would result in rejection of the shipment. Several cases of rejection have occurred for *arabushi*.

Tataki is also prepared from skipjack loins, either chilled or frozen. It is similar to sashimi, but the loins are heat-seared or grilled on the outside with inner portions remaining soft and raw. *Tataki* is then frozen in loin form and sold to supermarkets or restaurants, where it is prepared by making cross-sectional cuts across the loin.

Shiokara is a fermentation product to which salt is added. One of the major products comes from the viscera, stomach, or low-fat meat of skipjack. Household consumption since 1970 has been reasonably stable at about 400 g per household per annum.

Tsukudani (tuna pickle) is skipjack flesh that has been boiled and seasoned with a solution of soy sauce and other flavoring materials. Because many Japanese producers have entered this market, overproduction is occurring.

Trends in tuna industry boat numbers and catches

Longline (far seas). Total far-seas longline vessels numbered 823 in 1985, of which 628 were greater than 200 GRT. Of the major fishing companies, only Taiyo with 6 vessels was represented in the longline fleet. The total fleet size has been relatively constant in the last ten years, averaging 840 vessels, but it is gradually increasing in terms of average vessel tonnage because mid-sized vessels (100–200 GRT) have slowly increased from about 70 in the late 1970s to 109 in 1985, while the number of smaller boats (50–100 GRT) has decreased from over 150 in the late 1970s to 86 in 1985.

The far-seas longliner fleet in 1985 recorded its best year for volume in 12 years, with 233,000 tonnes of fish caught. The major species caught in 1985 was bigeye at 120,000 tonnes, followed by yellowfin (39,000 tonnes) and bluefin (23,000 tonnes), but a large number of species were caught, including albacore, swordfish, marlin, sailfish, and shark. Catches of both bigeye and yellowfin have tended to increase gradually over this period, while catches of albacore and the minor species have remained stable.

Pole-and-line (far seas). The Japanese far-seas pole-and-line fleet in 1985 comprised 129 vessels. Only 2 of these were owned by a major fishing company (Hosui). The fleet has declined in size by over 60 percent since 1974 as operations have become less economical due to rising operational costs and scarcity of baitfish. For medium-sized vessels of 100–200 GRT, the fleet of 84 vessels in 1974 reduced rapidly and stabilized at about 10 vessels by 1978. Large vessels in the fleet (greater than 200 GRT) dropped from 279 in 1975 to a low of 95 in 1985. The number of smaller, more efficient pole-and-line vessels increased from 6 in 1975 to 35 in 1982 but dropped to 25 in 1985.

Skipjack is the species sought by the far-seas pole-and-line fleet, and it represented about 85 percent of the catch in 1984 and 1985; albacore is the main by-catch and constituted about 11 percent of the total catch volume since 1985. The skipjack catch in 1985 was only 99,000 tonnes, down 37 percent from the 1984 catch and down 33 percent from the 1980–84 average.

Purse seine. Medium- and large-scale tuna and skipjack purse seiners in 1985 totaled 82, down from the 1983 peak of 95 vessels. The number of large-scale vessels (greater than 200 GRT) has grown steadily from 10 in 1974 to 35 in 1985, of which 8 were owned by the major fishing companies. The number of seiners under 200 GRT has remained stable throughout the 1974–85 period at about 50 vessels.

The total catch of all medium and large purse seiners in 1985 was 186,000 tonnes, of which the skipjack catch was 122,000 tonnes (66 percent). Increases in the skipjack catch have been dramatic, rising from about 30,000 tonnes in 1979 to a peak of about 131,000 tonnes in 1984. The catch of yellowfin has also risen steadily each year from 1975 (7,700 tonnes) to reach more than 53,000 tonnes in 1985. Conversely, catches of bluefin have declined each year from the 1981 peak of 21,500 tonnes to only 2,900 tonnes in 1985.

Total catches by industry category. Tuna longline catches (total of all species) were greatest at more than 400,000 tonnes annually in the late 1960s. Catches declined to less than 300,000 tonnes in the mid-1970s but have since increased to around 320,000–330,000 tonnes per year. The proportion caught by the far-seas category has been fairly constant at between 67 and 70 percent, with 25 percent being caught by offshore fisheries and the balance being taken from Japan's coastal waters.

Pole-and-line vessel catches, although varying considerably from year to year, peaked at 445,000 tonnes in 1974 and were greater than 400,000 tonnes in 1976 and 1978. Since 1978 the catches have fallen, with less than 300,000 tonnes per annum caught during the period 1981–83 and again in 1985. This fall reflects primarily a rapid decline in the far-seas pole-and-line vessel catches. The decline has been compensated for by increased purse seine catches.

Energy-efficient vessels. Because of the impact of higher oil prices on tuna fisheries in Japan, considerable effort has been devoted to research and development of more fuel-efficient vessels. Increasing numbers of these vessels have been deployed by Japanese fishermen. The total production of these vessels has increased from 4,927 GRT in 1981 to 14,700 GRT in 1984. About 98 percent of production tonnage in 1984 was by far-seas tuna and skipjack vessels.

Japanese overseas tuna fishing activities

Regions. Figure 1 shows the distribution of the main tuna and skipjack fishing areas, principal species, and best fishing months as perceived by the Japanese fishing industry. The figure shows a large area between Japan and the equator in which the principal resource is skipjack and albacore tuna. Yellowfin and bigeye tuna is found in warmer latitudes, especially in eastern Pacific regions and the Atlantic, while bluefin is fished in southern latitudes off South Africa, Australia, and New Zealand.

Of the far-seas longline catch in 1983, some 63 percent was caught in the Pacific, with 67 percent from the east/central Pacific (mainly bigeye and yellowfin). For medium- to large-scale purse seiners and the far-seas pole-and-liners, the dominant catch area was the central west Pacific (78 percent of catch for purse seiners and 52 percent for pole-and-liners). The next most important catch region was the northwest Pacific, with 19 percent of volumes for purse seiners and 37 percent for pole-and-liners. Little was caught by either vessel type outside these two regions.

No precise records are available in Japan of frozen tuna catches by Korean and Taiwanese vessels, but a high proportion of these catches is destined for Japan. To the extent that Korean fishing operations are financed by Japanese trading or fishing companies through their *maguro shosha* affiliates, these catches are effectively Japanese catches. During the 1970s the *maguro shosha* also often financed the capital cost of the vessels used by Korean operations. Since then the supply status for frozen tuna has changed: frozen tuna became plentiful—especially

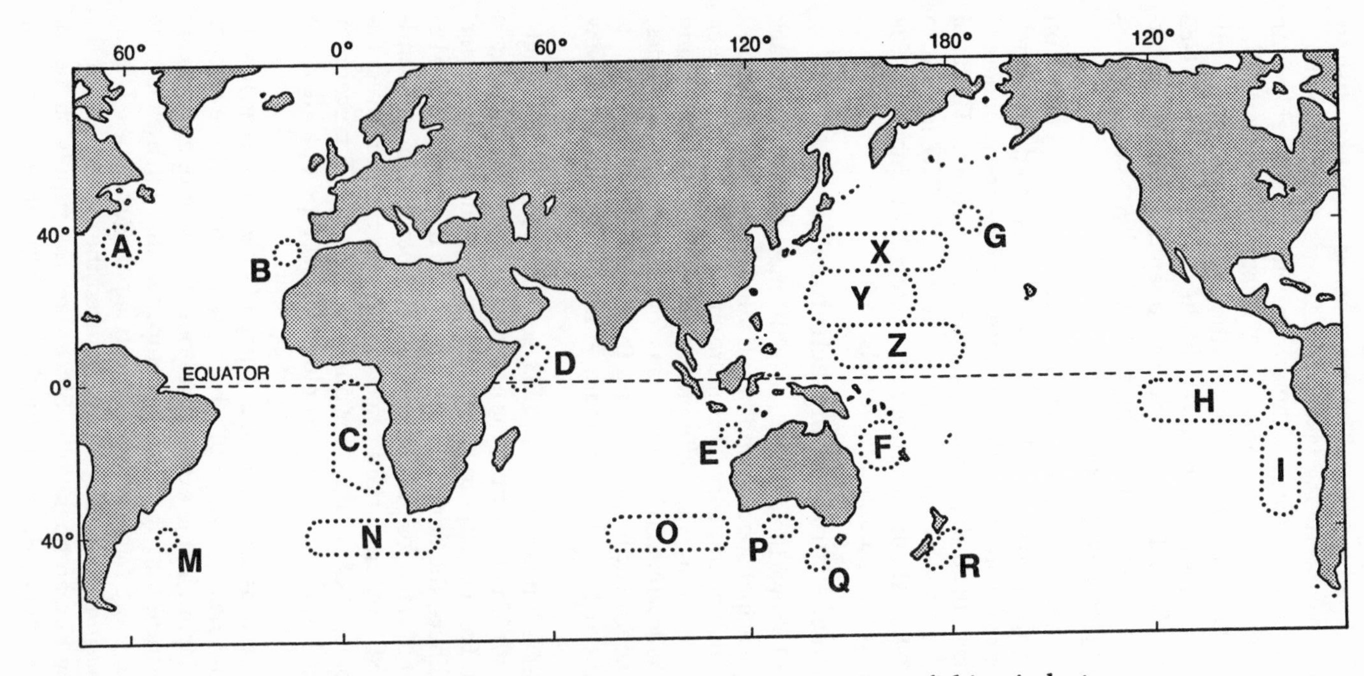

Figure 1. World tuna and skipjack resources as perceived by the Japanese fishing industry

Source: Based on data from Nippon Suisan.

Note: See legend for description of fishing areas, principal tuna species in each area, and best fishing months.

	Key area	Principal species	Months
A	New York	Bigeye and/or yellowfin, bluefin	November–January
B	Las Palmas	Bigeye and/or yellowfin	September–November
C	Angola	Bigeye	August–December
D	Indo (Somalia)	Bigeye and yellowfin	December–April
E	Java	Bigeye and/or yellowfin	All year
F	*Sangokai* (NE Australia)	Bigeye and/or yellowfin	October–December
G	*Kita-Oki* (Northern Sea)	Bigeye, albacore, and Japanese bluefin	September–March
H	*Seikei* (Ecuador) east/central Pacific	Bigeye and yellowfin	March–June
I	Peru/Chile	Bigeye and yellowfin	July–October
M	Montevideo (Uruguay)	Bluefin and albacore	August–December
N	Port Elizabeth	Southern bluefin	March–September
O	Southern Indian	Southern bluefin	July–October
P	Great Australian Bight	Southern bluefin	April–May
Q	Tasman	Southern bluefin	All year
R	New Zealand (N)	Southern bluefin	May–August
	New Zealand (S)	Southern bluefin	February–May
X	Eastern (from Japan)	Skipjack and albacore	August–February
Y	Central southern	Skipjack and albacore	—
Z	Micronesia	Skipjack	All year

yellowfin and bigeye—and Japanese tuna traders have had little incentive to provide financial assistance to foreign fishermen for capital purchases. The *maguro shosha,* however, can still assist vessels with operational finance.

Catching periods. The regions and timing of catches of tuna species have considerable significance because of the influence on quality, especially fat levels, and the general condition of water temperature and spawning times. As a consequence, the Japanese longline tuna industry has developed a distinct appreciation of appropriate fishing areas for different time periods.

Significant features for the Pacific area are evident. In the *Seikei* region (east/central Pacific), the catches are bigeye and yellowfin. In the southern part of the area (from 0° to 5° N), the fat content is regarded as too low because of higher water temperatures, but the bigeye and yellowfin tuna caught off Peru and Chile between May and August is considered to have the highest quality in the south Pacific region. The bigeye caught in the *Kita-Oki* region from October to March is also regarded as having a high quality, while the Indian Ocean resources, which are fished mainly by Korean and Taiwanese boats, are felt to have a lower quality. Similarly, in the *Sangokai* region of the western Pacific, off northeast Australia, the resource quality is not as high. But farther south off the northern coast of New Zealand, the quality of smaller yellowfin is regarded as good. The quality of bigeye and yellowfin to the north of Solomon Islands is lower because of the lower fat content of fish in these warmer waters. Yellowfin in Solomon Islands' area is regarded as best from July to October; during other months the fish tends to be thin and flabby.

Overseas ventures by fishing companies and trading companies

Fishing companies. As of 1985, six major Japanese fishing companies were involved in 109 overseas joint ventures or wholly owned operations in 31 different countries. Japanese general trading companies also were involved in 17 of these 109 investments. Other fishing companies were involved in an additional 13 overseas ventures.

Only one of the 122 overseas investments recorded in 1985 was specified as a tuna or skipjack operation; it is a joint venture in Solomon Islands.

Other overseas tuna investments by major fishing companies involve tuna trading only. This trading includes the Hoko operation, which purchases tuna from Maldives fishermen through a wholly owned local subsidiary, and Taiyo, which purchases bluefin in Spain, the United States, and Australia. Nippon Suisan has a range of tuna-purchasing operations similar to those of Taiyo.

Several smaller Japanese fishing companies have been involved in various *arabushi* operations in the Pacific region.

The Japanese tuna industry is involved in overseas tuna operations through various other structures. For example, the Japan Federation of Tuna Co-operative Associations (Nikkatsuren) through its commercial arm, Japan Tuna, provides servicing facilities for the Japanese tuna and skipjack fleet in six countries. Servicing involves provisions and oil supplies. Similarly, Japanese transshipment vessels make regular calls at many overseas ports, taking on frozen tuna from Japanese tuna and skipjack vessels.

Trading companies. Production-oriented investments in the tuna industry onshore have been made outside of Japan by companies other than the major fishing companies, primarily the large general trading companies. Mitsubishi Corporation owns Caribbean Tuna (a tuna cannery in Puerto Rico) and also has a major share in Mauritius Tuna Canning Enterprise, which operates both a cannery and a 499-GRT purse seiner. Similarly, C. Itoh, another large Japanese trading company, had a joint venture investment in a Fijian tuna cannery, and Mitsui had an albacore operation in Vanuatu.

A smaller fisheries trading company, Toshoku Suisan (part of the Toshoku trading company group), has recently established a longlining joint venture in New Caledonia. It is the first known example of a Japanese company taking advantage of the availability of small (90 GRT) and cheap secondhand longliners in Japan for overseas operations.

Toshoku Suisan (with its parent company known as Toshoku) has pioneered air-freighted imports of chilled sashimi bluefin tuna into Japan. At first, shipments came from the eastern seaboard of the United States, but subsequent shipments were bluefin from the Mediterranean. Toshoku has also been a pioneer in transportation of tuna loins rather than the whole fish. This business has been based on the large bluefin from the Mediterranean in both chilled and frozen form.

Japanese general trading companies are not necessarily constrained to partner only Japanese fishing companies in joint fishing enterprises offshore. The venture operated by Mitsui in Vanuatu, in which the Vanuatu government had a 10 percent share, initially used Korean longliners for catching albacore. The fishing operation has subsequently been transferred to Taiwanese longliners.

Tuna and skipjack exports and sales abroad

The greatest involvement of the Japanese tuna fishing industry in foreign business is in sales of tuna and skipjack to foreign vessels and canneries.

In 1984, when high catches occurred, 72,500 tonnes of fresh or frozen skipjack were exported from Japan, but of this amount about 48 percent was exported directly from Japanese vessels, the balance being first landed in Japan. In 1985 total skipjack exports fell by more than one-half to 33,000 tonnes. Major export destinations throughout 1985 were Thailand and the United States.

Most exports of albacore in recent years have been of fish previously landed in Japan. Total exports of albacore fluctuated widely during 1981–85 but were the largest in 1984 at 12,140 tonnes. The 1985 sales fell by 37 percent. The United States was the major market, taking 10,900 tonnes or 90 percent of exports in 1984 and 4,140 tonnes or 54 percent of exports in 1985.

Exports of fresh and frozen yellowfin have been modest over the four-year period from 1982 to 1985, with 7,000 tonnes shipped in 1985.

Exports at sea by Japanese longline vessels were small (less than 1,000 tonnes in 1984) and still declining. Most of this trade was destined for South Africa, and most was albacore.

There are no significant Japanese exports of the premier sashimi tuna species, bluefin and bigeye.

Processing sector

The tuna processing industries—canning and *fushi* production—are operated by relatively small companies, and the factory scale is generally small. The major fishing and trading companies have little direct involvement in these processing sectors. A large proportion of canning and *fushi* operations is based in the region of Yaizu, where most of the skipjack catch is landed.

Despite the fact that the seafood processing industries are relatively labor-intensive, nearly 60 percent of Japanese seafood canneries have fewer than 30 employees. The small scale of operation is even more pronounced for processing of *katsuobushi* and other *fushi* products, with 72 percent of factories having fewer than 10 employees and 95 percent having fewer than 30 employees.

Canneries. In 1972 there were 76 factories canning tuna and/or skipjack. Each year since then, factories have closed or merged so that by 1985 there were 44 tuna canneries operating; this number represented 19 percent of the 230 seafood canneries in production during that year.

The number of separate business entities operating tuna canneries declined steadily from 57 in 1972 to 36 in 1981 but has since stabilized at about 37. While the major fishing companies are not involved in the actual Japan production of canned tuna, three companies—Taiyo, Nippon Suisan, and Nichiro—exercise their strength in seafood distribution and contract the specialist tuna canneries to produce under their labels. The brands held by the fisheries majors are among the top ten domestic canned tuna brands, but the two companies with the major share of the market are both specialist tuna canners. These companies are Hagoromo, which has one-half of the brand market, and Inaba with a 13 percent share.

Several companies export canned tuna and skipjack. The top nine producers in 1985 held three-fourths of the market, and two companies held nearly 40 percent, namely, Sankyo Shokuhin with a 20 percent share and Hotei Canning with an 18 percent share. Neither Hagoromo nor Inaba exports its product; Hotei conversely had a 5 percent share of the domestic market in 1985.

In 1985 a severe decline occurred in canned tuna exports over the previous year, and the industry continued to lose ground in export markets in 1986. The decline in exports has resulted in pressure for sales in the Japanese domestic market, and competition has led to layoffs and severe financial difficulties in general for the industry. One company, Shimizu Shokuhin (SSK brand), recently went into receivership (SSK had a 5 percent share of the domestic market in 1985).

The domestic canned market could be further affected by competition from Thai imports. If these imports continue to be a commercial success, other Japanese importers/distributors will quickly follow suit, and imports from Thailand could increase rapidly and could involve the trading companies.

Fushi ***products.*** *Katsuobushi* is a traditional product in Japan, and small-scale production structures were established before the large fishing and trading companies became actively involved in tuna and skipjack. The large companies, therefore, have little involvement with either the production or the distribution of *katsuobushi* and other *fushi* products; the exceptions are where a major company has established *fushi* manufacturing ventures abroad, as is the case with Taiyo.

Processing into cuts (for sashimi). Several of the fishing majors and trading companies are actively involved in processing frozen sashimi tuna. This industry is dominated by the specialist tuna trading companies (*maguro shosha*) and wholesale market subwholesalers. It is concentrated in the large central markets and at Shimizu in Shizuoka prefecture, which is the major discharge port for frozen sashimi tuna, and also at Yaizu.

Several tuna traders own the supercold (–55°C) storage facilities that are required for sashimi tuna. Some specialist tuna secondary wholesalers operating in the landing port regions and central city markets are also called *maguro shosha* because of their trading activities outside the wholesale market system.

The processing of frozen tuna undertaken by subwholesalers and *maguro shosha* involves breaking down the bluefin, bigeye, and yellowfin tuna into loins or chunks for resale. The larger *maguro shosha* process tuna for retail outlets such as supermarkets and restaurants and large institutions, which generally then thaw it and prepare retail cuts on their own premises. The secondary wholesalers process the tuna into loins and chunks and also into retail cuts for small buyers at the wholesale market.

Although three of the major fishing companies exert influence over distribution through their *maguro shosha* and their strong involvement with wholesale companies in the central fish auction markets, the recent growth in frozen sashimi tuna distribution has been direct to supermarkets and restaurants and has been initiated mainly by those *maguro shosha* that are subsidiary companies of general trading companies.

Japanese management philosophy and consultative procedure

Japanese company behavior with respect to questions of social responsibility, industrial ethics, and philosophy of operation is a reflection in part of Japanese management systems and in part of industrial competition. For the large companies, long-term strategic planning and the lifetime-employment system are important.

Within the fisheries sector, these factors show up in strategies such as the deliberate attempts by the major fishing companies to move further into seafood and frozen-food product trading and distribution. At the same time, they are trying to optimize the adjustment period to their limited access to foreign resources by negotiating various licensed fishing agreements and joint-venture relationships with foreign parties.

Despite the intense competition among themselves, Japanese companies are able to discriminate readily between areas of corporate competition and areas of advantageous cooperation. Within the tuna fishing industry, cooperation usually is conducted through the various industrial associations; two examples are Nikkatsuren, which represents the majority of distant-water tuna and skipjack fishermen, and Zengyoren, which represents coastal fisheries.

GOVERNMENT INTERACTION

Fishing company relationships with foreign governments

Japanese tuna and skipjack far-seas fishing companies are relatively small, and their direct relationships with foreign governments are limited. The companies' interests in foreign waters are represented mainly by their trade association (Nikkatsuren) and the Japanese government. The prime objective in these relationships is to negotiate the most favorable terms possible for access to fish resources. Breaches of access terms by Japanese companies appear to be few, and overall the relationships are reasonably harmonious.

Closer relationships exist with foreign governments where large Japanese companies have established ventures in foreign countries. For the tuna and skipjack industry, these are few and may be of two types, joint ventures or 100 percent participation. The Japanese partner may be a large fishing company or a general trading company.

Where large Japanese companies are involved in offshore fisheries joint ventures or have established relationships through catching and trading with local companies, their direct involvement with foreign governments is usually limited. The large companies do not usually work through trade associations such as Nikkatsuren, and their relationship with the foreign government is usually through their local partner.

The large fishing companies also work closely in offshore ventures with the Japanese general trading companies, although there are no known cases of this within the tuna business. The general trading companies tend to take a leading role in formal government relationships.

Fishing company relationships with the Japanese government

In many instances, the Japanese companies, both the major fishing companies and the small operators, are supported significantly in their operations abroad by the Japanese government through various aid programs that are administered by organizations such as the Japan International Cooperation Agency and the Overseas Fishery Cooperation Foundation.

A major activity of the industry's own organizations (such as Nikkatsuren) is liaison with government, in particular the Fisheries Agency of the Ministry of Agriculture, Forestry and Fisheries (MAFF). The government's role, therefore, becomes that of an arbiter between the conflicting interests of different fisheries groups and between fisheries groups and political pressure groups outside the industry.

Because the Japanese tuna fishing industry is structured around small companies, its political influence is weak when conflicts arise or position-taking occurs within the overall Japanese fishing industry; this is especially the case when the tuna industry is in conflict or competition with the large fishing companies, which have a greater interest in tuna trade (and other seafood trade) than in tuna fishing or processing.

The major objective of licensing is to determine the number of vessels that can fish. Both the fishing ground and the period over which fish can be taken are specified in the license. In allocating the license, the Fisheries Agency of MAFF is subject to competing pressures, with different groups of tuna fisheries petitioning for favorable treatment, often at the expense of other groups. The Fisheries Agency is also petitioned by tuna and skipjack processors. The details of these petitions are relayed to a national council composed of experts, which recommends the number of boats to be licensed. The Fisheries Agency makes a final decision after reconciling the council's recommendations with the interests of foreign governments whose resources may be involved.

Government assistance programs to the tuna industry abroad

While Japan's tuna industry may be politically weak at home, its government does support the tuna fishing industry abroad. This support is direct in activities such as negotiation of licensed access to foreign fisheries resources, indirect through programs of assistance to develop foreign fisheries.

The number of experts sent to assist fisheries development in the Pacific islands region has generally ranged from 20 to 30 persons, with up to seven countries in the Pacific receiving such assistance. Ten fisheries trainees who were invited to Japan in 1983 represented three Pacific island countries.

IMPLICATIONS FOR THE PACIFIC ISLANDS REGION

The Japanese tuna industry in the mid-1980s is under extreme pressure. Increased costs and reduced catch rates have resulted in ongoing losses to skipjack pole-and-line fisheries and tuna longline fisheries. More recently, the dramatic upward valuation of the yen against most other world currencies has added to the financial troubles. The anticipated continuing strength of the yen could result in loss of competitiveness by the Japanese tuna fishing and processing industries.

The impact will be greatest on the skipjack industry—especially pole-and-line fisheries—because of its high level of exposure to international competition through (1) a substantial drop in the level of skipjack sales going to foreign canneries and (2) the exposure of Japanese tuna canneries to severe competition in both the export and domestic markets from foreign canneries, particularly the Thai canneries.

The impact on the raw tuna (sashimi) industry will be slower and less obvious. This industry has limited exposure to foreign competition because the Japan domestic market is the only significant world market for raw tuna, and the technologies associated with the catching and post-catch handling of sashimi tuna are highly specialized. There are indications, however, that the mystery that historically has surrounded this industry is beginning to disappear as the traditional technologies associated with the industry are challenged, not by foreign operations but rather by other Japanese operations.

The challenge has resulted from opportunities created as the market has matured in a way that places a greater premium on fresh chilled tuna than on frozen tuna. This development has seen a resurgence in growth of air-freighted sashimi tuna from both traditional and new sources, involving not only whole tuna but also tuna loins—which were previously considered unshippable by air.

The advantage of this type of fishery is the lower capital cost requirements. Instead of a supercold longliner valued at upward of $1 million, a 12-m trawler with adaptation is sufficient. For high-quality tuna the additional air-freight costs are easily covered by higher prices in Japan.

Another factor in question is the time period over which chilled tuna can be stored and the methods by which the fish can be transported. The opportunity exists to test and subsequently to commercially develop storage of chilled tuna in ice for shipment to Japan, not necessarily in specialized carrier vessels but possibly in chilled lockers on ordinary dry-cargo vessels.

The challenge to traditional longline fishing has come also from the use of small secondhand Japanese longliners operating on short fishing schedules from foreign ports and using transshipment technologies that have been developed and accepted over time. The practice of using foreign-based small longline vessels has made it possible for more foreign operators to consider participating in the Japanese frozen tuna market.

Operating on only 10- to 14-day fishing schedules, the locally based ships do not need to carry an extensive range of spare parts or specialist engineers. However, such vessels do need to carry a Japanese skipper and some skilled fishing supervisors. Shipment of cargoes to Japan is done by supercold carriers. Onshore supercold storage is not necessary if two longliners are worked in tandem with one vessel acting as the floating supercold store.

A Japanese government regulation prohibits the sale and export of secondhand vessels from Japan for operations such as this unless the vessels are older than ten years. These older vessels are available at relatively inexpensive prices. A 90-GRT longliner in good condition can be procured for about $120,000 to $190,000; an excellent 400-GRT vessel costs about $500,000. By comparison, a new vessel costs around $1.9 million.

Changes in competitive pressures and market demand within the Japanese tuna and skipjack industry have resulted in a steady decline in boat numbers. The Japanese government has encouraged this decline through its licensing policies, but the operators remaining in the industry generally have compensated by making a greater fishing effort in an attempt to revive profitability.

The implications of these changes for the Pacific island countries will depend largely on the reaction of Japanese tuna fishing and processing companies to the new circumstances. The fishing sector of the tuna and skipjack industry is based on a large number of relatively small companies, and their capacity to react is limited. Their probable reaction will be, as before, to concentrate primarily on the Japanese government, along with associated affected groups, in a lobbying effort for assistance through their industry associations.

Requests for assistance in restructuring (for the imposition of formal import restrictions for sashimi tuna and for the subsidization of canned exports) are likely to continue in the short term. There is no evidence, however, that the government will be any more receptive to these requests now or in the future than it has been in the past. Under the current climate of international trade liberalization—and with continued pressure from trading partners reacting to Japan's continued balance-of-payments surpluses—there is likely to be an even smaller probability of sympathetic response from the Japanese government. Nor do the politics surrounding the fishing industry provide any encouragement. The fishing industry has always been a "poor relation" in political influence compared with the agricultural sectors. Similarly, fisheries interests within MAFF are relatively weak. In addition, MAFF is likely to encounter considerable resistance from other ministries in the Japanese government in attempts to make concessions to the Japanese tuna industry.

As a result, the tuna- and skipjack-catching industries, as well as their associated processing industries, can expect little significant assistance from the Japanese government apart from assistance in minimizing disruptions as the industry winds down.

This lack of assistance has been tacitly recognized by the large-scale sector of the Japanese fishing industry. This sector has adopted a policy of winding down many fishing activities and promoting other strengths, especially in seafood distribution. The fishing majors (as well as the Japanese trading companies) have generally not been actively involved in Japanese tuna and skipjack fishing or in canned tuna processing or other skipjack products. Any involvement of the majors in this processing has been outside Japan. The tuna canning operations are now threatened by Thai competition, and thus their withdrawal from such foreign operations has been recent. This withdrawal has involved Japanese trading companies rather than fisheries companies (for example, C. Itoh from Fiji). Some operations apparently remain profitable (Mitsui and Mitsubishi in Puerto Rico and Mitsubishi in Mauritius).

The industry considers it highly unlikely that any Japanese company, large or small, will register interest in involvement with a tuna cannery operation in the Pacific islands region. The Japanese generally regard the commercial climate in the islands as unsatisfactory, the principal problem being the relatively high wage rates as compared with some parts of Asia and the low labor productivity. Possibilities may exist for further *arabushi* processing in the Pacific islands, although the large Japanese companies are unlikely to be attracted to *arabushi* processing as a stand-alone operation. They are also unlikely to be attracted to investment in skipjack pole-and-line operations that produce frozen skipjack for Thai canneries. Small purse seiners may be a more realistic option for such operations.

One development option for some Japanese companies is the preparation of skipjack loins in Pacific island countries for export in frozen form to Japanese canneries. The Japanese canneries could be favorably disposed toward this operation if it were to result in cheaper raw materials, given the fierce foreign competition. The canneries may not find support for the idea from the Japanese fisheries majors or general trading companies, which probably will find development of direct commercial and/or financial linkages with Thai canneries to be more attractive. Thus such a development will require considerable initiative from within the Pacific islands region. Nevertheless, the proposition does merit a more thorough investigation.

Development options are better for those Pacific island countries that have resources of high-quality sashimi tuna species (bluefin, bigeye, and yellowfin). These options include (1) satisfying the growing demand by the Japanese market for air-freighted chilled sashimi tuna, (2) participating in small-scale longlining using cheap secondhand Japanese vessels and then transferring the cargoes to supercold carrier vessels, and (3) developing chilled sashimi tuna shipments to Japan in the refrigerated lockers of dry-cargo vessels. Of these options, the third is speculative at this stage and involves considerable development risk, but the first two options are being carried out successfully by Japanese companies, including several major fishing companies. A major constraint for the islands is the quality of the tuna resource (as perceived by the market), with much of the Pacific tuna resource regarded by the Japanese as too low in fat to be prime-quality sashimi.

Sources in Japan suggest that for Pacific island countries with prime-quality sashimi yellowfin resources, the best approach to achieve higher returns in Japan would be to use chilled-storage technology to ship whole yellowfin to Japan. Loining of the fish in the islands is not recommended initially, but it could be started subsequently as experience in technology is acquired. This activity would require employment of Japanese skilled personnel (skipper, engineer, and master) and would be operated most efficiently using inexpensive secondhand longline vessels and possibly adapted trawlers. Joint ventures may subsequently be feasible.

In the event that Japanese fishing, trading, or processing companies are unwilling to participate in developments such as these, involvement by Japanese enterprises in the tuna industries of Pacific island countries is likely to become more concentrated on trading in Pacific island catches, particularly skipjack, with the major target market being Thailand.

16.
Japanese Trading Companies: Their Possible Role in Pacific Tuna Fisheries Development

Salvatore Comitini

ABSTRACT—This chapter examines Japanese trading companies and their activities in the international tuna market. The chapter emphasizes that the trading companies are primarily interested in providing intermediation services rather than investing in the production of the tuna industry.

INTRODUCTION

Japan has thousands of specialized trading firms, called *senmon shosa,* which function either in domestic distribution or in the export or import of a limited product line. They are generally representative of domestic wholesale distributors or the many export or import firms in the world.

The large trading companies, called *sogo shosha,* not only specialize in the export and import business but also link these activities with distributional activities in the various product and service markets in Japan. The Japanese Ministry of International Trade and Industry (MITI) differentiates the large trading companies from wholesalers, retailers, department stores, manufacturers, and others by their overwhelming weight in Japan's foreign trade.

In the words of one Japanese expert: "For a trading company to be a *sogo shosha,* it has to deal with many products (not concentrating in one product group, such as textiles or steel), engage in both export and import, have officers in various parts of the world, and wield considerable power in the spheres of marketing and finance. These criteria still serve as a rough guide to differentiate *sogo shosha* from ordinary trading companies."

He added, "In addition to meeting the criteria listed above, for a trading company to become a *sogo shosha*, it has to be engaged in offshore trade, undertake various activities (such as warehousing, transportation, resource development, manufacturing, etc.) to complement its trading activities, command large financial resources, play a leadership role in industrialization, etc."

There is another distinction between *sogo shosha* and ordinary trading companies. The latter are sometimes subsidiaries of manufacturing companies. Among the 30 top trading companies, 8 are manufacturer subsidiaries, and 3 function as either satellite companies or sales outlets of the other larger trading companies. Consequently, 19 companies out of the 30 top companies are truly independent trading companies. Among the 19 companies, the first 10 are classified as *sogo shosha* and the remaining 9 as ordinary trading companies. How are these two groups distinguished? As already indicated, one difference is size, a second is a diversified sales composition, and a third is the share of foreign trade in the overall sales of the particular company.

Nine general trading companies account for approximately one-half of Japan's export and import activities in terms of value. Their major sources of revenue are (1) exports of plants/machinery and iron/steel; (2) imports of fuels (petroleum, gas such as liquified petroleum gas (LPG) and alternative energy, e.g., coal and uranium ore); (3) imports of raw materials including ore of ferrous and nonferrous metals; and (4) imports of foodstuffs including tuna.

The task of defining a *sogo shosha* and of then categorizing its activities has proven to be an elusive one for researchers. According to a Japanese expert, this is because the large Japanese companies share patterns of business strategy, operations, and organization that are very different from those of other types of business firms and do not fit into any of the conventional categories. The vacuum in knowledge and understanding of the *sogo shosha* is the inevitable outcome of the nature of their commercial operations, where confidentiality and secrecy are often important to business success. But a much more important source of the knowledge gap is the complexity of the institution and the consequent difficulty in grasping its essential nature. Moreover, there is a lack of readily available models, concepts, or analogs with which simply and accurately to characterize the institution.

The *sogo shosha* differ from other companies and even multinational corporations in that they are primarily trade-oriented, not user- or manufacturer-oriented. Thus they function as intermediaries, or problem solvers, seeking to bridge the gap between supply and demand for goods and services in trade deals between parties.

CODE OF CONDUCT

A general code of business conduct generally acknowledged by Japanese general trading companies has been issued by the Japan Foreign Trade Council. It relates to corporate social responsibility, industrial ethics, and philosophy of operation.

According to this code, the Japanese general trading companies take both a private and a social view of their function in domestic and international society. That is, they view themselves as instrumental in promoting economic development, or affluence, in the national and world economies. Because of their historic mission and unique role in the changing social and economic environment, they should take measures to exercise self-control because of "the social influence of their business activities and the greatness of their responsibility."

The more specific functions and fields of activities that might relate to tuna trading and tuna industry development are given in Chapter 2 of the code. In section (1), for example, in the process of collecting and disseminating information on commodities all over the world, general trading companies have a social responsibility to attempt to stabilize commodity prices as a contribution to the general well-being of the people.

In sections (2) and (3), general trading companies are charged with having not only a distribution function but also an "organizer" function that could promote development of industries and enterprises in less developed regions and thus could "contribute to the harmonious development of the world and the improvement of welfare of domestic and international people."

Chapter 3 deals with relationships with other enterprises and industries. They should strive to "coexist and coprosper" with other enterprises, especially small businesses. In addition, their management standard should adhere to the "principles of orderly competition and fair trade."

Chapter 4 on self-control and regulation deals specifically with industrial ethics. General trading companies should adhere to "not only legal regulations but also to the international trust and business ethics... and are to make efforts in fairness and stabilization of business transactions."

Chapter 5 on establishing mutual trust with society deals generally with the philosophy of their operation. In order to gain public understanding and recognition of the impact of their business activities, "efforts are to be made to establish, one by one whenever possible, common measures concerning items which are desirable to be unified among public information like reports on settlement of accounts."

TRADING ACTIVITY WITHIN THE CORPORATE STRUCTURE

Range of operations

The four largest Japanese companies that have extensive dealings in tuna trading are Mitsubishi Corporation (Mitsubishi Shoji) with total sales as of March 1986 of $94.6 billion; Mitsui & Co (Mitsui Bussan) with total sales as of March 1986 of $100 billion; C. Itoh & Co, Ltd (Itochu Shoji) with sales as of March 1986 of $86.5 billion; and Marubeni Corporation with sales as of March 1986 of $77.8 billion (Appendix). According to *Forbes* magazine, these 4 companies are among the top 6 of the 200 largest foreign companies.

Although they engage in numerous types of business and diversified business ventures, their core business is still trading. Their activities include independent buying and selling, as well as functioning as trade intermediaries between buyers and sellers.

Each trading company handles between 10,000 and 20,000 products. They are mainly large-volume, first-stage wholesale traders of industrial raw materials, agricultural and fishery products, and standardized intermediate products (e.g., steel, synthetic fiber, and fertilizer). They conduct simple one-way trade and two-way seller-buyer transactions. They naturally prefer to handle the latter so as to maximize sales and trading for basic raw materials and provision of equipment and supplies.

They are, in effect, trading conglomerates, not manufacturing conglomerates. Although they have hundreds of small subsidiaries and large joint-venture operations both in Japan and in the world that engage in resource exploitation and development, manufacturing and processing, construction, financing and leasing, and subcontracting, their primary objectives in all these activities are to support the core business of buying and selling and to generate and promote new business opportunities. The trading companies do not function for purposes of development and manufacturing alone.

The broad range of services that the trading companies provide to customers and the vast resources that they control are targeted to support and expand their core trading business. Thus a majority of their subsidiaries and ventures are in sales, warehousing, transportation, and other service industries.

Because of dynamic shifts of imports and exports to and from Japan over time, the general trading companies are said to have "sensor tentacles" that stretch throughout the economy seeking exportable products and becoming aware of importable products (e.g., food items, industrial raw materials, and machinery). They are always on the lookout for expansion into commercial activities where they can turn a profit.

Tuna trading operations

The four largest Japanese trading companies have been involved in the tuna trade, among other activities, since at least the 1960s. The trading companies' intention to participate in fishing ventures initially was to expand their export trading of frozen tuna for marketing in the United States and Italy. Their involvement at this early stage was not extensive, but it involved trade credits as advances against purchases of fish in exchange for marketing rights (particularly for export sales of tuna).

As the growth of the Japanese economy proceeded and the demand for sashimi tuna rose in the 1970s, trading companies began to direct their trading activities to supplying tuna to the domestic sashimi tuna market. They eventually were able to increase their marketing share of the sashimi trade by negotiating whole boatload purchases of tuna from distant-water Japanese tuna companies and by acting as marketing agents for Korean and Taiwanese tuna fishing enterprises for exclusive sales rights to the Japanese market.

Because the vast majority of distant-water tuna vessels land their catches at Yaizu (Japan), each trading company set up operating subsidiaries or contracted with agents in Yaizu to conduct its tuna trading operations. These subsidiaries or agents operate cold-storage warehouses, do some processing, and also act as marketing agents on behalf of the parent trading company.

Processing by these marketing companies involves preparing tuna or skipjack from whole fish into small loins or steaks for the retail trade. Most of these companies have contracts with major supermarket chains in Japan to supply these retail packs.

Toyo Reizo Cold Storage (located in Yaizu) is the tuna marketing and processing subsidiary for the Mitsubishi Corporation. Yaizu Suisan operates a cold-storage facility and acts as the tuna marketing and processing agent for C. Itoh Ltd. Mitsui & Co does its tuna trading through Tokyo Commercial Co located in Yaizu. It also deals occasionally through a tuna trading company called Towa Suisan and located in Tokyo. Marubeni Corporation does its tuna trading through Maruko Suisan located in Yaizu. Maruko, on the other hand, contracts its cold storage from another company in Yaizu. Marubeni Reizo, located in Tokyo, also operates a cold-storage facility for Marubeni.

The trading companies view their basic function as providing efficiency and coordination through price and service. In the case of tuna trading, they view their service functions as (1) arranging transportation, (2) financing, (3) providing documentation formalities, (4) ensuring delivery, and (5) inspecting fish. Even the Federation of Japan Tuna Fisheries Co-operative Associations (Nikkatsuren) acknowledges some benefits of the trading companies in the tuna trade in that the offloading of fish at Yaizu is now done in one or two days, whereas formerly it took up to ten days. However, because they also hold frozen tuna in cold storage for speculative sales on the Japanese market, there is a feeling that the trading companies have some control over the market price of sashimi, which operates to the disadvantage of Japanese tuna fishermen.

The approximate percentages of their tuna trading, which change annually, are sashimi 70 percent, and processing (canned and *katsuobushi*) 30 percent. Tuna trading is small relative to the overall turnover of a large trading company. For Mitsubishi it is $150 million to $200 million; for C. Itoh it is $140 million to $150 million, compared with total sales of around $100 billion for each.

Profitability of tuna-related trading

According to the Japan Foreign Trade Council, the gross profit for all trading companies averages 1.3 percent (before tax), and the net profit averages 0.07 percent of net sales. For the major trading companies, according to Mitsubishi, the average net profit is 0.2 percent of net sales. Generally, the trading companies operate on high volume sales and very low profit margins (Appendix).

The profitability of their tuna trading operations depends on the market situation. C. Itoh is currently a marginal operation. C. Itoh indicates that it would be satisfied with a 1 or 2 percent return on net sales, but currently it is losing money on its tuna operations. Mitsubishi's margin is less than 1 percent of net sales on a volume of 100,000 tonnes of tuna per year.

Previously, there were more tuna traders than there are at present. The reasons for the reduction are that (1) traders now have higher risks and lower profit margins; (2) traders must advance working capital to vessel operators (e.g., for oil, general operating expenses, and wages); and (3) tuna has to compete with pork and chicken in the domestic market.

MARKET POSITION

Tuna trading arrangements and agreements

Given the keen rivalry of the major Japanese trading companies for market share and their constant vying for leadership positions in any particular activity, each company is quick to replicate and follow any successful venture or mode of contracting a supply source developed by any other company. The following section examines selected operations of the two most prominent trading companies in the tuna trade—

Mitsubishi and C. Itoh. The details are based on personal interviews conducted in 1986.

Mitsubishi. Mitsubishi was a shareholder with other private fishing interests in Japan and with Nikkatsuren in the establishment of Kaigai Gyogyo Kabushiki Kaisha (KGKK—Overseas Fishing Co Ltd), which operated from 1958 to 1983. KGKK was initially formed to promote development of overseas fishing operations by Japanese vessels especially in the Pacific, Southeast Asia, and Africa. This required the establishment of overseas bases for vessels to operate from and to land fish for processing and transshipment. KGKK's activities also included the establishment of joint ventures in fishing and fish canning/processing.

In Malaysia a joint venture operating under its ownership included (1) a cold-storage facility for transshipment of frozen tuna landed by fishing vessels of Japan, Taiwan, and Korea; (2) a cannery for packing canned tuna for export to the United States, the United Kingdom, and Europe; (3) a processing factory for frozen prawns, etc., for export; (4) a fish meal plant; and (5) a fleet of as many as 11 tuna longliners.

In Mauritius a joint venture operating under KGKK ownership included a cold-storage facility for transshipment of tuna landed by longliners from various nations, a cannery for packing canned tuna for export to the United Kingdom and Europe, a tuna longliner, and a tuna purse seiner.

In Papua New Guinea a company owned by KGKK operated a fleet of pole-and-line vessels. This venture was terminated in 1982 as a result of both taxation problems with the Papua New Guinea government and changes in the U.S. tuna market.

KGKK also had a venture that operated a fleet of ten pole-and-line vessels in Madagascar. This operation, which began in 1972, closed in 1976 due to the nation's economic and political difficulties.

In addition, as consumer demand for tuna increased in the early 1960s in Japan, the co-owners of KGKK saw an opportunity to finance Korean tuna longline operations. This was done by offering boats, technical know-how, and financial credits in exchange for rights to purchase their fish catches, initially for export trading and later for imports into the Japanese market.

Korean tuna longline operations actually started as early as 1963 with the Korea Marine Industry Development Corporation (KMIDC), which operated ten longliners based in American Samoa to supply Star-Kist's cannery. It was financed by a consortium that included Star-Kist and Italian and French tuna interests. The arrangement with the consortium provided long-term capital to KMIDC for the purchase of tuna longliners and advances for working capital for trip expenses. These loans and advances were to be repaid through the supply and sale of fish to the consortium.

Mitsubishi provided mainly debt capital in KGKK, using the same type of arrangement with the Korean vessels as used by the consortium that financed KMIDC.

The financial structure of KGKK was such that only 2 percent of its capitalization was equity investment while 60 percent of its debt capital was provided by Mitsubishi. Mitsubishi also provided technical assistance for handling and preparing tuna for the Japanese sashimi market by the Korean vessels.

Five Korean tuna fishing companies were financed by KGKK in 1966 and 1967 (in the form of long-term capital for purchasing fishing vessels and operating capital in the form of advances). The plan was to provide an additional supply of fresh tuna for the sashimi market in Japan through Korean imports. Nikkatsuren eventually sold its interest in KGKK because the company's operations were at cross-purposes with the interests of Nikkatsuren. Mitsubishi then became a major equity holder in the company. The five Korean companies eventually failed in the 1970s after the first oil shock, the subsequent high interest rates worldwide, and the economic downturn in the Japanese economy.

In 1983 KGKK ceased operations. However, Mitsubishi established a company under the same name to take over and carry on the Mauritius joint-venture tuna cannery and now plans to increase the number of purse seiners based there from one to three. Its canned tuna sales are mainly to the European Economic Community (EEC) countries.

C. Itoh. C. Itoh, unlike Mitsubishi, as far back as 1970–71 actually built longline boats and chartered them to Korean companies, as well as provided financial and technical assistance.

C. Itoh saw this arrangement as a source of inexpensive and efficient labor. Since then Korea has established its own fishing companies as a result of the repayment of credit extended by C. Itoh. These companies are still closely tied to C. Itoh because they need sales commitments for their catches, vessel financing, and operating expenses due to Korea's exchange control restrictions.

Tuna supply sources

Tuna trading by the Japanese trading companies is now divided into two components. One is to deal with tuna for sashimi in the Japanese market, and the other is to deal with tuna for canning and processing.

Trading companies buy raw tuna on a spot basis from Japanese tuna fishing vessels. They also contract with Korean and Taiwanese boat owners for tuna deliveries. As for the source of tuna supply to the Japanese market of 400,000 tonnes (50,000–60,000 tonnes fresh), the Japanese longline fleet accounts for 70 percent, Korean vessels for 18–20 percent, Taiwanese vessels for 6-7 percent, and other foreign countries for the balance.

Sashimi tuna. Yaizu Suisan, a subsidiary of C. Itoh, buys tuna for sashimi from Japanese longliners and also from Taiwanese and Korean vessels. Most dealings are arranged under long-term contracts, and C. Itoh will extend to its contractual parties certain loans and advances. Such financial assistance is commonly given to Korean vessel owners because of the tight foreign exchange regulations still in effect in Korea.

The supply of sashimi varies by grade. The top grade is southern bluefin tuna. Because of the catch quota imposed in this fishery, the catch is limited and the market strong. Bigeye tuna caught in the higher latitudes is considered high-grade sashimi and a substitute for bluefin tuna to some extent. Medium-grade sashimi has less fat content than the higher grades, has a good red color, and is in relatively plentiful supply. The middle latitudes of the central Pacific and Indian Oceans yield medium-grade bigeye and yellowfin tuna, which is caught primarily by Korean and Taiwanese longliners. These catches always find their way to the Japanese market.

Canning/processing tuna. Mitsubishi buys processing tuna mainly from Japanese vessels, but because of the yen revaluation, purchases from Korea and Taiwan have been increasing. American purse seiners are not generally competitive because their operating costs are higher.

Mitsubishi currently uses Tinian as a transshipment base for canning/processing tuna purchased from Japanese, Korean, Taiwanese, and American vessels. The fish is transshipped to reefers of Korean or Panamanian registry at transport costs to American Samoa (Star-Kist and Van Camp) of $132/tonne and to Thailand at $80–$90/tonne. The costs of transportation favor transshipping to Thailand.

C. Itoh also deals in the trading of tuna for canning/processing use. The company buys fish from Japanese, Taiwan, and Korean purse seiners for sales to Japanese canners or to Thai canners, with transshipment at Tinian or Guam.

Four or five years ago C. Itoh started buying tuna (skipjack and yellowfin) from Indonesia and the Philippines for shipment to the United States and Japan. They also buy from U.S., Taiwanese, and Korean boats and sell to canners in Thailand and the Philippines and assume the risk of non-payment by the cannery buyers.

Market share

In terms of the total tuna trade in Japan, the market share of each company obviously can change from year to year. Mitsubishi is the acknowledged leader in purchases from both Japanese vessels and foreign vessels. By its own estimate, Mitsubishi has a 30 percent share of the total tuna trade in Japan and a 40 percent of the total tuna marketing from the Pacific Ocean region. An estimate by a competitor trading company has put Mitsubishi's share at 50 percent of the sashimi trade.

Approximately 35 percent of the total tuna trade in Japan is shared equally by the other major trading companies, 10 percent is done by medium or small traders, and 5 percent is handled by very small traders. Some of the very small traders act as agents for foreign fishermen and occasionally do a small trading business, for example, cold-storage warehousing, sales on the auction block, or sales to other traders or fishmongers through direct negotiation.

The share of the imported tuna trade by the Japanese trading companies' subsidiaries and/or agents is as follows: Toyo Reizo (Mitsubishi) 45–50 percent, Yashima (independent) 10–15 percent, Hassui (Kyokyo) 10 percent, Yaizu Suisan (C. Itoh) 4–5 percent, Maruko (Marubeni) 3–4 percent, Nippon Suisan 2–3 percent, Kanetoma (independent) 2–3 percent, Marumi (independent) 2–3 percent, and Tokyo Shosha (Mitsui), Toshoku Trading (Mitsui), and others 1 percent each or less.

The market share for the tuna trade with Thailand's canners is as follows: Interpral (French brokers) 25–30 percent, Mitsubishi 20–25 percent, C. Itoh 20–25 percent, Thai vessels 10 percent, Marubeni negligible, and Taiyo (Solomon Islands) negligible.

The share of Japanese trading companies in tuna imports changes from year to year. From one-half to two-thirds of the catches of Taiwanese tuna vessels are handled by Taiwan-sponsored companies registered in Japan. On the other hand, almost 100 percent of Korean tuna imports is handled by Japanese trading companies.

IMPACT OF JAPANESE TRADING COMPANIES IN THE TUNA TRADE

The basic function of a general trading company is to buy commodities from producers or suppliers and to sell commodities to manufacturers, processors, and wholesalers, or even to consumers. For purposes of facilitating procurement and trading commodities, there are many ways that a trading company can extend assistance to the producers or suppliers or to the end chain of buyers.

Initially, the trading companies study the requirements of producers or suppliers for capital, infrastructure, technical assistance, or market information, as well as the needs of manufacturers or processors for raw materials or the needs of wholesalers and consumers for merchandise to trade. When they see that the needs of manufacturers or processors are growing and when they see that these needs are not being adequately met with a sufficient supply of raw materials, they may extend assistance to producers and/or suppliers of raw materials to raise their production capacity. Or when they see that the supply of raw materials is growing, they may extend assistance to manufacturers, processors, and wholesalers, or even to consumers, to increase their purchases of raw materials or merchandise.

In assisting both ends of the trade line, a general trading company at times can become involved in equity holding or finances, but its real intention is to increase its trading activity with products that have market potential. Its position is always as an intermediary, and a general trading company never really wishes to be a part of production or manufacturing.

In the early 1970s, when the Japanese trading companies saw the demand for tuna rapidly growing while production remained on a low level, they assisted the producers in increasing their production capacity. With the increased production, the trading companies hoped to increase their trading business. As cases in point, Mitsubishi and C. Itoh, as well as other trading companies, extended financial and technical assistance to Korean and Taiwanese tuna producers (fishing companies and/or tuna boat owners). Moreover, at times they encouraged principal officers of existing companies to leave and become independent owners by offering them tuna vessels on lease arrangements and credit advances for meeting operating expenses.

Almost all the trading companies that dealt with Korean partners built tuna vessels and leased them to the Koreans on the condition that the lease payment (actually an installment on the vessels' purchase price) would be paid with fish catches and that ownership of the vessels would be transferred to the operators after the necessary number of years of installment payments. Under this arrangement, the trading companies were assured that the fish catches would be delivered to them.

Today, however, the supply of tuna is readily available, or even rather excessive, and the trading companies can easily purchase the commodity without involving themselves on the production side.

There also appears to be no current interest on the part of trading companies in vertical integration in tuna production and marketing of the joint-venture type (either equity or loan investment). This is borne out by projections of the companies' financial performance. Because of the rapid appreciation of the yen, overall sales and profits declined in 1986, and projections for any immediate improvement are not considered promising.

The trading companies are in an advantageous position to arrange the efficient and inexpensive freight for transportation of sashimi-grade tuna. Although fishing vessels do land their catches at Japanese ports on completion of a fishing trip, because tuna fishing is now carried on in waters that are more and more distant, tuna catches are increasingly being transported to Japan by refrigerated carrier vessels. Because sashimi tuna is now expensive, not all buyer-distributors have the financial ability to buy up a whole boatload from a tuna longliner. The trading companies can intermediate these purchases.

The market is currently speculative, and tuna prices tend to fluctuate widely. Sometimes heavy inventories of stock have to be held, and tuna trading, particularly for sashimi, has become risky for the medium and small trader. Only those with financial backing can sustain themselves in this trade, and the market is now dominated by fewer than ten buying-distributing companies.

CONCLUSION

The activities of general trading companies are related primarily to international marketing and procurement of commodities together with trade-supportive activities such as trade financing, market research, warehousing, and shipping. Thus they essentially provide the infrastructure that supports independent or semi-independent productive and extractive activities. The features that characterize the trading companies' network of affiliated ventures and trading arrangements are now recognized as a "new" (or unconventional) form of international business organization.

Their primary interest in direct and indirect investments in overseas resource development ventures with foreign partners is the creation and protection of *shoken* (the commercial right to intermediate), whereby, either through an equity or a debtor-creditor relationship, foreign producers are obligated to sell the product through the intermediation of a particular trading company to whom they pay a commission.

In the early days of the trading companies' entry into tuna trading, especially during the late 1960s and early 1970s, they saw an opportunity to provide an additional supply source of tuna to the foreign processing markets. Later it became an alternative supply of tuna to the Japanese domestic sashimi market, which was faced with excess demand as a result of the rapid economic growth of the economy. In engaging in these activities, the real motive of the trading companies was to generate commissions through an extension of their trading business.

The culmination of these early activities involved them in commercial commitments and joint ventures of various types. The current tuna market situation, either for the sashimi trade or for the canning/processing trade, is not conducive, nor does it currently lend itself, to either direct or indirect investment interest of the types that were prevalent in the earlier growth period.

The immediate prospect of entry into the Japanese market is not realistic on any account. The sashimi market is currently saturated. The increasing imports and sagging prices of red-meat sashimi is having the effect of stimulating efforts toward imposing some restraint on additional tuna imports from all countries.

The trading companies may finance new fisheries development projects if they seem promising, but the current recession in the tuna industry, which is felt to be structural and profound, would make few, if any, projects financially attractive. In addition, the trading companies are giving priority to those existing project commitments that require infusion of new capital investment, for example, the Mauritius joint venture by Mitsubishi. To be sure, the Lome Convention countries (e.g., Solomon Islands, Fiji, and Papua New Guinea) would have preferred status in any future investments on the basis of their qualifying for preferential access to the EEC market.

The general pessimism in the tuna industry is pervasive and has not escaped the sensitive feelers of the trading companies. Under current market conditions, the only solid interest expressed by Japanese trading companies in any extension of Pacific islands tuna-trading activities is in the canning/processing transshipment trade. The obvious reason is that this type of trading precludes any long-term capital commitment on the part of the trading companies and allows them to perform their real role as intermediaries in tuna trading. If tuna prices eventually do recover, the companies will probably again look favorably to various forms of long-term capital involvement in the production aspect of the tuna trade.

APPENDIX: Sales and Financial Ratios of Four Largest Japanese General Trading Companies

Mitsubishi Corporation	
Sales[a] (March 1986)	$94.6 billion
Sales breakdown	
Fuels	31%
Metals	22%
Machinery	22%

Foodstuffs	10%
Chemicals	8%
Textiles, industrial materials, and other	7%
Trade ratio[b]	61%
Equity ratio	6.8%
Gross profit/sales (March 1985)	1.23%
Net profit/sales	0.14%
Net profit/equity	7.49%
Mitsui & Company	
Sales (March 1986)	$100 billion
Sales breakdown	
Steel products	14%
Machinery	19%
Chemicals	11%
Foodstuffs	12%
Petroleum, gas	22%
Nonferrous metals	13%
Textiles	4%
Other	6%
Trade ratio	62%
Equity ratio	5%
Gross profit/sales (March 1985)	1.52%
Net profit/sales	0.03%
Net profit/equity	2.62%
C. Itoh & Co.	
Sales (March 1986)	$86.5 billion
Sales breakdown	
Energy, chemicals	30%
Metals	14%
Machinery, construction	28%
Foodstuffs	12%
Textiles	11%
Lumber, other	5%
Trade ratio	55%
Equity ratio	3.7%
Gross profit/sales (March 1985)	1.37%
Net profit/sales	0.04%
Net profit/equity	6.03%

Marubeni Corporation	
Sales (March 1986)	$77.8 billion
Sales breakdown	
Metals	18%
Machinery, construction	29%
Energy, chemicals	28%
Textiles	9%
Foodstuffs	10%
Commodity transactions	6%
Trade ratio	66%
Equity ratio	3.7%
Gross profit/sales (March 1985)	1.20%
Net profit/sales	0.06%
Net profit/equity	6.75%

Source: Japan Company Handbook 1986.

[a]Sales figures given in yen were converted to dollars at 185 yen = one U.S. dollar.

[b]The trade ratio is the ratio of export, import, and offshore trading to sales.

17.
U.S. Tuna Processors

Robert T. B. Iversen

ABSTRACT—This chapter describes the organization and operations of the major U.S. tuna processors, including the problems that they have overcome in restructuring the production and harvesting components of the tuna industry. Also discussed are the measures taken by the processors, which are characterized by flexibility and mobility, to return to profitable operations.

INTRODUCTION

The U.S. tuna processing industry is in the process of emerging from one of the most chaotic periods of its existence. Faced with aggressive foreign competition and a buildup of modern purse seine fleets by other countries, U.S. tuna processors have taken drastic action to ensure survival.

The industry has closed all but one of its processing plants on the U.S. mainland where labor costs are relatively high, and it has consolidated almost all of its productive capacity in offshore U.S. territories where labor costs are lower and tax incentives are generous. Current indicators are that the operations of U.S. processors are profitable and show signs of remaining so.

The presence of the U.S. fleet in the Pacific islands region and the subsequent seizure of several U.S. flag vessels promoted an adversarial relationship between U.S. fishermen and island countries. These difficulties led to a series of protracted negotiations between the United States and Pacific island nations that concluded in October 1986 with agreement on a tuna treaty providing for U.S. vessel access to the exclusive economic zones (EEZs) of island countries.

INDUSTRY OVERVIEW

Principal U.S. tuna processors

There are four principal wholly U.S.-owned tuna processors: H.J. Heinz Company, which processes tuna through several subsidiaries bearing the name of Star-Kist; Ralston Purina, through its subsidiary Van Camp Seafood Company Inc; Bumble Bee Seafoods Inc; and California Home Brands Inc, through its division called Pan Pacific Fisheries.

Two other secondary processors of tuna are located in the United States but are owned by Japanese interests. They are (1) Mitsubishi Foods Inc, a subsidiary of Mitsubishi International Corporation and Mitsubishi of Tokyo, and (2) Ocean Packing Corporation, a subsidiary of Mitsui and Co (USA) Inc, which is owned by Mitsui and Co, Ltd, a large trading firm. Mitsubishi processes tuna in Puerto Rico through its subsidiary, Caribe Tuna Inc, and Ocean Packing processes through its Puerto Rico subsidiary, Neptune Packing Inc. This chapter deals predominantly with the four major U.S.-owned tuna processors.

Star-Kist. The principal tuna operating company of Star-Kist is Star-Kist Foods Inc, founded in 1917 and acquired by H.J. Heinz Company in 1963. Its corporate headquarters are located in California. Through its several wholly owned subsidiaries, Star-Kist is the largest U.S. tuna processor and producer of canned tuna and tuna-related products. It has over a one-third share of the domestic market, selling canned tuna under the Star-Kist label. Tuna and tuna-related products processed by Star-Kist account for the largest share of the total product sales of H.J. Heinz Company. Its principal tuna processing plants are located in Mayaguez (Puerto Rico) and Pago Pago (American Samoa). The Mayaguez plant is the world's largest tuna cannery, and its Pago Pago cannery is the world's second-largest. Star-Kist also has a tuna processing plant and cannery in Canada.

The Puerto Rico plant has an annual capacity of approximately 136,100 tonnes and is operating at close to full capacity. The plant employs about 4,000 persons. The American Samoa plant has an annual capacity of about 101,600 tonnes and is operating at about 90 percent capacity. The average employment in Star-Kist's American Samoa plant for the months of February, May, August, and November 1986 was 2,360.

Van Camp. Van Camp Seafood Company Inc is a wholly owned subsidiary of Ralston Purina Company Inc, which acquired Van Camp in 1963. Both companies have their corporate headquarters in Missouri. It is the second-largest U.S. tuna processor and producer of tuna and tuna-related products in the United States with about one-fifth of the domestic market. Van Camp produces canned tuna and salmon under the Chicken of the Sea label and operates tuna processing plants and canneries in Puerto Rico and American Samoa.

Van Camp does not operate tuna processing plants outside a U.S. territory, nor does it engage in any tuna processing joint ventures with foreign partners. One company has a transshipment/collection station,

Ghana, which also transships through Trinidad. Tuna purse seiners delivering fish for Van Camp caught in the western Pacific are transshipping in Guam and Tinian.

Bumble Bee. The new Bumble Bee company is the third-largest U.S. tuna processor and canner, selling under the Bumble Bee label. It holds about 16 percent of the U.S. domestic market. Bumble Bee's present facilities include a tuna processing and canning plant in Puerto Rico and a processing plant and cannery in Ecuador. It also has a subsidiary known as Bumble Bee International Inc in Tokyo, where it trades in tuna, including tuna for the Japanese sashimi market.

In 1985 when Castle & Cooke was divesting itself of its Bumble Bee subsidiary, tuna operations contributed the largest share of Bumble Bee's combined sales of tuna, salmon, and shellfish. In 1984 tuna revenues were $214 million out of the total Bumble Bee seafood sales of $301 million. The present Bumble Bee tuna processing and canning facility is in Puerto Rico and has an annual capacity of about 64,000 tonnes. Recently, it has been operating at about a 93 percent capacity and employs about 1,300 persons. The processing and production capacity, as well as the number of employees for its Ecuador operation, is not known.

Pan Pacific Fisheries. Pan Pacific, a division of California Home Brands Inc of California, is the fourth-largest tuna processor and canner in the United States, selling tuna under several private labels for supermarket chains and also under its own house labels of C.H.B., Top Wave, Lucky Strike, and others. Pan Pacific's contribution to the company's overall revenues is presumed to be considerable because in 1984 Pan Pacific's fish products (which also include mackerel) provided 30 percent ($88 million) of C.H.B.'s total revenues of $291 million. The company operates one tuna processing and canning facility in California, which employs about 500 workers. It has no overseas operations.

Secondary U.S. tuna processors

Mitsubishi Foods, headquartered in California, employs about 700 people at its Puerto Rico plant and has a processing capacity of about 27,000 tonnes per year. The company's canned tuna is sold under the 3-Diamonds brand, which is primarily distributed in the eastern and midwestern parts of the United States.

Canned tuna sales make up about 50 percent of all of Mitsubishi Foods sales, the remaining 50 percent being other types of canned foodstuffs, primarily fruits and vegetables. Mitsubishi Foods does not own or operate any tuna fishing vessels.

The Ocean Packing Corporation employs about 500 persons in Puerto Rico, but its processing capacity is unknown.

Relations between harvesters and processors

Relations between tuna purse seiners (harvesters) and the four principal processors have undergone significant changes since the late 1970s. Prior to 1980, tuna processors often owned outright, or had large financial controlling interests in, purse seiners that supplied frozen tuna. This type of relationship helped the processors to have a guaranteed source of supply. But as the high carrying cost of investment in vessels increased, U.S. processors began to divest themselves of seiners. For example, Bumble Bee purchased 12 seiners from the Gann fleet in 1975 for $31 million in cash and notes, only to resell them in 1982. This coincided with the reduction in raw tuna prices on the world spot markets in the early 1980s. By 1986 Van Camp owned or operated 17 tuna seiners, of which it had a 100 percent interest in nine, a 50 percent or less interest in six, and a long-term lease on two. Star-Kist has an equity interest in about 15 or 20 seiners, but at one time it had an equity interest in almost 50 purse seiners. Bumble Bee owns two seiners of 180 and 220 tonnes carrying capacity based in Ecuador. In 1984 Pan Pacific owned 11 seiners but recently sold its remaining 6 vessels (2 to a U.S. firm, 3 to Venezuelan interests, and 1 to Ecuador). The end result of this divestment in fleet by the processors was a shift of financial risks associated with vessel operations from the processors and harvesters entirely to the harvesters.

Decisions by processors to divest themselves of ownership and interests in seiners caused major disagreement between the processors and vessel owners. Relationships between the two groups took a turn for the worse in 1985 when a group of vessel owners sued the three largest U.S. processors (Star-Kist, Ralston Purina, and Castle & Cooke).

Supply and source of raw tuna for processing

U.S. tuna processors obtain their raw tuna from an international market composed of landings in the United States from domestic vessels—primarily purse seiners—and from imports of frozen tuna. However, the U.S. share of world raw tuna imports has steadily fallen, from 49 percent in 1979 to 30 percent in 1984. In 1985 U.S. imports of raw tuna were 5 percent less than in 1984. In 1985 most of the catch of the U.S. fleet was produced in the western Pacific (117,417 tonnes or 52 percent of domestically caught cannery receipts and U.S. direct exports).

In 1985 the eastern Pacific area produced 102,430 tonnes of tuna for U.S. canning. In 1976, when the number of U.S. flag purse seiners and catches was at its peak, the fleet produced nearly 300,000 tonnes of tuna.

In 1984 and 1985 the United States imported frozen tuna from 33 countries—245,354 tonnes in 1984 and 233,682 tonnes in 1985.

The methods by which U.S. tuna processors arrange for the purchase, acquisition, and delivery of frozen tuna are diverse and have undergone several changes in recent years. Details of contract terms with vessels, spot-market brokers, and long-term supply arrangements are confidential. However, as world production of frozen tuna increased (lowering frozen tuna prices), processors have tended to purchase on the spot market or from individually owned tuna vessels.

Tuna processors in American Samoa have an advantage over those in Puerto Rico in the delivery of frozen tuna. Because American Samoa is outside the U.S. customs district, foreign flag fishing vessels can unload their catches directly into the canneries in the territory, while in Puerto Rico the tuna caught by foreign fishing vessels must be transshipped outside Puerto Rico for delivery to the canneries.

Share of tuna processing activity within overall corporate structures

In recent years (1984–86) sales of Star-Kist's tuna and tuna-related products have made up 19 percent of the total sales of its corporate parent, H.J. Heinz Company. Van Camp's sales of seafood products (which include a small amount of non-tuna products such as canned salmon and oysters) averaged 6 percent of the total net sales of Ralston Purina Co during 1981–86, ranging from a high of 9 percent in 1981 to a low of 5 percent in 1985. Because Bumble Bee is now privately held, tuna sales are assumed to make up the majority of its seafood product line, which also includes canned salmon, smoked and whole oysters, and Figaro cat food. When Bumble Bee was a subsidiary of Castle & Cooke Inc, its revenues from tuna sales averaged about 14 percent each year as a percentage of the total revenues from Castle & Cooke in 1982, 1983, and 1984. Pan Pacific produced the highest share of revenues for its corporate parent. During 1980–84 Pan Pacific's share of C.H.B.'s total revenues was never less than 30 percent (1984) and was as high as 35 percent (1980).

Since Bumble Bee became privately owned, its annual sales have reportedly increased to the $250 million range. Star-Kist appears to be the most successful subsidiary in terms of the parent corporation's recent earnings. For the period 1984–86 Star-Kist's domestic operations contributed 22 percent (1984), 23 percent (1985), and 23 percent (1986) to H.J. Heinz's total international earnings, and industry sources report that Star-Kist has consistently been one of H.J. Heinz's most profitable subsidiaries. Van Camp's share of operating income to Ralston Purina for the period 1982–86 fluctuated from a low of minus 1 percent in 1982 to 3 percent in 1985. A negative figure for 1982 was associated with an operating loss of about $16 million on tuna vessels. Bumble Bee's tuna contributions to Castle & Cooke during 1982–84 were negative in all years, with a net loss of $12 million in 1982, $7 million in 1983, and $1 million in 1984. The operating profits of Pan Pacific Fisheries fluctuated widely during the period 1980–84, ranging from a high of $7 million in 1980 to a low of minus $3 million in 1984.

Employment and wage structure

The combined number of persons employed by the four major U.S. tuna processors in Puerto Rico, American Samoa, and California is approximately 11,300.

Current wage rates for the Pan Pacific processing plant in California range from $6.82 per hour for general labor to $8.34 per hour for a retort (skilled) operator. However, if benefits are also taken into consideration, the hourly dollar amount is higher.

According to the U.S. Department of Labor (USDOL), the average hourly earnings for the 3,318 workers in American Samoa engaged in tuna processing and canning activities in November 1985 were $2.94 per hour. At that time 2,481 workers (75 percent of the total cannery work force) were paid a Fair Labor Standards Act (FLSA) minimum rate of $2.82 per hour. Workers in American Samoa are exempt from receiving the minimum wage paid elsewhere in the United States (i.e., $3.35 per hour). Hearings held by USDOL in March and April 1986 resulted in a ruling to raise the minimum wage in American Samoa to $3.35 from $2.82 per hour. The American Samoan government, Star-Kist, and Van Camp opposed the ruling. The two canneries together are the largest single employer in the private sector in American Samoa, a territory with a population of only 35,000. A final ruling to raise the minimum wage was published by the USDOL on 20 June 1986. However, this issue became moot when the Omnibus Territories Bill (1986) contained a provision that overturned the ruling to raise the minimum wage in American Samoa. Therefore, the minimum wage in American Samoa is still $2.82 per hour.

In American Samoa the minimum wages paid by Star-Kist and Van Camp are the highest minimum wages in the territory. Minimum wages in other industries range from a low of $1.46 per hour (for laundry and dry cleaning workers) to a high of $2.82 per hour (for fish cannery and processing workers).

Processing technology

The processing of frozen or fresh tuna into its final canned product form has not changed for decades. The process is relatively straightforward and involves receiving the frozen or fresh tuna, thawing if necessary, butchering, pre-cooking the fish fillets, cooling, and loading onto conveyor belts where cleaners separate the flesh into various components (white- or lightmeat for human consumption and red- or darkmeat for pet food).

The stations, which are occupied by the fish cleaners, are called lines, and they may have dozens of workers on each side of the conveyor belt. Bones, viscera, and fish heads are converted into fish meal. Next, the meat is packed with water or oil in hermetically sealed cans. These cans of tuna are then placed in a retort for a second cooking and sterilization, cooled again, and sent through machines for automatic labeling and boxing.

A technological improvement, which is the subject of much discussion but which to date has not been perfected, is an automatic butchering machine. The problem is that tuna is of different species and comes in a variety of sizes, ranging from 1.5-kg skipjack tuna to 55-kg yellowfin tuna. A recent development has been the trial use of a "Danish" line to automatically butcher tuna. This technique is supposed to produce tuna fillets ready to be packed, thereby eliminating the precooking and cleaning phases and thus saving money. The Danish line has not been adopted, however, because of the different sizes of fish and because of the low yield of tuna for canning.

Production of tuna loins. One proposal to lower production costs is to have the tuna processed to the loin stage in an area where labor costs are low and then have the loins transported to another area where the final cooking, canning, labeling, and packing are done. Because about 70 percent of the typical processing plant's work force is employed in getting the tuna to the loin stage, processing the product in such a two-stage method should result in savings to the processor.

Loining as a separate operation, however, may not be undertaken for several reasons. First, direct labor costs are a relatively small part of the overall costs of tuna processing and canning, accounting for about 5 to 15 percent. Therefore, loining operations may result in only relatively small savings. Furthermore, the cost of transporting tuna loins to the final processing destination is considerable because tuna loins have a short shelf life. Another important reason is that certain species, especially skipjack, have quality-control problems.

Company management policies in developing countries

There seems to be no written public management policy by U.S. tuna processors that bears on their relationships with developing countries. The only statement resembling a corporate policy is one by Star-Kist: "Whenever and wherever appropriate, Star-Kist is prepared to make commitments in any part of the world that will maintain its position of leadership and will aid in local fishery and economic development."

GOVERNMENT INTERACTION

Relationships between U.S. processors and foreign governments

Relationships between the processors and foreign governments can be broadly categorized into two different groups: direct and indirect. In the direct relationship, the tuna processor negotiates directly with a foreign government for the right to participate in certain activities. In an indirect relationship, the processor often interacts with the foreign government through a third party, such as the U.S. government, or perhaps a tuna industry-related organization in which the processor participates or has an equity.

Tax incentives and repatriation of profits

Puerto Rico. In Puerto Rico, fiscal incentives cover relief from corporate income taxes, property taxes, and municipal taxes (sales tax). Puerto Rico's Industrial Incentive Act (1978) allows those tuna processors and commercial fishing operations that supply Puerto Rican canneries tax exemptions of up to 90 percent of industrial development income for periods ranging from 10 to 25 years, depending on the industry's location.

Under the 1978 law all tuna processors, except Neptune Packing, have a 90 percent exemption on corporate income taxes with a maximum rate of 45 percent on taxable income. The present exemptions for the five processors are effective until the years 1993–2000, depending on the company. Neptune Packing's present tax exemption is 85.5 percent. Bumble Bee, Star-Kist, and Van Camp are in a zone providing maximum tax benefits, and their income tax exemptions will be reduced from 90 percent to 65 percent by the year 1995 (1993 for Star-Kist). Neptune Packing's benefit will be reduced to 35 percent by 1994, but Caribe Tuna's income tax exemption will be reduced to zero by 1993. Under the Industrial Incentive Act, however, the processors will have an opportunity to renegotiate their present level of tax exemptions.

Puerto Rico also has what is termed a toll gate tax (effectively a dividend withholding tax), which is a tax levied on dividends declared by Puerto Rico-based companies that normally would be distributed to the parent corporation or outside Puerto Rico. The regular toll gate tax is 10 percent, but the average effective rate paid by most tuna processors is estimated to be in the range of 4 to 6 percent. A lower toll gate tax is provided if processors agree to keep some of their profits in Puerto Rico in what are termed "eligible activities."

American Samoa. American Samoa grants exemptions from the payment of certain categories of taxes for periods up to ten years for the establishment or expansion of qualifying industrial or business enterprises under the Industrial Incentives Act. The exemptions may be extended to encourage new businesses or significant expansion of existing businesses. Other benefits include the non-taxation of dividends paid by wholly owned subsidiaries of U.S. parent companies operating in American Samoa, as well as the carrying forward of business losses for tax purposes for up to seven years. American Samoa has no taxes on sales or gross receipts, property, exports, or value-added items. American Samoa also permits 30- to 60-day free port storage for transshipped freight through Pago Pago (where the two tuna processors are located).

Both Samoa Packing Company (Van Camp/Ralston Purina) and Star-Kist Samoa Inc have agreements with the government of American Samoa covering a wide variety of tax benefits. The agreements also call for the two companies to make investments to increase processing capacity and employment, improve the environmental conditions in Pago Pago harbor, and train American Samoans for positions as managers in the processing plants and as crewmen on purse seiners. Both agreements cover the same general types of benefits, but they differ with respect to the actions that must be taken by the processors so as to continue their tax benefits.

Exemptions from income taxes on activities related to the delivery of tuna to the two processing plants are provided for tuna longline and purse seine fishing vessels, as well as for related vessels owned by corporations incorporated in American Samoa. Each corporation organized under American Samoa laws that owns or operates purse seiners is also exempt from all corporate income taxes on its income from tuna fishing if at least 20 percent of its annual tuna catch is delivered to American Samoa for processing. Tax rates imposed by American Samoa on corporate income are the same as the U.S. government's tax rates on corporate income.

Relations between U.S. processors and the U.S. government

Numerous major pieces of U.S. legislation either directly or indirectly affect relations between the processors and the U.S. federal government. The legislation includes the Marine Mammals Protection Act (1972), the Nicholson Act (1793), the Fishery Conservation and Management Act (1976), and the Fisherman's Protection Act (1967). In addition, there have been (1) investigations into the U.S. tuna industry conducted by the government, (2) rulings on foreign trade preferences, and (3) cases of financial assistance by government or quasi-governmental agencies to the industry.

Industry investigations

In 1984 and 1986 the U.S. government conducted two extensive investigations into conditions in the U.S. tuna industry. One, known as a "201 investigation," was undertaken after the U.S. International Trade Commission (USITC) received a petition for import relief on behalf of the United States Tuna Foundation (USTF), C.H.B. Foods Inc, the American Tunaboat Association (ATA), and three fishermen's unions under Section 201 of the Trade Act (1974). A determination was made that tuna canned in water and in oil was not being imported into the United States in such increased quantities as to be a substantial cause of injury to the U.S. domestic industry.

The other investigation, known as a "332 investigation," was a comprehensive study of conditions in the U.S. tuna industry under Section 332 of the Tariff Act (1930). The investigation was requested by the U.S. trade representative, but the results did not include any policy determinations or recommendations. Its primary purpose was to update the data collected by the USITC in 1984.

According to industry sources, the 201 investigation caused some disharmony between U.S. processors; Bumble Bee and Van Camp withdrew from the USTF in 1984 because they were not in favor of the petition. However, in 1986 they rejoined USTF. Both investigations caused U.S. tuna processors to supply the USITC with a large amount of data, much of which was considered proprietary, in response to government questionnaires.

Some U.S. government rulings have helped U.S. processors, as did a 1983 ruling by the U.S. International Trade Administration (USITA), which concluded that all manufacturers, producers, or exporters in the Philippines of canned tuna had subsidies from the Philippine government that constituted bounties or grants within the meaning of the countervailing duty law. As a result, the USITA ordered that a countervailing duty of an additional 0.72 percent be added to the duty normally charged on canned tuna imported from the Philippines.

MARKET POSITION AND SHARES

Location of sales

The principal market for canned tuna from U.S. processors is the domestic U.S. market. The export of canned tuna is negligible compared with total U.S. production. For this reason, data are not collected separately on U.S. exports of canned tuna. According to official statistics, canned tuna is the most commonly consumed fish product in the United States. Sales and consumption of canned tuna occur throughout the country but tend to be concentrated regionally, especially in the metropolitan areas along the U.S. coasts—the northeast, southern California, and the Pacific northwest. There are geographical preferences for different varieties of canned tuna, with the east coast market preferring whitemeat (albacore) tuna and the west coast preferring lightmeat (yellowfin and skipjack) tuna.

U.S. processors sell their brands at all locations throughout the United States. However, one of the secondary processors, Mitsubishi Foods (which markets its canned tuna under the 3-Diamonds label), sells its product primarily on the east coast, with little tuna being sold beyond the U.S. midwest.

Relative market shares of various processors

The retail market shares of the different processors are shown in Table 1. It does not indicate the relative market position of Pan Pacific Fisheries

Table 1. Market shares of U.S. tuna processors, September 1986

Brand	Processor	Market share (%)
Star-Kist	Star-Kist	36.0
Chicken of the Sea	Van Camp	20.0
Bumble Bee	Bumble Bee	15.5
3-Diamonds	Mitsubishi	3.7
Geisha	Mitsui	2.8
Empress	Mitsui	1.4
Private labels		17.0
Other		3.6
Total		100.0

Source: SAMI 1986.

because its products are sold by California Home Brands under a variety of private labels. Industry sources indicate that its share makes up a large proportion of private label sales and that it is fourth in the U.S. market.

Product price history

Domestic whitemeat tuna was $25.81 per case in 1975, reached a peak of $47.44 in 1981, and declined to $39.90 in 1985. Domestic lightmeat tuna was $23.61 per case in 1975, $35.58 in 1980, and $26.00 in 1985. Canned imports sold for $17.34 per case in 1975, reached a high of $30.38 in 1981, and by 1985 sold for $19.06.

The total wholesale value of all canned tuna sold in the United States also varied widely during 1975–85. It was $699 million in 1975, rising to $1,337 million in 1978, then generally fluctuating downward, and reaching a level of $1,030 million in 1985.

Brokers

The four major processors use brokers to sell their canned tuna. There are reportedly more than 200 brokers selling U.S. canned tuna. They are organized on a regional basis, and each broker usually sells only one brand of tuna. This is required by processors. Imported canned tuna is generally marketed by the importing firm, which may also act as a broker for some domestically produced canned tuna. Institutional food brokers also distribute imported canned tuna because imports are concentrated in this sector.

Profitability of tuna processors

Net sales increased strongly from 1979 to 1981, from $961 million in 1979 to $1.2 billion in 1981, a 27 percent increase. Sales then began to decline, and by 1985 they were down to $1 billion—a drop of 15 percent. As tuna sales began to decline in 1982, the processors reported net income losses of $61.7 million (-5.5 percent) in 1982, $50.4 million (-4.7 percent) in 1983, $4.6 million (-0.4 percent) in 1984, returning to profitability in 1985 with net income of $57.9 million (+5.6 percent). However, even though the processors in the aggregate showed losses in net income only for the years 1982–84, in every year except 1980 at least one processor reported operating losses. The worst year was 1982, when four processors reported operating losses, but in every other year between 1979 and 1984, either one or two processors reported operating losses.

The net losses in income during 1982–84, however, should not be attributed solely to declines in net sales because gross income was positive in all years from 1979 to 1985. During the period 1981–84 several firms wrote off large costs of closing their California plants, which were reflected as net losses between 1982 and 1984.

In 1985 the U.S. canned tuna industry enjoyed a marked increase in profitability with a positive net income of $57.9 million, even though net sales were down from 1984. This improvement in profitability was partly due to a decrease in the costs of production and divestments in fishing vessels. Preliminary data submitted by processors for 1986 show a continuation of the trend toward profitability.

INDUSTRY RESTRUCTURING

Since the mid-1970s, and especially since the early 1980s, events on the international and national scene for U.S. tuna processors have combined to bring about fundamental changes in the world tuna industry. These changes included (1) a rapid buildup of the U.S. purse seine fleet to a peak of about 140 vessels in 1975, followed by a rapid decline in the 1980s; (2) a worldwide recession starting in 1981; (3) an increase in foreign tuna purse seine fleets that increased the world's tuna supply and reduced prices; (4) changes in the location of U.S. fleet operations; and (5) the emergence of large-scale tuna canning in foreign countries, notably Thailand. Foreign imports of canned tuna to the United States have now captured almost 35 percent of the U.S. market for canned tuna. Faced with relatively high domestic labor costs and falling domestic production, the U.S. processors closed all their mainland canneries except one and moved offshore.

External factors

The recession of 1981 and its associated rise in interest rates began to cause financial difficulties for the U.S. purse seine fleet. By 1986 the number of active U.S. purse seiners had declined to 72, down 51 percent from the fleet's peak in 1975.

At the same time, foreign tuna purse seine fleets were undergoing expansion. In 1980, 14 Japanese purse seiners were fishing in the western Pacific, but by 1985 the number had increased to 33. There were also 53 other foreign seiners fishing the western Pacific in 1985, in addition to the U.S. fleet. European fleets of purse seiners began fishing in the Indian Ocean, and the Mexican government began a construction program of modern tuna purse seiners.

The expansion of purse seine fleets led to more frozen tuna being available on the international market, which depressed prices for frozen raw tuna. As a result, U.S. tuna processors started to change their tuna procurement policies and began divesting themselves of vessels in which they had a financial interest. As a result, many seiners today are fishing on an "open ticket," have no guaranteed market for their fish, and must negotiate prices with the processors on a per-trip basis, often when they return to port.

Changes in the dietary habits of the U.S. consumer have also played a role in changing market conditions. Canned tuna packed in water (which consumers prefer for its lower calorie content) has captured the largest U.S. market share, rising from 45 percent in 1979 to 72 percent in 1985.

The lack of access arrangements between the ATA and the Pacific island countries after 1984 caused the fleet some problems. Without agreements with the better-producing areas, such as the Federated States of Micronesia, Kiribati, and Papua New Guinea, seiners have had their mobility restricted.

Internal factors

From 1980 to 1984 domestic tuna landings at U.S. ports and domestic tuna processing fell dramatically. In 1980 domestic landings were 181,436 tonnes, but by 1984 the landings were only 96,161 tonnes—a drop of 47 percent. In addition, the volume of canned tuna produced domestically fell from 14.8 million standard cases in 1980 to 6.5 million cases in 1984—a decrease of 56 percent. U.S. processors were faced either with continuing their processing activities both offshore and in California (where they operated expensive canneries) or with consolidating all of their processing offshore where costs were significantly lower.

Another contributing factor to the offshore move was the change in relationship with the U.S. fleet that supplied the mainland canneries. Most of the vessels were homeported in California, and so long as the processors either owned the vessels or had large equities in them and maintained their canneries in California, the vessels probably exerted pressure on the processors not to move. By 1980–83, however, most seiners were operating independently of the processors and presumably exerted less influence over the decision making and policies of the processors.

IMPLICATIONS FOR PACIFIC ISLAND COUNTRIES

The U.S. tuna processors succeeded in overcoming serious problems connected with resource location, abundance, and price, as well as increased costs of production, rapid changes in consumer demand for canned tuna, intense competition from foreign processors, and large-scale changes in both the U.S. and foreign purse seine fleets. That they have succeeded in overcoming these problems may be due as much to their being part of large corporate entities as it is to decades of experience in a changing industry. Historically, the Pacific island countries have not had to deal with these issues.

The activities of U.S. processors in foreign countries over the past two or three decades have shown that they must maintain flexibility and mobility if they are to succeed in business, but their involvement with local government and business interests can lead to economic development in countries with relatively undeveloped tuna fishing and processing industries.

Due to the increase in the world's population, the demand for canned tuna can be expected to increase over the long term, and this should present new opportunities for Pacific island countries to become a more significant part of this worldwide industry.

Impact of U.S. processor operations on the Pacific islands

U.S. processors will probably need to repair relations that have eroded over the past five years with Pacific island nations so that future business discussions may proceed in a positive manner. There are indications that this is already beginning. Tuna processors are expecting a continued annual increase in the U.S. consumption of canned tuna. If it should reach levels beyond the productive capacities of their plants in Puerto Rico and American Samoa, these processors may consider processing operations in areas closer to the tuna resources supplying their plants.

Because the processors are likely to continue to reduce their holdings in vessel ownership and operations, they will probably seek business arrangements that allow them to purchase frozen tuna direct at the lowest possible prices. If U.S. purse seine owners can enter into profitable joint ventures with island governments (in which the government is willing to subsidize some operational costs of the vessel in return for the social and economic benefits of basing a purse seiner in a developing area, leading to lower frozen tuna costs), processors could be expected to endorse such an arrangement.

Processors can also be expected to contribute to technology transfer under the terms of the U.S. tuna treaty, as well as make substantial contributions in this area.

Processors are in the position to offer marketing skills to those Pacific island nations that are producing canned tuna so that they can sell their products on the U.S. market. Because some Thailand processors are already exporting canned tuna bearing the labels of major U.S. processors, U.S. processors might consider similar arrangements with island processing plants, provided the costs of the canned tuna were competitive with the Thai product, even if they are not involved in the ownership or management of the processing plants.

Tuna development alternatives

If interested Pacific island countries could provide a labor force whose wages would be sufficiently low to encourage investment in processing plants, or if they could provide potential investors with sufficient tax incentives or other concessions, they are then faced with the problem of ensuring that the cost of raw tuna will not jeopardize the processing operation. If the potential investor does not have the ability to obtain low-priced frozen tuna on the world market, these countries may have to resort to operating their own fleets of tuna vessels to supply the processing plant. The capital needed for enough catcher boats to supply the processing plant may be beyond the capability of governments. Thus they might consider joint ventures with, say, a U.S. purse seine owner who is seeking a way out of financial difficulties. If an island government is willing to merely break even on vessel operations or perhaps even contribute a subsidy toward vessel operations so that the processing plant can receive a steady supply of reasonably priced fish, the effort might be worthwhile, based on the assumption that the general social and economic benefits from an operational processing plant would outweigh the subsidy.

Medium- or small-size seiners might work in countries such as Papua New Guinea or Solomon Islands that are close to the fishing grounds. They probably would not be suitable for operations in Kiribati, where fishing is spread over greater distances. A key factor is that of travel time relative to fishing time. If some U.S. purse seiners based in the western Pacific are operating at a profit, Pacific island countries might contract with them to supply the fish needed to maintain a processing operation.

Instead of a processing operation, Pacific island nations also could consider developing transshipment facilities in locations closer to the center of fishing operations, with the view that if these proved successful, they could lead to the further development of a processing plant. However, the establishment of large transshipment facilities from scratch may be financially prohibitive.

One other possibility is that Pacific island countries might obtain their own reefer vessels to transship tuna. If world tuna production continues to increase, there should be a need for additional reefer vessels, and if Pacific island countries can figure out a way to operate reefer vessels more cheaply than the foreign reefer vessels now used, there could be a market for their services.

CONCLUSION

U.S. tuna processors in the past three years have changed from being vertically integrated industries. They have taken drastic actions to cut production costs in order to meet foreign competition and to regain profitability.

U.S. processors realized that in order to be profitable, they would have to operate in areas where labor costs were lower than in California and where they could compete with the increasing amounts of imported canned tuna, particularly from Thailand. They therefore consolidated their U.S. processing operations offshore in Puerto Rico and American Samoa, except for one medium-size cannery still processing tuna on the U.S. mainland.

Faced with additional losses due to ownership or equity in purse seiners, U.S. processors divested their interests in the unprofitable seiners. This has led to a fundamental change in their relationships with vessel owners and has contributed to the legal action being taken by a large number of seiners against the processors.

U.S. processors may have concluded that even though their processing operations in Puerto Rico and American Samoa are sufficient for them to meet the demands of the U.S. market, future increases in U.S. consumption of canned tuna or a decrease in foreign imports if the tariff on tuna packed in water is raised may require an expansion of their processing capacities. As a result, they will continue to investigate the possibilities of processing in the Pacific islands region close to the fishing grounds.

Island countries have already indicated an interest in developing their own tuna fisheries; thus the time appears appropriate for the U.S. processors, as well as other segments of the U.S. tuna industry, to seek new business arrangements with Pacific island countries.

18.
A Summary of International Business Operations in the Global Tuna Market

Linda Lucas Hudgins and Linda Fernandez

ABSTRACT—This summary chapter reviews companies engaged in global tuna transactions. These companies fall into three general categories: (1) small companies within a domestic industry whose activities are confined to a single country; (2) relatively large specialist companies (e.g., Japanese fishing companies) that may have international operations; and (3) large multinational corporations (e.g., StarKist, Safcol, and C. Itoh) that are often divisions or subsidiaries of large food production companies. This chapter examines general business activity in tuna markets, particularly the operations of multinational corporations that dominate global tuna transactions.

INTRODUCTION

The restructuring of world tuna markets essentially began when multinational corporations moved out of their own countries to conduct tuna business in foreign countries.[1] Before 1980 tuna production and consumption were usually integrated in the same country. That is, the U.S. tuna fleet caught the tuna that was ultimately canned by U.S. processors and consumed by American consumers. The Japanese fleet likewise caught most of the tuna consumed by Japanese consumers. Since 1980, however, tuna exchanges have taken place in a more sophisticated international environment. Today, the tuna consumed by U.S. consumers may have been harvested in the western Pacific, transshipped to Thailand, canned in a plant leased to a multinational corporation, labeled, and distributed by yet another multinational. Tuna harvested by the Japanese fleet may be shipped or traded worldwide by *maguro shosha,* specialist tuna-trading companies.

The causes of this shift in location of production are varied. They include the declaration of exclusive economic zones (EEZs) by a number of countries with tuna resources, the exchange rate fluctuations, and the world recession in the early 1980s that caused corporations to seek lower wages and production costs in foreign countries.

In addition to the change in location, there have been concurrent changes in activity by corporations. The previously vertically integrated processors have moved away from harvesting and, in some cases, away from processing toward distribution/trading. In addition, pressure from lower-priced products (e.g., the Thai canned tuna product) has forced various domestic markets to open up, resulting in shifts in market shares within domestic markets.

Although the tuna is now caught and processed in locations different from those in the past, the major tuna-consuming countries (United States and Japan) remain the same, and the corporate headquarters of major tuna companies remain the same (United States, Japan, and Australia). European markets are either relatively self-sustaining or import through multinational corporations.

While new industries have been developed on the international tuna scene (e.g., Thailand, Mexico, and Pacific island countries), the same corporations dominate transactions in harvesting, transshipping, processing (canning/drying/freezing), distribution, and trading. Almost 70 percent of all tuna and tuna products today is caught, processed, traded, or distributed by ten multinational corporations. These corporations operate on every continent and at every level of the market.

This chapter describes general business operations in global tuna markets and examines the activities of the largest multinational corporations: Bumble Bee (formerly a Castle & Cooke subsidiary), C. Itoh, John West Foods, Marubeni, Mitsubishi, Mitsui, and Safcol (Thailand, Philippines, and Australia), Star-Kist (H.J. Heinz subsidiary), Van Camp (Ralston Purina subsidiary), and H.J. Heinz (Australia and United States). This chapter briefly reviews corporate histories and finance, while concentrating on recent corporate objectives and capital mobility in response to the worldwide restructuring of the tuna industry.

GENERAL BUSINESS ACTIVITY

Companies engaged in global tuna transactions fall into three major categories: (1) small companies within a domestic industry whose activities are confined to a single country; (2) relatively large specialist companies (Japanese fishing companies) that may have international operations; and (3) large multinational corporations (e.g., Star-Kist, Safcol, and Japanese trading companies) that are often divisions or subsidiaries of food production parent companies. In the third category, the parent company may trade or produce hundreds of products besides tuna. The subsidiaries usually can take advantage of the parent company's distribution systems or communication networks.

Small domestic-oriented tuna companies

The small domestic-oriented tuna company is most obvious in Japan (harvesting), Thailand (canning), and Australia (harvesting and canning). These firms have found their niche in a particular domestic market and do not participate directly in any international activities. They are often formed into federations for government lobbying efforts or coordination of production and sales. There is little opportunity for Pacific island countries to interact with this type of firm, specifically because they are domestically oriented.

Specialist companies

Specialist companies are characterized by both domestic and international operations. A key distinction from the small type of firm is participation in foreign joint ventures and operations at more than one level of the industry. For example, the six major Japanese fishing companies (Taiyo, Nippon Suisan, Nichiro, Kyokuyo, Hoko, and Hosui) were involved in 109 overseas joint ventures or wholly owned operations in 31 different countries as of 1985. These ventures predominantly involve tuna trading (purchases), provision of services to the Japanese fleet throughout the Pacific, or transshipment. (Under Japanese law, transshipment must be made in port rather than at sea.) The Taiyo joint venture in Solomon Islands is an example. These companies provide a wide range of potential interaction for Pacific island countries.

Multinational corporations

This group includes about ten corporations (or Japanese trading companies) that have consistently been involved in world tuna markets at every level of production. Table 1 lists corporate activity by selected locations. Although individually unique, the corporate strategies reflect some common orientations. For example, the corporations are increasingly divesting themselves of vessel equity of any sort, including vessel financing, which once allowed fleets to develop (e.g., Korean vessels were financed by Japanese, and U.S. processors held equity in U.S. flag vessels). As competition increased, many vessels lost their contracts when processors went into the spot market to buy tuna at lower prices. This divestiture will inevitably lead to further declines in some established fleets while opening the door to the entry of others.

It is difficult to align a particular company with a particular specialization in production or even with a particular location in which each production phase is undertaken. This difficulty merely highlights the

Table 1. Major corporate entities in international tuna markets by past and current activity and location

	Bumble Bee	C. Itoh	H.J. Heinz (Australia)	John West	Marubeni	Mitsubishi	Mitsui	Safcol	Star-Kist	Van Camp
American Samoa									1,2,3	1,2,3
Australia		1,2,3,4	1,2,3,4	4	1,2			1,2,3,4		
Canada									3,4	
Ecuador	1,2,3									
Fiji		1,4								
France		2							2	
Ghana									2	2
Guam									2	
Indonesia		1,2							2	
Ivory Coast									2	2
Japan	4,5	1,2,3,4,5			1,2,3,4,5	1,2,3,4,5	1,2,3,4,5			
Mauritius						2,3				
Philippines		1						3		
Puerto Rico	1,2,3					3	3			1,2,3
Reunion Island									2	
Senegal									2	
Thailand								2,3		
Trinidad										2
United Kingdom				4						
United States	1,3,4					4	4	5	1,3,4	1,3,4

Key: 1 = harvesting, 2 = transshipping/cold storage, 3 = processing, 4 = distribution, 5 = trading complexity of multinational corporations that operate widely and within a number of subsidiaries. In addition, as the production phases have changed in relative importance within the world tuna market, companies have responded by moving into these new operations. For example, transshipping has become more important as increasing supplies of tuna are caught in the Pacific and need to be transported to major processing areas in Puerto Rico, American Samoa, the Philippines, and Thailand. This procedure requires increased coordination and perhaps even capital from tuna market participants. Table 2 shows past and current activity by level of market involvement.

A major point is that these corporations have both mobility and international linkages, and they use them. Safcol, for example, has six plants in Australia and seven plants in Thailand and previously operated plants in Indonesia and the Philippines. Mitsubishi now leases plants in Puerto Rico, and Star-Kist outsources from canneries in Thailand. The lease and outsource arrangements give the company more flexibility to respond to changes in markets, resource availability, or tax/tariff considerations. All but one processing plant has closed on the U.S. mainland, with production predominantly shifted to American Samoa and Puerto Rico. Bumble Bee has recently begun production in a plant in Manta, Ecuador.

These corporations are increasingly competing with one another at some levels of production, while at the same time cooperating and trading. Pacific island countries can expect that any interaction with these corporations will be characterized by profit motives that, in turn,

Table 2. Major corporate entities in international tuna markets by past or current level of market involvement

	Harvesting	Transshipment	Processing	Distribution	Trading
Bumble Bee	x	x	x	x	x
Mitsubishi	x	x	x	x	x
Marubeni	x	x	x	x	x
Mitsui	x	x	x	x	x
Safcol	x	x	x	x	x
Star-Kist	x	x	x	x	
H.J. Heinz (Australia)	x	x	x	x	
Van Camp	x	x	x	x	
C. Itoh	x	x	x	x	x
John West				x	x

Sources: Ashenden and Kitson 1987b; Comitini 1987; Crough 1987a; Iversen 1987.

depend on market conditions and institutional considerations discussed in the following section.

INSTITUTIONAL FACTORS

The corporations plan their tuna industry activities around certain policies and regulations that (1) directly affect their access to harvesting, processing, and marketing resources and (2) enhance their profit position in the market. These international enterprises take into consideration domestic regulations, as well as those in the overseas country in which they choose to operate.

In Australia, for example, all tuna harvesters recognize the limitations imposed by the national management plan for southern bluefin tuna. The harvesting quotas are said to have influenced processing corporations to shift out of vessel operations toward financing.

The operations of Safcol in Thailand have been influenced by the favorable position of the Thai Board of Trade with respect to export-oriented firms. The Star-Kist and Van Camp operations in American Samoa and Puerto Rico are clearly influenced by favorable tax and tariff concessions, as well as by special legislation that allows landing of fish by foreign flag vessels in these ports. The American Samoa tuna pack is also allowed tariff-free entry into the U.S. market.

The Japanese specialist tuna-trading companies, attempting to increase profit margins, are showing less and less interest in joint-venture activities, preferring to perform as intermediaries rather than producers. These companies will continue to be influenced by exchange rate fluctuations.

The corporations overall seem to be affected more by political and economic risk factors than by ownership barriers, for example, requiring a certain percentage of ownership by nationals. If all else is equal, the corporations appear more willing to locate in known legal situations such as American Samoa and Puerto Rico. Australian firms have longer-term relations with Asian markets and therefore locate there. Exceptions to this pattern exist, but they are still experimental , like the Manta, Ecuador, plant.

Pacific island countries can enhance the possibility of conducting business with these multinationals through careful targeting of incentive programs.

CORPORATE DESCRIPTIONS

Tuna production companies

Bumble Bee. This corporation is the third-largest tuna processor in the United States, operating a tuna processing and canning plant in

Mayaguez, Puerto Rico, and another in Manta, Ecuador. Bumble Bee also maintains a subsidiary, Bumble Bee International Inc, in Tokyo, which trades tuna in Japanese sashimi markets and others. The Puerto Rico plant operates at 93 percent capacity and processes about 64,000 tonnes of tuna annually.

This plant employs 1,300 persons at wages of $3.97 per hour. No data are available on the operations in Ecuador. As of 1985 Bumble Bee became a private corporation, after a long history as a subsidiary of Castle & Cooke, Inc. Statistics indicate that gross revenues from tuna operations in 1984 were $213.5 million. Bumble Bee owns only two purse seiners operating out of Ecuador.

H.J. Heinz Company Ltd (Australia). This large tuna producer handles Greenseas tuna products for human consumption, 9-Lives pet food, and Epicure/Golden Days health food products. Heinz has made many attempts to expand in Australia with the purchase of brand names of Seahaven and Frelish tuna from West Ocean Canning Pty, Ltd. Heinz, like Safcol, provided aerial spotting and shore facilities to enable Australian vessels to increase domestic tuna catches during the 1970s. Established in 1935, Heinz (Australia) is independent of its parent company, H.J. Heinz (USA), and employs about 1,500 residents. The total company earnings in 1985 from all products were $179 million.

Safcol Holdings Ltd. Safcol is Australia's largest tuna processing and trading group, maintaining six tuna processing plants in Australia and seven plants in Thailand. Safcol (Australia) earned about $83 million in 1985, and Safcol (Thailand) earned $63 million in 1985.

Safcol is a subsidiary of the Southern Farmers Group, the third-largest food processor in Australia, which handles food and non-food processing. The Southern Farmers Group, which first was a fishermen's cooperative, is a part of Industry Equity Ltd, the fourth-largest company in Australia.

Safcol imports all raw material for its Thai operations that process for human consumption and pet food. A trading company, Interpesco, based in San Diego, California, was recently purchased by Safcol. Interpesco purchases raw/frozen tuna (primarily skipjack) on the world market from Pacific Ocean fleets. This company, first contracted by Safcol in 1985, also handles distribution of Thai tuna products in the United States.

In mid-1986 Safcol entered the Japanese canned tuna market in response to the yen's appreciation. Negotiations were conducted directly with Japanese supermarket chains that sold the tuna under their own labels. This endeavor bypassed the usual wholesale market and avoided elaborate trading company arrangements and fees.

Safcol now centers its Australian operations in Victoria after closing its Port Lincoln and Melbourne plants. For harvesting, Safcol charters the New Zealand vessel *Daniel Solander*, which processes and freezes lower-grade sashimi caught by smaller boats for export to Japan. In 1985–86 this vessel processed approximately 300 tonnes. Prior to 1984 Safcol tried to remain competitive in vessel operations by using aerial spot surveillance and radio communications for fleets. The quotas on southern bluefin tuna, however, have reduced domestic catches and the profits of vessel operations. Safcol sells Sirena brand tuna domestically and has a contract to supply a competitor, John West Foods, with tuna for distribution.

A brief look at the short-term Safcol operations in the Philippines and Indonesia provides a background for Safcol's international operations as a competitive, profit-seeking, financially mobile entity.

Safcol (Philippines and Indonesia). In 1977 Safcol entered a joint venture for Philippine tuna canning with Judric Seafoods Co of Hong Kong, a company favored by the Philippine Board of Business. By 1980 the plant operated 24 hours per day processing 200 tonnes daily, an amount equal to one-half of the country's tuna exports. These production levels led to a second venture between Safcol and Judric called the Philippine Tuna Canning Corporation, which operated several canneries. The canned tuna supplied institutional markets in the United States and represented the largest percentage of any cannery in the world supplying the United States. A deterioration in international tuna market prices and poor domestic government policy made the Philippine operations vulnerable to serious financial losses. Finally, in the early 1980s, competition and worker union protests increased the financial and political deterioration. By 1984 the Manila plant had closed, and subsequent labor disputes closed the plant in Zamboanga. Total losses for Safcol in the Philippines were estimated at $4.9 million in 1984.

Safcol also lost money on its operations in Indonesia during the early 1980s. The Indonesian losses are attributed to remote site location and to corruption on the part of management. This plant closed in 1984. Safcol then expanded operations in Thailand.

Star-Kist Foods, Inc. Star-Kist is a subsidiary of H.J. Heinz Company and has one-third of the tuna market share in the United States. In 1986 Star-Kist earned $829.6 million, which was 19 percent of the total H.J. Heinz earnings in 1986. Star-Kist operates the world's first- and second-largest tuna processing operations in Puerto Rico and American Samoa. The Mayaguez, Puerto Rico, plant operates at 100 percent capacity, processing 187,077 tonnes of tuna per year. The plant in Pago Pago, American Samoa, operates at 90 percent capacity, processing

101,604 tonnes of tuna per year. The Puerto Rico plant employs 4,000 people at the minimum wage of $3.35 per hour. The American Samoa plant employs 2,358 persons at $2.82 per hour.

Star-Kist operates cold-storage facilities in France, Ghana, Ivory Coast, Reunion Island, and Senegal, as well as a processing plant in Douarnenez, France. The site in Ghana also processes tuna for sale to local markets. The company has reduced its equity in purse seiners from 50 to 20 in an attempt to avoid financial risk at the harvesting level.

Van Camp Seafood Co, Inc. Van Camp represents the second-largest U.S.-headquartered tuna processing company with about 16 percent of the U.S. tuna market. The best estimates of production indicate that the Mayaguez, Puerto Rico, plant processes 41,000 to 55,000 tonnes annually, and the American Samoa plant processes 70,000 tonnes annually. The Puerto Rico plant employs about 1,244 persons at $4.40 per hour and another 1,226 persons at $2.82 per hour. Van Camp maintains cold-storage facilities in Ghana, Ivory Coast, and Trinidad. The company sells under Chicken of the Sea label and became independent from the Ralston Purina food distribution company in 1985. Its total earnings in 1985 were $270.8 million.

Tuna trading companies (*sogo shosha*)

The companies described in this section are members of the specialist tuna traders network in Japan (*sogo shosha*), as well as international businesses in their own right. In general, the four top companies, Mitsubishi Corporation, Marubeni Corporation, Mitsui and Company, and C. Itoh are among the top 6 of the 200 largest companies in the world. They have been involved in the tuna trade since the 1960s with the central objective to trade in tuna products and earn commissions. The *sogo shosha* view their function in the tuna industry as fivefold: (1) arranging transportation, (2) financing, (3) providing documentation formalities, (4) ensuring delivery, and (5) inspecting fish.

With a noticeable decline in the profit margin along with substantial financial risks from advancing capital to tuna harvesting vessels, the number of tuna traders has declined. All have effectively increased their market share of the sashimi trade (collectively holding about 70 percent of the Japan sashimi market) by negotiating whole boatload purchases of tuna from Japanese boats operating in distant waters, as well as by acting as marketing agents for Korean and Taiwanese tuna fishing enterprises. All have cold-storage warehouses and agents at the main landing ports of Yaizu, Japan.

Mitsubishi Corporation. This *soga shosha* has several subsidiaries in Japan (Mitsubishi Shoji) and the United States (Mitsubishi Foods Inc) that ship tuna and manage tuna distribution. Mitsubishi's main port operations in Yaizu, Japan, are handled by Toyo Reizo cold-storage company, which provides processing and marketing. Mitsubishi ships approximately 10,000 to 15,000 tonnes annually for processing from the western Pacific and another 10,000 tonnes annually of frozen skipjack tuna from Japan to canning operations in Asia (Thailand and Taiwan). Mitsubishi's role as a shareholder in the Pacific Ocean fishing endeavors of Kaigai Gyogyo Kabushiki Kaisha (KGKK Overseas Fishing Company Ltd) (1958–83) has generated past joint-venture fishing and transshipment arrangements in Malaysia, Mauritius, Papua New Guinea, and Madagascar. Political and economic conditions in these locations reduced Mitsubishi activity; now it includes only the Mauritius joint-venture tuna cannery.

Tinian serves as Mitsubishi's transshippment base for tuna purchased from Japanese, Korean, Taiwanese, and U.S. vessels in the Pacific. Mitsubishi maintains a cannery, Tuna Inc in Ponce, Puerto Rico, through a U.S. subsidiary, Mitsubishi Foods Inc. Its capacity is 27,000 tonnes per year, and it employs 700 people. Mitsubishi markets its product under the 3-Diamonds brand, much of which is sourced from Thailand. Total earnings of Mitsubishi's tuna operations were $150 million to $200 million for 1986.

Marubeni Corporation. This corporation arranges many tuna vessel shipping contracts through its sashimi market agents for Australian supplies of raw tuna for distribution. Its subsidiary, Marubeni Australia Ltd, had earnings of about $88 million for 1985–86, contributing to overall tuna trading profits of $49 million. Once tuna supplies arrive in Yaizu, Marubeni's subsidiary, Marukai Suisan, subcontracts cold storage prior to distribution. The Tokyo subsidiary, Marubeni Reizo, provides cold storage for tuna products distributed to sashimi markets in Japan.

C. Itoh and Company. Although C. Itoh built several tuna purse seiners in the past, it now purchases tuna only from fleets operating in the Pacific and Indonesia (usually Korean and Taiwanese vessels). These transactions are characteristically risky. C. Itoh ships these purchased tuna loads to and from its Japanese cannery and out to Western European markets. C. Itoh's subsidiary, Yaizu Suisan, provides cold storage, processing, and marketing in Japan. Its earnings for tuna in 1986 were $140 million to $150 million.

Mitsui and Company. Two subsidiaries, Tokyo Commercial Company in Yaizu and Towa Suisan in Tokyo, handle tuna trading strictly in Japan. Mitsui owns Mitsui and Co (USA), which in turn operates a wholly owned subsidiary, Ocean Packing Corporation, headquartered in White Plains, New York. Ocean Packing Corporation operates a tuna processing and canning facility, Neptune Packing Inc, in Mayaguez, Puerto Rico, which employs 500 people.

CONCLUSION

In the past ten years the production and consumption of tuna have been separated by location. This restructuring has opened some levels of the tuna industry, particularly harvesting and processing, to new entrants. Established corporations have moved away from harvesting toward trading and distribution. The corporate activity is clearly driven by the profit motive.

The ten largest corporations use their mobility and international linkages to respond quickly to market or institutional changes. Those Pacific island countries intending to participate in the global tuna market will find strong competition from profit-seeking multinationals.

NOTE

1. This chapter synthesizes the chapters by Ashenden/Kitson, Comitini, Crough, and Iversen for PIDP's tuna project.

V. DEVELOPMENT OPTIONS AND ISSUES

19.
Financing a Tuna Project

Boris Skapin and William S. Pintz

ABSTRACT—This chapter explains in a non-technical way the nature of, and the problems associated with, the financing of large-scale tuna fishing and/or canning projects in the Pacific islands region. After a general introduction, the chapter outlines the components of a financing plan, which is followed by a review of finance and risk. Sources of debt are then discussed along with the financing environment for industrial projects. International mechanisms for encouraging risk taking by lenders and investors are evaluated and the terms and conditions of loans reviewed. The nature of tuna fishing as viewed by lenders is examined, followed by a discussion of the consequences of project collapse. The second half of the chapter focuses on issues associated with raising funds to finance ventures. The conclusion notes that no financial wizardry can substitute for strong project prerequisites.

FRAMEWORK FOR FINANCE

Background

The successful financing of a tuna venture is a substantial and complex undertaking. In addressing such an undertaking, the principal project sponsors must understand both the framework within which the necessary funds are to be obtained (or made available) and the implications of different financial structures. Identifying and maintaining an appropriate financial structure in a timely manner, while combining other project elements, can determine success or failure of the fishing operation. However, finance, in itself, can never substitute for the fundamental elements of the project. If sound marketing arrangements are not made, if efficient catching and/or canning with rigid cost control and qualified management do not exist, or if such a project is not internationally competitive, the project will fail irrespective of its financial structure. Indeed, the laborious effort and documentation required to raise funds are intended to ensure that the venture is fundamentally viable.

Whether or not it is a direct financial participant, a government often has substantial policy interests in the financial structure of commercial fishing ventures. Although it would probably be imprudent to attempt to regulate or control the financial strategies pursued by purely private tuna operations or operations with foreign partners, the government has, if nothing else, direct tax and foreign exchange concerns to protect. These concerns suggest that the government should take an active interest in how the funding for a new fishing project is assembled.

Where the government is the principal sponsor or a major equity participant in a joint-venture tuna project, additional policy questions become important; these relate to the sharing or leveling of risks, the liability exposure, and the preemption of lender quotas. In joint ventures, the interests of the parties involved coincide on some points and diverge on others. Where divergence of interests occurs, any project sponsor must be prepared to vigorously present its viewpoint, defend its position, and negotiate the best arrangement. The first step in organizing finances to start the project operations is to understand the concerns and objectives of the participants: project equity investors, lenders, public agencies, and private entrepreneurs.

Financing plan

The standard structure of a finance plan for any project, including a fishing project, is outlined in Table 1. The design of the finance plan must begin with the estimated project costs, which include expenses for all fixed and other assets required to start operations. The costs (including provisions for contingencies and escalation) must be carefully estimated because once the financing plan is structured, additional funds are difficult to raise. In project financing, it is uncommon, unusual, and unprofessional to return to the financiers seeking additional financial assistance after they have already been convinced and have signed contracts based on project feasibility studies. The original cost estimate is always used to determine the projected profitability of a venture. A sample cost estimate is given in Table 2.

Finance and risk

One of the first financial distinctions to be made is related to the question of ownership. Ownership is always associated with the notion of investor and equity investment, which in turn is associated with the entity that must ultimately bear the main risks of the project. In general, financial structuring has two components: (1) the adequate

Table 1. Outline of a project finance plan

	Local currency	Foreign currency	Total ($000)
Equity			
Share capital			
Sponsor 1			
Sponsor 2			
Sponsor 3			
Other forms of equity or quasi equity (subsidized loans)			
Cash during construction or expansion			
Subtotal equity			
Loans			
Source 1			
Source 2			
Source 3			
Subtotal loans			
Total			

Table 2. Sample cost estimate

	Local currency	Foreign currency	Total ($000)
Fixed assets			
Vessels			
Machinery			
Other			
Other costs			
Pre-feasibility costs			
Finance charges during construction			
Working capital			
Escalation provision			
Total			

tailoring of debt and the raising of money and (2) the sharing or shifting of project risks. Whatever the structure, ultimately the equity owners must bear the risks that others are unwilling to assume.

Institutions are normally willing to bear risks in the expectation of compensation (usually interest or profits). The greater the risk, the greater the expected compensation. Risk and compensation may be thought of as mirror images of one another; financial agencies range from aggressive risk-taking entrepreneurs (seeking substantial equity profits) to bilateral aid donors (seeking humanitarian or political rather than commercial benefits). Somewhere between these extremes lie the multilateral developmental institutions like the World Bank (WB), the Asian Development Bank (ADB), and other commercial and merchant bankers, as well as the specialized financial service agencies that mobilize funds in the international money centers through a variety of financial instruments.

The risks vary over the life of a project. In addition, the nature of the risk changes as a project moves through its feasibility study stage, its implementation/construction stage, and its ultimate operational stage. Although project risks tend to decline or become more controllable as a tuna project enters its operational stage, uncertainties and external forces can still occur. These include an above-average increase of fuel costs, the disruption of supplies, the depression of selling prices, or even an unexpected climatic change like El Niño (which may cause the relocation of fishing grounds). Thus a distinction must be made between those risks that are temporarily associated with a particular project phase and those that are likely to continue throughout a project's life. Table 3 lists typical risks that can be encountered by a Pacific island tuna operation.

Table 3 shows that considerable uncertainties can face any new project during implementation and operational stages. For this reason, Clement-Jones (1987) argues against majority or 100 percent equity ownership by government in new fishing projects. The same applies for any other form of ownership, for example, a 100 percent privately owned company.

From the point of view of bankers or other non-equity financiers, the relevant risk period is only the period until the debt is fully repaid. Equity owners, on the other hand, must face uncertainty throughout the project's entire life.

Risk—spreading, sharing, and hedging

Because the total risk of a fishing venture is determined by several institutional, market, technical, and biological factors, the risk must be considered as fixed or defined (controllable) at any given moment. Risks do exist, and if one party to the project wants to avoid risk, then some other party must assume additional risks. The assumption of risk usually carries a price. This price may increase by higher interest rates, insurance premiums, price discounts, loan guarantees, or a combination of these and other devices.

There are two main categories of risks: a "credit risk," which involves the lending of money to a project, and an "equity risk," which is taken by the owners or equity sponsors of the project. Credit risk is conservatively defined by lenders and is often underwritten by mortgages, liens, pledges, or other purchases of physical assets, associated insurance of the sponsor or project, or loan guarantees by project

Table 3. Financing risks in fishing projects

Category	Characterization
Resource risk	Biological yield of fish stock
Operating risk	
Technical	Can fish be efficiently caught?
Cost (including fuel)	Is inflation of major costs likely to make operation uneconomical?
Management	Can company manage operation?
Climate	Will climate affect fishing ground?
Infrastructure	Will infrastructure meet project needs?
Environmental	Will environmental resources like processing water or baitfish be depleted or polluted?
Marketing	Can catch be sold at economical price?
Political	Will local/national policies affect economics of project?
Participant	Do participants have common interests or objectives in project?
Completion	Will project be completed as planned (time, cost, performance)?
Legal	Are legal complications likely to jeopardize project (implementation or operation)?

owners. However, even a conservative appraisal of the project and tangible loan security or guarantees are sometimes insufficient to cover project uncertainties. For example, in the last decade the U.S. and Mexican purse seine fleets have been particularly hard hit by bankruptcy and loan defaults. This experience has not only left several commercial banks with bad loans and unsellable repossessed assets but is also undoubtedly affecting the attitude of commercial lenders toward making new loans to the tuna industry. In other words, the credit risk of commercial tuna fishing ventures is currently estimated to be quite high.

Empirically, the distinction between credit risk and equity risk can be defined in terms of the premiums that lenders charge to make risky loans. Most international commercial lending is based on a reference interest rate, often the London Interbank Offered Rate (LIBOR) plus a risk premium or a spread or a margin. Commercial banks are usually unwilling to lend money at interest rates above LIBOR plus a certain percentage, which could be as high as 3 percent. The rationale for this reluctance is that the risk levels that would justify a risk premium higher than 3 percent are more properly classified as equity risks rather than as lender or credit risks. Other hard currency loans bear different cost or spread margins that tend to be considerably higher if local currencies (particularly in countries with high inflation) are involved.

In contrast with the variable nature of commercial lending (e.g., interest rates that vary with LIBOR), borrowing from multilateral organizations like the WB, the International Finance Corporation (IFC), and the ADB is often on a fixed interest rate basis. In itself, a fixed interest rate has the effect of reducing project uncertainty because financing costs can be more precisely and directly calculated before a commitment is made to undertake the project. In effect, any lender offering a fixed interest rate is accepting the risk that its cost of money will not exceed the fixed, on-lending rate to the borrower. Of course, unless the investment agreements specify otherwise, by taking a fixed rate loan the borrower agrees to pay the same price even if money becomes less expensive (i.e., if the interest rate falls).

Commercial banks will sometimes form loan syndications in which each bank assumes a fraction of the overall loan and thus shares the risk in the event that the loan is not repaid. On the borrower's side, risk sharing may occur when each partner in a project agrees to guarantee a portion of the project's debt. Such sharing is common in large or risky projects but occurs to some extent in almost all financing strategies or structures. The basic idea behind risk sharing is that the overall risk exposure is more manageable if it has a large number of small components than if the entire project pivots on a few major obligations or commitments. The wisdom of this portfolio management strategy has been repeatedly verified by empirical studies.

Sources of debt

Because the inclination of lenders varies over time, each country's tactical decisions should make optimal use of the potential finance within the context of a particular economic sector or its overall development strategy. Five sources of financing are representative:

- Bilateral aid agencies
- Multilateral lenders
- Export and supplier credit institutions
- Commercial banks
- Loans from sponsors or shareholders

Bilateral aid agencies. Aid funds from bilateral donors at concessionary interest rates are available for a broad range of projects in developing countries. But even when these funds are obtained, they usually do not represent a substantial portion of the project's financial plan. In other words, funds from bilateral agencies tend to be a marginal contribution, and often funds are tied to specific purposes: for example, conducting pre-feasibility studies and training personnel. Historically, many donors have tried to focus on social or infrastructure projects and to avoid lending money for commercial activities. This non-commercial focus has been partly justified on philosophical grounds; however, of growing concern is the criticism by the recipient country that this aid competes unfairly with its commercial financial institutions. This non-commercial orientation may be changing with the current emphasis of many donors on privatization, which has led several aid agencies to modify their programs as adjuncts or catalysts to commercial financing. Thus project assistance might be favorably viewed if it were part of an expansion of existing vessel fleets or shore-based operations. However, irrespective of their motives, bilateral donors are often reluctant to provide funds for projects that directly compete with their own domestic industries. In other words, if the distant-water fishing industry is to be affected, the project might qualify for supportive aid funds, but if the project is to replace distant-water fishing, this source may not be appropriate.

Multilateral lenders. The WB, together with its affiliates, is the major development institution serving developing countries. Several of the region's countries (Papua New Guinea, Fiji, Western Samoa, Vanuatu, and Solomon Islands) are WB members. An even larger number of island countries participate in the lending activities of the ADB, which is an equivalent regional financial institution. These multilateral financial institutions make project loans that must be government-guaranteed. For countries with a particularly low gross national product (GNP) per capita, the WB's International Development Association (IDA) provides soft loans with long repayment periods and symbolic interest rates. Where private investors are involved, a WB affiliate—the IFC—may participate as both lender and equity owner.

Loans from the WB or the ADB are generally for longer terms (from 12 to 20 years) with a two- or three-year grace period for first payments of principal, and interest rates are often below commercial levels. However, the project appraisal is detailed and the approval process lengthy. In addition, the denomination of the loans in foreign currencies is sometimes criticized as carrying substantial currency risk, particularly if local components are financed with these funds. Nevertheless, these multilateral lending institutions are often the only possible sources of foreign-currency financing for developing countries, and they are highly regarded by the international financial community. The participation of these organizations is seen as an endorsement of the creditworthiness of a project because of their careful and professional evaluation of the technical, financial, and economic viability of a proposed project. Sometimes developing nations are given technical assistance grants (ADB) or loan support (WB) for project preparation. In contrast to the tied procurement requirements of many bilateral donor countries, multilateral lenders insist on international competitive bidding procedures (except for the IFC, where selection of machinery and equipment is entirely determined by the project sponsor).

For Pacific island countries, two other agencies deserve mention for their increasing role in the region: the European Investment Bank (EIB) and the Commonwealth Development Corporation (CDC). The EIB, which serves as a financial arm of the European Economic Community (EEC), has several specialized lending programs that should be investigated in cases where European trade is involved. (For example, EIB financing should be explored in connection with tuna marketing or processing in Italy.) The second international financing agency is the CDC, which has been quite active in palm oil and other agricultural activities in Pacific island countries. Although the CDC has yet to lend or invest in tuna projects, the organization's mandate is sufficiently broad to permit this kind of participation.

Export credit and supplier credit institutions. Many industrial countries have export-financing agencies that seek, through loan guarantees and subsidized interest rates, to promote national exports. In theory, competition among these agencies is to be controlled under an agreement called the Berne Union, but competition is, in fact, often intense. Because not all countries with export credit agencies produce suitable fishing vessels or cannery equipment for the tuna industry, advance research into vessel or equipment sources is necessary. In addition, care must be taken that equipment secured under export credit arrangements is internationally competitive and that the benefits of attractive finance terms are offset by any expensive purchase prices. Good procurement and financial advice are particularly useful when countries deal with export credit agencies.

Commercial banks. Commercial bank lending to developing countries has been sharply curtailed as a result of the current world debt crisis. Nevertheless, this source of lending continues to be quite important in some regions and for certain projects. As previously noted, commercial lending to commercial tuna fishing ventures has had a poor record in recent years, and the major banks are likely to be extremely cautious—and expensive (in terms of LIBOR plus the margin interest rate). Nevertheless, borrowers with sound tuna projects, strong management, and solid sponsor guarantees should be able to obtain commercial bank finance. Funding from this source will normally cost more and have shorter repayment periods than finance from the previously mentioned sources. The borrower should pay particular attention to negotiation of the most favorable terms and conditions because commercial banks have flexibility in determining interest rates.

Loans from sponsors or shareholders. Large natural resource companies or trading companies are often in a position to lend directly to commercial projects in which they have an equity interest. This lending frequently takes the form of shareholder loans and may be an adjunct to other lending sources. Shareholder lending could be manipulated and even abused to circumvent exchange restrictions on repatriated profits or to avoid government taxation. Because this form of financing is by its nature less than arm's length, it must be approached with caution, and probably it should not be a first or preferred financing option.

The financing environment

Potential financiers are understandably concerned primarily with the economics of the proposed project. The appraisals of a fishing project can sometimes vary among the types of lenders. In many instances, the appraisals address the overall physical and economic context of the project, as well as the narrow debt-repayment questions. The purpose of such appraisals is to identify as many risks and uncertainties as possible and, wherever feasible, to quantify the range within which project variables may fluctuate.

The broadest appraisals are usually undertaken by international financial institutions like the WB and IFC. Commercial banks tend to be more narrowly focused in their appraisal procedures, although they are expanding their perspective as a consequence of the world debt crisis. Bilateral aid funding tends to have the least rigorous appraisal and documentation procedures; however, where donor funds are tied to procurement of vessels or equipment from the donor's country, additional non-financial procedures may be involved.

Bilateral or multilateral lenders are somewhat insulated from political risks by international treaties or agreements. However, commercial lenders are highly sensitive to political risks in developing countries and often differentiate their concerns into two groups: "country risk" and "sovereign risk."

Country risk generally refers to the risk that the recipient country, for economic or political reasons, may not permit the transfer of currency to overseas lenders to cover interest or principal payments. In contrast, sovereign risk refers to projects where the government (i.e., the sovereign nation) is an equity participant or sponsor.

In addition, commercial bankers are also concerned with the risks that the government might nationalize the fishing venture or that political instability might interfere with the commercial operation of the fishing or cannery project. Needless to say, the appraisal of such broad political and economic concerns is not a precise science, and although many attempts have been made to quantify such risks, the task remains fairly subjective. For this reason, commercial lenders often are attracted by arrangements that link their lending to broad international political treaties and agreements.

International mechanisms for encouraging risk taking by lenders and investors

As international financing has grown in recent decades, lenders and investors have increasingly sought mechanisms through which risk exposure might be managed. The industrial world and multilateral institutions have perceived this need to overcome certain types of risk uncertainties as a barrier to increased private sector involvement in developing countries. Three such mechanisms are now examined to see how they might be used in compiling a financing package for a tuna fishing project in the Pacific islands.

Cofinancing. The cofinancing of a project implies the linking of funding from several types of financial agencies in a complementary fashion. Often this linking involves the merging of funds from several sources (bilateral aid agencies, export credit institutions, multilateral banks, and commercial banks) through parallel lending to the project. Because each financial agency has its own lending priorities and restrictions, cofinancing must be specifically tailored to the needs of the fishing project and those of the individual donors. A hypothetical, but not unlikely, cofinancing scheme for a fish harvesting/cannery operation might involve an export credit agency's financing the purchase of the fishing fleet, with the cannery development using funds from both commercial bank credits and multilateral or bilateral agencies for shore-based infrastructure such as the wharf and power supply. Of course, sources and uses may vary, particularly if the private sector is involved.

The attractiveness of such cofinancing schemes is twofold. First, because each participant is providing only a share of the overall financing, sometimes attractive overall interest rates, repayment terms, and procurement conditions can be obtained without preempting finance that might be needed for other purposes. Second, lenders take comfort in participating in such a broad scheme because they recognize that any unfavorable policies by the government can have far-reaching consequences. Frequently, the multilateral lender will act as a collection agent for other cofinancers. Table 4 summarizes the recent cofinancing activities of the WB. The IFC has also been active in arranging cofinancing and participation schemes.

Cross-default provisions. Cofinancing is often associated with another risk-sharing mechanism, cross-default provisions. Cross-default provisions are agreements among lenders that if one project loan is in default, all other loans are considered to be in default. The obvious attractiveness of such provisions to lenders is that they contractually assure that the borrower will suffer only the broadest possible consequences of any repayment default.

In the case where cross-default provisions exist between commercial lenders and other lenders (e.g., between a commercial bank and the WB), the project sponsors, including the government, would face default consequences that have a magnified impact and that might extend well beyond the project itself. Such magnified default consequences are the result of the international status of multilateral financial institutions and their position in world financial affairs. Such arrangements could be generally considered as political risk hedging.

Cross-default provisions might be disadvantageous when developing country governments are direct project participants. Conversely, if

Table 4. World Bank cofinancing operations, 1975–84 (billions of dollars)

		Cofinanciers' contribution				Bank contribution		
Fiscal year	Number of projects with cofinancing	Commercial banks	Export credit agencies	Other official sources	Total	IBRD	IDA	Total project costs
1975	51	0.1	1.0	0.9	1.9	1.0	0.3	8.8
1976	67	0.3	0.9	1.1	2.2	1.6	0.4	9.6
1977	78	0.7	0.2	1.5	2.4	1.9	0.7	10.0
1978	79	0.2	0.5	1.8	2.5	1.7	0.8	11.4
1979	105	0.5	0.3	2.0	2.8	3.0	1.1	13.3
1980	86	1.7	1.6	2.6	5.9	3.0	1.6	20.3
1981	72	1.1	0.5	1.5	3.1	2.6	1.5	15.1
1982	98	1.2	1.8	2.2	5.3	4.1	1.2	20.0
1983	84	1.1	3.0	1.8	5.7	3.3	1.1	20.8
1984	98	1.1	0.9	2.0	4.0	4.6	1.3	21.7

Source: World Bank 1985, 123.

Note: Components may not add to totals because of rounding. These amounts represent private cofinancing as reflected in the financing plans at the time of board approval of A category loans. They do not represent private cofinancing loans actually signed in the fiscal year. An analysis of cofinancing operations can also be found in World Bank Annual Reports.

they provide sufficient lender comfort to enable the financing of a fishing project that otherwise would require an increased cash contribution from the government, then they may be worthwhile. Thus the acceptability of such mechanisms ultimately depends on the attractiveness of the fishing project. A marginal venture with a low debt-service ratio, whose major justification is secondary economic benefits, is probably better financed in a manner that avoids the necessity for cross-default provisions.

Investment insurance. Just as lenders use cross-default provisions to minimize their concerns over political risk, investors from industrial countries can often directly purchase investment insurance. Normally, such political insurance covers three types of investor risks: expropriation, inability to convert local currency, and political violence (e.g., civil war and revolution).

In exchange for insurance by a quasi-public agency of the government, investors normally pay an insurance premium. In recent years the extent of the risks covered under these quasi-public investment insurance schemes has been gradually expanding.

Such investment insurance has obvious application to the structuring of foreign equity holdings in Pacific island tuna ventures. However, issuance of the insurance is dependent upon the existence of a comprehensive agreement between the government and the foreign insurance agency. Although some Pacific island countries are signatories to a large number of such agreements, other island countries have been slow to endorse such arrangements. While such arrangements may imply some limitations on national sovereignty, the mere existence (or lack of existence) of such an agreement sends a positive message to prospective investors.

In early 1981 the WB began discussions about the establishment of a Multilateral Investment Guarantee Agency (MIGA). These discussions led to concrete proposals that were presented in 1985 and that resulted in an international convention in mid-1986. MIGA is legally and financially separate from the WB and offers coverage to investors from member countries. Four categories of non-commercial risk are covered: (1) the transfer risk resulting from government restrictions on conversion and transfer, (2) the risk of loss resulting from legislative or administrative action or omission of the government, (3) the risk resulting from the repudiation of a contract by the government, and (4) the risk of civil disturbance and war.

RAISING THE FUNDS

Looking for money—the prerequisites

Before any prospective financier is approached, a thorough feasibility study of the proposed venture must be undertaken. In most instances, suitable feasibility studies are beyond the capability of the government (and sometimes the project sponsors) and require contracting with an outside consultant firm. However, occasionally, if the funding is to be provided by a concessionary foreign aid donor, assistance in the feasibility study can be obtained directly from the donor or through an agency like the United Nations. If commercial or multilateral funding is to be sought, then the consultant should have international experience and reputation. The consultant identification process can be usefully started by asking commercial banks or multilateral financial institutions for a list of reputable firms. This approach can increase both the credibility and the reliability of the resultant feasibility study.

The heart of any feasibility study (i.e., the evaluation of all possible alternatives for project implementation) is the cash flow analysis of the project that quantifies a range of technical, marketing, financial, and taxation issues. While the government for its own purposes may also desire to undertake a broad economic assessment of the project, lending decisions are almost always made on the basis of a much narrower financial analysis.

If the focus is on the estimates and rationale of the cash flow analysis, it is easy to lose sight of the myriad assumptions that lie just below the surface. Collectively, these assumptions form the basic risk parameters of the project, and a good lender appraisal will carefully probe the major risk exposures. In general, the greater the lending risk assumed by the lender, the more complete, thorough, and convincing the feasibility study must be.

Nevitt (1983) presents a comprehensive picture of typical lender appraisal questions (Table 5). These questions must be addressed by the feasibility study, and ideally they should be specified in the project manager's or consultant's terms of reference.

For tuna ventures the greatest project uncertainties have historically related to the cost of fuel and to the market price of the catch. Any lender appraisal can be expected to pay particular attention to these topics. In addition, for many Pacific island countries, assumptions about the cost and productivity of the shore base or cannery labor force will be carefully scrutinized by financiers. Because the cannery operations in American Samoa are the oldest and most profitable tuna operations in the Pacific islands region, lenders inevitably will use these operations as a guide in their appraisal review.

If the feasibility study and resultant cash flow analysis have been adequately prepared, the lender will evaluate the project on a range of financial ratios and on an assessment of the riskiness of the proposal, country, industry, and so forth. In this assessment the experience of the lending institution, both with the industry and with the particular borrower, is likely to play a significant role.

Where finance is being sought from a multilateral source like the ADB or the WB, the appraisal process is likely to be lengthy and involve the visit of several appraisal missions. Because multilateral lenders

Table 5. Criteria for project financing

Assured costs of supplies and raw materials
Assured energy supplies at reasonable costs
Existing market for the project's commodity or service
Transportation of product to market
Adequate communications
Availability of building materials
Experienced and reliable contractor
Experienced and reliable operator
Management personnel
No new technology
Contractual agreements among joint-venture partners
Political environment, licenses, and permits
No risk expropriation
Country risk
Sovereign risk
Currency and foreign exchange risk
Adequate equity contribution
Project as collateral
Satisfactory appraisals
Adequate insurance coverage
Force majeure risk
Cost overrun risk
Additional capital by sponsor
Standby credit facility
Fixed-price contract
Completion guarantee extension to debt maturity
Takeout of lenders
Sponsors' escrow funds for completion
Delay risk
Inflation risk
Realistic interest rate projections

Source: Nevitt 1983.

do not, as a rule, lend to commercial ventures (the WB's IFC is an exception), money from these sources will often be directed at the associated (cofinanced) infrastructure and may be linked (through cross default) to other financial sources. Invariably, such integrated financial arrangements will consume additional time, but they will assure the local sponsors of the project's viability. Moreover, the process will ensure that the project is adequately structured and creditworthy.

Bilateral donors are usually less rigorous in their project appraisal but are seldom any faster. Again, the non-commercial policies of aid agencies mean that different elements of the project will have to be simultaneously processed through parallel funding agencies. Export credit agencies, although more attentive to the financial risks of a project, will share with aid agencies a home-country bias toward the project. This bias will be concerned with questions such as the importance of the project to the home-country market, the provision of capital goods (vessels, machinery, etc.), construction services to the project, and the strengthening of trade ties between the donor and the recipient country.

Such non-financial lending objectives, when combined with government loan guarantees and a less rigorous loan appraisal, can lead to a lending environment that encourages marginal projects that otherwise would be rejected on purely commercial grounds. There is nothing intrinsically wrong with non-commercial development projects undertaken on broad social and development grounds so long as the inherent risks and potential ongoing subsidy requirements are clearly recognized and politically accepted.

Finally, shareholder loans to the fishing project may occur expeditiously and with minimal appraisal because a parent company is simply lending to its own subsidiary. Such shareholder loans are often made to offset cost overruns and sometimes in lieu of a contingent liability such as a loan guarantee.

Terms and conditions

The interest rate charged to a fishing loan will be dependent on the type of lender, the risk assessment of the project, the loan security, and the lender's cost of money. Mention has already been made of commercial loans that are, if coming from foreign sources, often pegged to LIBOR. The cost for such loans varies throughout their life as LIBOR varies with changing conditions in the money markets. In contrast, some multilateral lending, together with repayable foreign aid credits, is made on a fixed interest rate basis.

Because wide differences exist in the creditworthiness of developing countries and because not all countries borrow every year, the spread differential fluctuates significantly from year to year. Nevertheless, developing countries as a group generally borrow from commercial sources at higher interest rates but have roughly comparable repayment periods.

The disadvantageous terms of developing-country borrowers in seeking commercial sources of debt are partly offset by their access to funds from multilateral development institutions. In general, loans from such institutions are available at interest rates that may be below commercial rates. In addition, designated low-income countries are eligible for highly concessionary WB credits from the IDA. Those Pacific island countries that qualify for IDA concessionary finance include Solomon Islands, Vanuatu, and Western Samoa.

A particularly attractive feature of the WB and the ADB lending programs is the longer maturities over which the loans are repayable. These loans normally carry up to a 6-year grace period with a repayment schedule extending for some 10 to 15 years thereafter. WB loans can be extended up to a total of 20 years from the approval date, and IDA credits have a total of 50 years. The IFC, which is also trying to provide between 15 and 20 percent of equity (while not directly participating in the management of a company), grants loans with repayment periods of up to 10 or 12 years, including a 1- or 2-year grace period dating from project completion. Such lengthy repayment schedules are extremely important in indebted fishing ventures, where substantial annual variations in revenues are to be expected and where the heavy debt burden associated with short loan maturities increases the likelihood of cash flow problems.

Export credit agencies provide debt finance that is intermediate between commercial and multilateral lenders. These institutions normally offer interest rates below commercial levels that sometimes approach or are lower than the interest rates charged by the WB's IFC. On the other hand, the usual loan maturity is likely to be from 8 to 10 years, which, although longer than commercial terms, is not as attractive as the 15- to 20-year period offered by the WB.

Export credit agencies make fixed interest rate loans in their home currencies. Occasionally, where a project does not generate revenue in the currency needed to repay the loan, a currency or foreign exchange risk is created. In addition, care must be taken that the substantial advantages of export financing are not offset by overly expensive or inappropriate processing equipment or fishing vessels. In this regard, it is important that the feasibility study be conducted in sufficient depth to identify procurement sources for major capital imports to the project. Finally, the organizing of export credits is often a lengthy and tedious process that, in extreme cases, may delay a project. Nevertheless, the terms offered by these agencies are sufficiently attractive to warrant careful consideration by Pacific island countries.

Shareholder loans may sometimes be used as a substitute for direct equity contributions or investor guarantees. While the hazards of this type of debt finance have previously been noted, mention should be made about the likely terms and conditions. Shareholder loans are often mandated by some contingency condition like a major cost overrun or a project delay. When loans are made for such purposes, they are likely to be subordinate to the project's "senior debt," which is raised from other financial sources.

In addition to project shareholders, subordinated debt may also be provided by equipment suppliers or product purchasers. Although each of these lenders may have different objectives, all are likely to seek the shortest repayment period consistent with the structure of the senior debt of the project. While shareholder loans may sometimes carry attractive (below commercial) interest rates, subordinated credits from other lenders are likely to be at or near the rates charged by commercial institutions. One of the advantages of international financial institutions' participation in a project is that it often leads to the participation of other partners or lenders.

The nature of tuna fishing as viewed by lenders

Tuna fishing is a rather uncertain business. The catch is under the sea and is therefore only partly controllable. Its financing experience in recent years has not been good, and some sectors of the market are unlikely to see even moderate expansion over the medium term. In addition, any new cannery in the Pacific islands region should be internationally competitive. As a commercial venture, tuna fishing for the high-volume cannery market is considered to be a high-risk, low-reward business.

On the other hand, for many Pacific island countries, tuna is the largest natural resource and offers a major developmental opportunity. For these reasons, the harvesting and processing of tuna cannot, and should not, be ignored. In addition, Pacific island countries do have a competitive advantage in this sector. Within this setting, projects should be encouraged, but the selection of a financial structure needs to be carefully considered if operations are to survive over the long term.

Traditionally, tuna finance has made extensive use of debt finance (in financial jargon, it has been heavily leveraged). This leveraging means that both fishing and processing ventures must face substantial annual costs regardless of catch rates, product prices, or fluctuations in expenses such as fuel. The obvious strategy under such circumstances is to decrease the amount of debt leveraging by increasing the owner's equity, to decrease annual debt-service requirements by seeking long-term, low-interest loans, and/or to accept the need for some sort of government subsidization.

Because the availability of equity or subsidization funds is determined by external commercial or developmental policy, the key financial question addressed here is the structure of the debt-service obligations of fishing ventures.

To illustrate the importance of the terms of debt financing to the tuna industry, an analysis was made of the cost structure of the U.S. purse seine industry using the most current estimates available. These data are presented in Table 6 for selected operational cost items in 1985.

Table 7 demonstrates the effect of the interest rate and loan maturity on annual capital payments. The matrix shows the combined effects of commercial interest rates and terms (7 years) versus interest and

Table 6. Cost structure of U.S. tuna purse seine fleet in 1985

	$000	Total cost (%)
Crew costs	621	18.9
Fuel costs	533	16.3
Repairs	261	8.0
Insurance	267	8.1
Vessel capital charges[a]	910	27.8
Other costs	687	29.9
Total	3,279	100.0

[a]Based on $4.5 million vessel commercially financed at 10.5 percent over seven years.

Table 7. Effect of financing terms on purse seine vessel costs (expressed in percentages of annual mortgage charge)[a]

	7 years	10 years (%)	17 years
Commercial	Base	-4.0	-25.1
Commercial -0.75 percent	-2.4	-7.0	-29.5
Commercial -1.5 percent	-4.8	-10.0	-31.8

[a]Assumes vessel cost of $4.5 million with commercial rate of 10.5 percent.

terms for a typical export credit loan (commercial -0.75 percent for 10 years) and a WB loan (commercial -1.5 percent for 17 years).

To put these figures in some perspective, the U.S. International Trade Commission (1986) financial survey estimated that over the period 1979–85, the average U.S. purse seiner lost about $347,000 per year. The effect of a longer mortgage repayment period would have reduced this annual loss by 85 percent (at concessionary interest rates such as those offered by multilateral lenders).

Again, it should be noted that because the WB does not finance private ventures, such attractive loan terms would probably not be available for the purchase of tuna vessels in the Pacific islands. Nevertheless, the above example does demonstrate the substantial impact that favorable loan terms may have on tuna fishing economics.

To the degree that the overall venture could be structured to permit participation of, say, export credit finance for purely commercial activities and multilateral finance for associated infrastructure expenditures, a fairly robust financial structure would be created. In addition, such a venture might advantageously use debt and equity funds from the WB affiliate, the IFC.

This blending of equity funds (to reduce overall fixed capital charges) with debt raised from a variety of sources on different terms is essential to project viability. This ingredient is particularly important in tuna fishing and processing industries where product prices and important input costs like fuel are highly volatile. A key ingredient in successful financing is to strike a balance between the amount of fixed debt a project can carry (debt leveraging) and the terms under which that debt is obtained (e.g., interest rates and loan maturities). There is no simple formula to apply here, but the recent experience in the U.S. and Mexican tuna industries clearly shows the disastrous consequences of financial structures with short loan maturities and relatively high interest rate margins.

Risk and project collapse

The bankruptcy of a major natural resource project in a small country can be a traumatic occurrence. Often governments place their political prestige on highly visible development initiatives and, for this reason, feel compelled to support clearly uneconomical ventures. In the developing world the most visible examples of this national prestige phenomenon is the endless subsidization and restructuring of national airlines. But beyond such nationalistic concerns are substantive issues involving the loss of employment, an inappropriate allocation of government services or infrastructure, and a reduction in secondary or support industries.

In contrast to the broad political and economic consequences of a collapsed project, the financial costs can be readily quantified. The matter of who will bear these financial costs is, of course, directly related to the sharing of risks, which is implicit in the financial structure of the project. For heavily debt-leveraged projects, assets may be seized by senior creditors and sold to offset outstanding loans. If assets are inadequate or if contractual arrangements make provision for contingent liabilities, creditors may seek to recover funds from various loan guarantors. With multilateral lenders, who provide funds only to governments, the ultimate loan guarantor of government-borrowed funds is the national tax base. In purely private ventures, lenders may seek recourse from the parent company.

Investors undertaking natural resource projects in Pacific island countries often seek several types of financial concessions from the government. In many cases, the intention behind such negotiating positions is not immediately clear, nor is it quantifiable. Typical investor financial concessions include tax-relief measures, provision of infrastructure, loan guarantees by government, and access, through government, to concessionary financial sources.

With the exception of fiscal concessions relating to tax or tariff relief, most financial concessions directly increase both the government's risk exposure and its vulnerability to the demands of creditors in the event of project collapse. On the other hand, as has been shown, the longer loan maturities and concessionary interest rates available through the government can play an important role in the viability of the fishing project. Thus increased government involvement, particularly in the first years of project operations, simultaneously increases risk exposure and decreases the project's vulnerability to the short-term market or to cost uncertainty. Government incentives are thus vital in encouraging project sponsors to initiate the venture.

When the advantages of financial risk are balanced against its disadvantages, loan guarantee arrangements are preferable to direct borrowing by government. This trade-off may be deceptive and should be carefully evaluated. Although concessions can sometimes be obtained on loan terms with a government guarantee, the usual circumstance is that such concessions are unlikely to be as significant as those that the government itself could obtain directly. Too often the government simply guarantees the risk that the investor would otherwise have borne, without a measurable impact on the vulnerability of the fishing project.

If the fishing venture proceeds smoothly, of course, this discussion of risk exposure and default liability is academic. But projects do not always turn out as planned, and loan agreements are primarily designed to deal with liability conditions when projects fail.

CONCLUSION

The financing of a tuna fishing or processing venture is a major and complex undertaking for any individual sponsor or group. For Pacific island countries, whose main developmental potential may be their marine resources, this financial undertaking assumes added significance. The determination of an appropriate capital structure, an optimal financial strategy, and adequate funding is vital to a successful tuna venture.

All participants bring thier own objectives and concerns to financial decision making, and inevitably some of these objectives are in conflict. Nonetheless, these objectives have to be coordinated throughout the life cycle of a project. Many objectives deal with risk: how to spread it, how to shift it, how to reward it, and, if all else fails, how to manage it and to absorb it. The government is often the least-sophisticated risk manager in the venture, and its need or desire to mount a high-visibility, employment-providing, and particularly profitable activity may lead it to assume a disproportionate share of the project's inherent financial, technical, and economical uncertainties.

While sound financial planning can substantially contribute to the likelihood that a tuna operation will succeed, it can never substitute for the fundamental prerequisites of the project. These prerequisites are

- Strong management with demonstrated competence and experience in the technical or sector area (e.g., fishing and canning) and the financial/marketing area, together with (if necessary) a strong technical or market partner from abroad.
- Realistic initial assessment of revenues, capital requirements, and operating costs that demonstrates financial (from the company's point of view) and economic (from the country's point of view) feasibility, with particular attention being paid to international competitiveness.
- Access to good and proven fishing grounds and to tuna markets and (where necessary) ability to organize food distribution networks for canned tuna or pet food or to acquire strong marketing expertise from abroad.
- Utilization of modern, proven technology that is cost-effective and combined with skilled labor (particularly in fishing) with an efficient reward system.

Without fundamentally strong project prerequisites and an economically accessible natural tuna resource, the best financial wizardry will be rendered ineffective.

20. Public Investment and Taxation: Approaches to Investment

Robert Clement-Jones

ABSTRACT—This chapter provides guidelines for Pacific island countries that want to become more actively involved in their tuna industries through public investment. Issues discussed include the rationale for investing in tuna industries, alternative fiscal instruments, and direct public investment.

INTRODUCTION

Because investment considerations are relevant to all types of industrial and natural resource development, the overall message of this chapter can be applied to the general economy rather than to the tuna industry alone. Thus this chapter presents guidelines and preliminary analysis rather than detailed prescriptions, overall approaches rather than precise negotiating positions.

No amount of planning can substitute for careful financial and economic analysis in determining the viability of a particular investment, nor can any general position substitute for the careful drafting and negotiation of contracts and project agreements. These tenets reflect the premise that investment (both public and private) is not a precise science. While the components of an industrial enterprise (machinery, production lines, time management, etc.) can be said to be "scientific" in design and operation, the initiation of an enterprise depends on many considerations.

Decisions to invest must take into account both implicit and explicit assumptions about how the bottom line (i.e., profitability) is reached. The preparation of projected cash flows, profit-and-loss statements, and balance sheets requires interpretation of these assumptions. Similarly, the negotiation of contracts and project agreements requires not only legal skills but also creativity, perseverance, and awareness of purpose and reality.

Pacific island nations, many of which became independent in the 1970s, wish to benefit from one of their most important natural resources—the sea. In addition, these nations must contend with the changes, actual or prospective, in international and domestic laws (e.g., creation of exclusive economic zones and the Law of the Sea) and the changes in the relationship between north and south (e.g., the Lome Convention). These changes were designed to provide greater access for several developing countries to markets in the European Economic Community, while the Caribbean Basin Agreement was intended to give preferential access to U.S. markets.

All these changes are fundamental to the underlying force behind the increasing emphasis on interisland cooperation (for example, the Nauru Group) and to the growing interest of these new nations in assuming control of their resources for revenue, employment, regional development, and environmental self-determination. Historical patterns of exploitation, such as distant-water purse seining by Japanese or U.S. vessels, still represent a large proportion of all tuna fishing in the region, but island governments are aiming for (and achieving) higher revenues, more information, and greater emphasis on local ownership of fishing vessels and processing facilities. The success of some countries (e.g., Solomon Islands) in establishing local processing facilities and others (e.g., Papua New Guinea) in developing revenue systems (and licensing arrangements) more beneficial to the country obviously tempts other countries in the region to follow suit. Furthermore, island governments may feel frustrated about their seeming lack of control over their only natural resource and therefore look to their own active participation in its exploitation to derive adequate benefits.

This chapter poses some of the questions that Pacific island governments should ask prior to any investment in the tuna industry. The questions do not have easy answers; the informed interpretation of a particular set of circumstances facing individual (or collaborating) governments will ultimately determine the direction taken.

WHY PUBLIC INVESTMENT?

At a time when the role of the public sector vis-à-vis the private sector is under worldwide political and academic debate, it is useful to review some of the salient features of the debate. By definition, public investment in revenue-earning ventures includes all uses of public monies (from all sources) applied to capital formation for directly productive activities. This definition includes

- Investment by government (or government-owned financial institutions) in joint ventures or wholly owned government enterprises;

- Assumption of an equity share in an already functioning commercial enterprise (whether by forgoing royalties—"carried equity"—or by making cash contributions);
- Investment in infrastructure, which is directly related to a particular enterprise and for which the enterprise pays a negotiated user charge that reflects at least the opportunity cost of that capital to the government.

In addition, the term "public investment" should include other transactions such as government guarantees for loans taken by ventures or other financial transactions that represent contingent liabilities on sovereign debt. The definition is purposely broad.

The following reasons are often used by governments to justify public investment (as broadly defined) in revenue-earning or commercial enterprises:

- To raise more revenue,
- To capture rent for a natural resource,
- To assert rights in an exclusive economic zone (EEZ),
- To foster investments that would not otherwise take place,
- To gain knowledge and understanding of the industry,
- To overcome transfer pricing.

This list is by no means exhaustive. Ideological or political stances may also favor public over private interests, particularly in those nations that are trying to reduce their economic dependency on former colonial powers. Nor are the elements of this list mutually exclusive: policymakers may use a multiplicity of arguments to justify their plans and projects; in addition, revenue-raising objectives are entirely compatible with and mutually reinforcing of natural resource rent-capturing objectives.

The essence of policy formulation and decision making is to achieve a clear understanding of why such policies need to be formulated and why such decisions have to be made. The "whys and wherefores" of policymaking are the objectives: the broad goals by which the results of implementation of the policies can be measured. Objectives may be used to justify certain policies or decisions, but they should not be confused with the means or instruments used to achieve them.

The attraction of public investment as an instrument of public policy and objectives lies in its direct nature. The government can use the means that it has at its disposal to directly influence investment patterns, increase employment, and increase export earnings. Even though it is a direct instrument, it is neither the only nor the most effective way of achieving desired objectives. The following section reviews the principal alternative instruments.

Licensing

The simplest and most basic instrument used in the region's tuna industry has been licensing. It is being used to satisfy several objectives: raising revenue, capturing rent from a natural resource, asserting rights within an EEZ, gaining knowledge about the exploitation of the resource, and even promoting investment. If, within a region, its terms can be similar for all the nations concerned, licensing is a powerful tool to assert the rights of small nations in their negotiations with the tuna industry.

Licensing has been the first and most fundamental step toward the establishment of industries based on the long-term sustained exploitation of the tuna resources in the Pacific islands. Although an essential first step, by itself it may not satisfy the objectives of all nations in the region. Revenues may not be sufficient, and the rent that can be captured may be limited. Knowledge about the industry may be inadequate, and fishing observers may not be able to collect accurate information, even when they are actually on board the vessels.

Taxation regimes

Taxation policies are important tools that governments have at their disposal. In many cases, they can substitute for direct investment to achieve specific objectives. Taxation can have an impact on raising revenue, collecting rent from a natural resource, and promoting investment.

Taxation regimes are an important factor in investment decision making. From a government's point of view, they represent a balancing act between the need to acquire revenues and the impact that taxation has on economic activity (including investment). From an investor's point of view, these policies represent important determinants of the returns that can be expected from investment. As an overly harsh taxation regime will encourage investors to go elsewhere (or to engage in tax avoidance), so will an overly generous tax system involve unnecessary economic or financial costs to a country.

Governments seeking to encourage investment in their countries are often well advised to examine their tax systems to ensure that they do not discourage serious investors. Taxes based on profits (i.e., income net of all expenditure, including financing costs) are preferable to taxes based on output. Processing companies should pay taxes on their profits and not, for example, on how many tonnes of fish are exported. At the same time, governments must ensure that companies do not make unreasonable deductions; there should be regulations on the proportion of debt and equity used to finance a company.

The higher the proportion of debt, the higher the financing costs (interest) relative to profits. If, as is generally the case, financing costs are a tax-deductible item, then there will be lower profits on which to levy taxes. This is particularly important with loans raised offshore because interest payments on loans will be a net outflow of foreign exchange from the country. In practice, this procedure is predicated on "reasonable" debt:equity ratios such as 3:1 or less.

Rent. Taxation is also the principal way by which a government captures the rent from a nation's resources. Licensing of fishing boats is taxation in its most basic form. More advanced types of taxation are specifically designed to ensure that adequate returns are made by the owner of a natural resource in its exploitation. Good examples of such tax regimes are those that often apply to mining and petroleum development companies. The basic concept is that companies engaged in natural resource development should receive rewards commensurate with the risks and uncertainties involved. At the same time, they should not receive all the benefits of extraordinary changes in external conditions (such as the world market price for the particular commodity).

Investment promotion. An important determinant for investors in making decisions about whether and how to invest is the expected rate of return on their equity. Regardless of the economic or financial return on the project as a whole, if the investor perceives that the tax rates are too high or are unfairly applied, then that investor will either look elsewhere to invest or negotiate new tax rates. A government can encourage investment in many ways through the judicious use of tax incentives. These may range from tax holidays, whereby companies pay little or no tax in the early years of the investment, to extra tax deductions for capital investment, employment, export performance, and so on. Similiarly, tax exemptions may be granted to the various inputs (capital equipment, materials, and fuels) that are required for the production process.

Most countries have developed investment policies that explicitly incorporate tax regimes aimed at encouraging investment. Thus countries wishing to promote investment must ensure that their own tax regimes do not act as disincentives to investment. This applies both to direct and indirect taxes. Direct taxes affect the returns to an investment by taxing net income (i.e., income less expenditure). Indirect taxes affect returns to an investment by increasing the cost of an input (e.g., fuel import tax) or by decreasing the price of the output (export tax on a primary product). Any assessment of the value of an investment should take into account the overall tax regime.

Numerous publications have been devoted to the costs and benefits of particular tax regimes. It is commonly accepted that tax rates are a determinant of the overall level of economic activity. At the same time, governments need to recognize that tax incentives are only one of several active components in investment promotion. If a tax regime is too generous in the form of tax holidays, accelerated depreciation allowances, or allowances related to export performance, significant costs to the government may result, with little additional investment. Tax incentives that favor certain sectors of the economy (e.g., industry) may lead to underinvestment in other sectors that are more profitable for the economy as a whole. Incentives relating to investment in capital (i.e., machinery and equipment) also tend to promote more capital-intensive and less labor-intensive investment than would otherwise be the case.

The general objective of tax incentives should be to promote increased overall investment and not to divert investment from one sector of the economy to another.

Knowledge. One major problem faced by countries that have exploitative industry is transfer pricing. Transfer pricing is a mechanism used by companies to reduce their overall levels of taxation. Companies undervalue their products in sales to parent companies (often located overseas), thereby reducing their earnings in the country where the product is derived. Typically, the product is a raw material for the parent company's production process.

Transfer pricing is doubly damaging for developing countries because it reduces the value of foreign exchange earnings from their own natural resources and because it reduces government revenues by decreasing the tax liability of companies involved.

DIRECT PUBLIC INVESTMENT

The basic concept of direct public investment is that government makes the conscious decision to use its own funds (tax revenues, aid funds, royalties, etc.) to invest in commercial enterprises. Explicitly or implicitly, government considers that (1) it is a better judge of how and where the funds should be invested than the private sector (which would otherwise keep the taxes), or (2) the particular investment at the margin will yield a better return than alternative investments.

Before the different forms of public investment are discussed, a few general considerations are reviewed:

- The principle of equity is that it is risk capital. Because all commercial investment involves an element of risk, the funds applied to this type of investment by government should be written off as budgetary expenditure. Returns to equity are entirely dependent on the outcome of the project and constitute the last claim on the resources of the company, should it fail.

- In general, governments should not borrow to take equity in a company because the loan will require repayment while the equity provides no certainty of a return with which to repay the debt.
- Equity in a company entitles the holder to participate in the profits of the company, but new projects often have substantial lead times before the equity yields a return; major projects may have a two- to three-year period before they are fully operational, and at least another four or five years are required before they start to show any profit.
- In many cases, the shareholders (particularly in new companies with only a few shareholders) are required to provide separate guarantees for the loans taken by the company.

Preconditions

The following conditions should be met before governments contemplate investment in a commercial venture such as a tuna fishing, marketing, or processing project:

- The investment must appear to be robust; that is, it must show a good financial and economic return, irrespective of *financing arrangements*.
- If the government's primary goal for the project is to provide revenues, then the financial analysis of the project must explicitly balance the expected (projected) returns to equity with (1) an assessment of the risks of the project, (2) the cost of any loans or other financing that the government will have to provide for the project, and (3) the tax revenues that would have been received in the absence of the project.
- The project company must be established on the basis of a properly structured project agreement that clearly sets out the rights and responsibilities of all parties involved.
- In principle, all inputs to and outputs from the company should be traded at "arm's-length prices."
- When the costs of inputs for the project are reviewed and compared with those from different sources, the cost of financing must be taken into account. For example, many countries offer export credits to promote the sale of their equipment. Such credits are, in fact, subsidized loans and insurance (which insures the exporter against

non-payment by the importer). The interest rates that apply to these credits can vary, and the total financing package must be reviewed for the goods provided.

Forms of public investment

One hundred percent government shareholding. In this case, governments assume total responsibility for the company; they provide all the necessary financing to establish the company, directly with equity, or to guarantee the debt from commercial or publicly owned banks. This is a common structure for government corporations around the world, particularly public utilities such as electricity and telecommunications. In general, the directors are drawn from the civil service, and the company is often treated as an extension of government, answerable directly to the minister in charge. The government normally uses such companies to fulfill objectives that are not completely commercial.

The management structure of wholly government-owned companies varies from quasi-civil services (typically with overstaffing and inefficient management systems) to more commercially oriented structures with professional management companies providing managerial leadership and know-how. Except in only a few specialized cases, this type of financial structure and ownership is the least desirable of all types of government investment in commercial enterprises. The conflicting objectives mentioned above often lead to inefficiencies far in excess of the actual cost of, say, providing the subsidy. On the other hand, the use of outside management companies to provide managerial expertise is often a costly proposition because rewards to the management are usually established irrespective of the level of performance. In the case of the tuna fishing and processing industry, this company structure should not be considered, even if the government has sufficient financial capability.

Majority government shareholding. By "majority" is meant greater than 50 percent of the total share capital of the company. The rationale often given for this kind of ownership structure is that government should control the company because of its strategic or economic importance. The argument is that additional shareholding is required to attract resources such as

- Finance,
- Management expertise,
- Industry-specific technology,
- Marketing channels for the product.

However, the government, by nominating the chairman of the board together with a majority of the board members, will be able to control the operation of the company and ensure that it is consistent with the national interest.

While this structure is preferable to that of a wholly government-owned company, it is still prone to the same problems of inefficiency and political interference. Certainly, it allows the inputs mentioned above to be brought in; in particular, it gives the managing party (who should have a sizable shareholding in the company) a vested interest in making the company a success. At the same time, the notion that government is somehow controlling the operation of the company through its position on the board of directors is somewhat erroneous.

First, there are the practical aspects of running the company. These are undertaken by professional managers who are hired for the purpose. Directors often do not have a sufficient grasp of the day-to-day operation of a company to provide more than overall guidance. This is especially true when the directors and chairman are civil servants with little or no commercial experience. Typically, the permanent secretary (head of department) in the relevant ministry is appointed to be chairman of the company's board. Such civil servants are often unable to do more than attend board meetings and approve the proposals made by the managers. Furthermore, where the directors are appointed from the junior ranks of the civil service, they are often unable to understand the workings of the company fully because they do not have sufficient experience and therefore do not fully appreciate the implications of the decisions to which they are a party.

In addition, there are the legal requirements. In most company acts under which companies function, the responsibilities of directors are set out. In particular, even though the directors are the legal representatives of the shareholders, they must act in accordance with the best interests of the company itself. For example, when the company declares a dividend, the directors must make certain that the financial requirements of the company are satisfied first. Such decisions may cause conflicts of interest, for example, where government-appointed directors are mandated to declare a dividend that would leave the company short of funds for reinvestment. Finally, instances occur when governments appoint private citizens who are knowledgeable in the particular field to the boards of companies in which the governments have a shareholding. In principle, this is a sound strategy, but it can lead to a conflict of interest in which the appointee is engaged in activities similar to those of the company.

Based on the above information, the following conclusions can be drawn:

- Majority shareholding by a government, as in the case of wholly owned companies, places the main financial burden on the government.
- Other shareholding parties (such as in joint ventures) can attract managerial and technical inputs and thereby promote commercial objectives for the company.
- Government control, for various reasons, may be neither possible nor desirable: the functions, responsibilities, and experience of government-appointed directors may limit their role as instruments of government policy and direction.

Minority shareholding. Less than 50 percent of the total equity in a company constitutes a minority shareholding. Of all types of government ownership, minority shareholding represents the least costly method of satisfying the political objectives of government involvement in projects such as tuna resource developmemt. There may also be legitimate instances where the participation of the government will help an overseas investor reduce its perceived risk in the project—a risk that is more political than financial. Government financial commitment to a project may satisfy the potential investor that the government will continue to provide its support to the company through the implementation of the project. At the same time, the fact that it is a minority shareholding may help assure the investor that government does not intend to interfere in the running of the company to the detriment of its profitability and the returns to the investor's capital.

In general, the less government equity at the start of the project, the safer it is for government. Governments should aim to minimize their exposure in commercial companies but should retain the right to buy into the company at a later date. The highest risks in a project occur during the early stages of its implementation and operation. There may be increased costs of construction (of a factory, for example) or other inputs (such as fishing vessels). The early years have greater uncertainties in the operation of production lines or equipment, and the marketing channels for the outputs are less likely to be fully established. Hence, because of risk aversion, governments should await the smooth operation of a project before they increase their shareholding.

In terms of revenue, the choices are not always so clear-cut. Once a project is operational, it is easier to judge the future earnings of its shares. However, any determination of the value of additional shares bought by government should reflect a greater certainty about their earnings. Thus governments should negotiate "options" to purchase additional shares at rates that are based on the historical value of the shares adjusted for earnings (i.e., the equivalent amount of capital that could have been earned in the international financial markets). In other words, if the project earns better returns than the international returns on capital, the government will be able to acquire additional shares and earn greater income than it would with its capital invested in the international financial markets. On the other hand, because the purchase is an "option" to buy and not an obligation, if the project earns less per share than would an alternative investment, the government need not exercise its option.

Other types of public investment

Two other types of public investment (according to the broad definition used here), which relate to commercial projects such as tuna fishing and processing, are as follows:

Project-related infrastructure. In some cases, there may be bottlenecks to the development of a natural resource due to the lack of infrastructure such as a road or a port. Even though infrastructure may be included in the government's long-term development plans, the project's implementation would be greatly facilitated by the earlier construction of the facilities; moreover, the project company might be unable or unwilling to finance the infrastructure. This may be an instance where the government, by bringing its plans forward to develop the infrastructure, could enhance conditions for private investment.

In return for government financing of the road or port, the project company would be expected to pay a "user charge" for the facilities. In essence, this procedure reduces the up-front expenditure that the project must make to become established, while at the same time it gives the government a return on its investment. In cases where such an infrastructure investment is made, governments may use a number of options to finance it. The preferred option is to establish a limited liability company (say, Company A), which is wholly owned by government, with a normal structure of debt and equity. Company A may or may not actually construct the facilities itself, but it provides the financing to do so.

For income, Company A receives payments from the project company (say, Company B) for the use of the facilities constructed. In some cases, the operations of Company A may be financed with only a minimum of financial exposure of the government. This could be achieved by negotiating a contract with Company B that guarantees payments to Company A to cover the cost of repaying the loans taken by it to finance the facilities. The government should also expect some return on its equity in Company A. This kind of cost-effective arrangement could help satisfy several government objectives—for example, by raising revenue, promoting investment, removing investment bottlenecks, and reducing risks for investors.

Government guarantees. Of all the options for encouraging investment, the least beneficial is the provision of government guarantees for commercial debt. A government guarantee means that the government undertakes to repay a loan if the person or company defaults. In many cases, a guarantee is claimed to be a costless way to encourage investment or development. The first principle is that if a commercial project requires a government guarantee in order to proceed, then there are reasons to believe that the project is not a viable one. The corollary to this principle is that a good project should not need a guarantee; it should stand on its own merits.

A government guarantee represents all the risks for government and none of the responsibilities, few of the rewards if the project works out well and all of the costs if the project fails. All parties concerned should understand that government guarantees represent contingent liabilities on sovereign debt without any control of the company that is using the loan. Furthermore, the loans are usually unbudgeted, and if they need to be paid by government, resources will have to be taken from other areas of public expenditure.

21.
Packaging Tuna Projects

Ian D. Richardson

ABSTRACT—To maximize the benefits accruing from its fishing zones, a developing country must acquire the fishing, processing, and marketing skills within locally based companies. Joint ventures offer one solution, but many tuna joint ventures have failed to meet the expectations of the Pacific island governments. This chapter reviews some of the problems of tuna joint ventures and suggests an alternative approach wherein the government takes the initiative rather than merely responds to unsolicited proposals from foreign interests. The government identifies potential partners with proven technical and commercial records in fishing, processing, or marketing tuna that are willing to participate and invest in a locally registered company, established and operated in a manner clearly determined by government. The government does not itself take a major financial stake or become involved in the direction of the company. Instead, the government identifies and encourages local entrepreneurs and bilateral and international funding agencies to provide equity and loans. The government also ensures that the long-term interests of the nation and its people are safeguarded and that there is strong commercial direction through its selection of the major equity holders and board members. Thus, unlike many tuna ventures, the government's financial and management involvement in the joint-venture company is strictly limited.

INTRODUCTION

This chapter highlights some of the issues raised and problems experienced when Pacific island nations have attempted to establish tuna fishing and processing ventures. The difficulties arise principally because of the need to harmonize the purely commercial objectives of the foreign private-sector investors (particularly in a joint venture) with the long-term responsibilities and objectives of the nation. To illustrate the problems and suggest some solutions, reference is made to existing ventures and accounts of the actions taken by governments such as Papua New Guinea to establish tuna ventures.

The desire to maximize the benefits from the natural resources of the country has posed problems to both developed and developing countries. The acquisition of rights to the ownership and exploitation of resources within the exclusive economic zone (EEZ) is a case in point, particularly when internationally tradable species such as tuna and prawn are involved.

An obvious solution is to license the access to and the exploitation of the resource. Although countries may not have an immediate alternative to licensing, they cannot realize the full socioeconomic benefits unless local industries are established.

To establish a country's capability to extract and exploit the resource, to process it (preferably in an added-value form), and to market it on the world market, the country must acquire technical skills, managerial skills, and marketing expertise, as well as the finance to meet the capital and operating costs of the project. All these factors are clearly recognized by the Pacific island nations. The problem is how to acquire these skills and assets in a manner compatible with the long-term interests of the country.

The creation of joint-venture companies or corporations offers one solution. Indeed, joint ventures have been successful, particularly in the manufacturing and agricultural industries, but fisheries joint ventures have not experienced the same high degree of success.

If private capital and entrepreneurial initiative are limited, the government may provide the major part of the required finance; vessels and equipment may be purchased and operational expertise acquired through agreements with individuals or foreign companies. Experience has demonstrated that although such arrangements offer 100 percent control by the government and theoretically only minimal dilution of the financial and socioeconomic benefits to the country, they are rarely effective in the long term and do not result in a cost-effective enterprise.

Given the need for external financial and commercial involvement, the question is by whom and under what terms it is to be provided so that the benefits to the nation are maximized in the long term.

Origin of tuna joint ventures

A salient feature of past tuna joint ventures is that the initiative has invariably come from foreign sources, either commercial or governmental; of course, this initiative is coupled with the vested interest of the foreign company proposing the venture.

Interested external parties fall into the following two categories:

1. A commercial company wishes to gain access to the source of supply of raw material, specifically the right to fish the tuna within the country's EEZ. The objectives are to supply the canneries in the home country or canneries controlled by the foreign partner and to service the traditional markets. The profit center is offshore.

   The interest of the nation is to maximize benefits in the long term, implying a transfer of technology and skills leading to a wholly locally operated fishing, processing, and marketing venture in which the full value of the product is returned to the Pacific islands country indefinitely.

   These two objectives are basically incompatible, and a failure to recognize the objectives of the foreign commercial concern as distinct from those of the nation, as well as to make appropriate arrangements, has been the origin of dissatisfaction in many past ventures.

2. A commercial venture in the recipient country frequently receives an offer of a joint venture in which the foreign company offers vessels or equipment as its contribution to equity. The vessels may no longer be able to fish the traditional grounds economically. Some Pacific island countries have been approached by foreign fishing companies or shipbuilders under the guise of foreign government bilateral aid. The vessels on offer may or may not be of appropriate design and size, and they are often offered without the essential operational and maintenance backup. Invariably, the recipient government finds itself in an embarrassing situation, having acquired vessels or equipment without the capability to operate them and having assumed a large debt repayment obligation.

   In many cases, agreements to train and to employ locals have been made with a foreign partner in an attempt to ensure localization and technology transfer. Similarly, management and marketing agreements can be and have been made between foreign interests and the Pacific island country.

   Invariably, with the passage of time, differences of opinion have arisen, resulting in dissatisfaction on the part of the recipient country with the performance of the foreign interests. Examples are a poor or non-existent training program and questionable marketing arrangements leading to a belief that the anticipated revenues have not accrued to the recipient country.

Where the initiative is taken by a supplier of vessels or equipment, the possibility is that the government, as the signatory to the contract, has accepted a considerable financial commitment, either as principal or as guarantor. The financial commitment may appear to be reduced by offers of favorable funding terms from the supplier or its government or by apparent capital cost reductions, although frequently the vessels and equipment may be unnecessarily expensive in relation to the operational requirement. On occasion, the profit on the sale of vessels and equipment has been used as the foreign partner's equity stake. Favorable financing terms may ease the financial burden without eliminating government financial exposure or guaranteeing the ability of the venture to generate repayments.

Technical operating, management, and marketing agreements can, in theory, reduce the risk and generate confidence in a return on investment. However, the supplier of vessels and equipment does not necessarily have operational experience. Even if it does, its commitment to use that experience over the long term is unlikely to be maintained unless repayments are spread over a long period and are directly related to the success of the venture.

Where the initiative is taken by a foreign marketing and/or processing company, the motivation is clearly to obtain supplies and maximize profits. The foreign interests usually insist on a marketing contract that permits the foreign partner to accept or reject the raw material on the basis of quality. In addition, the foreign partner usually requires exclusive rights to buy at prices that are not necessarily competitive. Such arrangements reduce the potential benefits to the country. A failure of the foreign partner to strengthen local capability through training may further reduce the benefits.

A Pacific island country must, of course, accept some loss of benefits in return for the technical, managerial, and marketing skills acquired from outside it. However, an acceptable level of "cost" must be determined from the outset. The venture must then be structured in such a manner as to ensure that there is no excessive long-term siphoning off of profits outside the country and that ultimately it can operate as a local enterprise.

Role of government

Many developing countries lack finance and entrepreneurial spirit in the private sector. Even where they do exist, the government may have to play a catalytic role, particularly in the primary producing ventures such as agriculture and fisheries, where risks are high, where there is a long initial period before full production, where infrastructure costs are high, and where there is a shortage of local people with the necessary expertise.

As a result, the government finds itself the prime point of contact for foreign interests, playing a leading role in setting up the venture and having a significant financial involvement and participation in the ongoing management, at least at the board level. This situation usually occurs when the initial approach is made by foreign private or governmental interests or when development agencies or bilateral donors that normally operate on a government-to-government basis are involved.

In many past joint ventures, the long-term success or failure can be correlated with the level of government involvement in financing and managing the venture. Furthermore, the extent of the government's continuing financial and management involvement is the major determinant in the success of the venture.

In both the developed and developing countries, government-run enterprises have not always been as cost effective as private enterprises. Nevertheless, in certain situations, the government must accept the responsibility for operating these ventures, particularly when no other alternative exists or when social services are involved. However, the mode of management differs significantly from that of a purely commercially oriented venture. Understandably, because of their different objectives and operating philosophy, a joint venture in which government and private interests are involved together in management is likely to experience difficulties. If the necessary safeguards of the interest of the nation and people can be assured, the government should encourage the establishment of the commercial venture but not participate in its ongoing direction.

AN ALTERNATIVE APPROACH TO ESTABLISHING TUNA JOINT VENTURES

One approach that recognizes all these problems, as well as many others in fisheries and agriculture joint ventures, was adopted in 1984 by Papua New Guinea. It had experienced several unsatisfactory arrangements with foreign tuna fishing and processing companies and suppliers of vessels and equipment, which were sometimes supported by overseas governmental or statutory agencies. The principal features of the Papua New Guinea strategy can be summarized as follows:

1. The government acts as catalyst in establishing tuna fishing and processing ventures for the purpose of exploiting and benefiting from the tuna resource within the country's EEZ.

2. One or more joint-venture companies are based in Papua New Guinea and are registered as Papua New Guinea companies operating within the existing legal and fiscal regulations and subject to all taxes, controls, and benefits available to local companies exporting at least part of the product.

3. Those technical and marketing skills that do not exist locally can be acquired by attracting companies with a proven track record to participate in the joint venture.

4. These companies are expected to subscribe between them up to 49 percent of the equity as a demonstration of commitment to the ongoing success of the venture.

5. Local entrepreneurs/companies are encouraged to participate in the company and to take an appropriate share of the risk.

6. The government's financial exposure should be severely limited, but the government will attempt to secure and pass on the benefits of bilateral and international development funds.

7. The government will rely on existing legislation to ensure that the company operates in the interests of the nation, and the people will, with certain safeguards, permit management to be in the hands of the equity holders.

8. The government will provide the infrastructure and services, as well as all the incentives and assistance as promulgated in the laws of the country with regard to new industries.

Essential to this strategy is the government's lead in defining the terms and operating strategy of the joint-venture company and in acting as the catalyst to attract companies to "bid" for participation in a venture run strictly on commercial lines according to the fiscal and legal regulations of the country. In addition, the government, acting on behalf of the nation and the people, ensures that the company is structured so that any "hidden" benefits or advantages cannot accrue to any one partner to the detriment of the company or the country. The government's assistance should be limited insofar as possible to providing infrastructure and loan funds to the company on favorable terms.

The above description summarizes one possible strategy, but other alternatives could be adopted to ensure that the benefits are equitably distributed among all partners to the venture.

Action

The capture, processing, and marketing of fish are highly specialized, and companies that have been involved in joint ventures are highly experienced in negotiating the terms of joint ventures. It is therefore advisable for the government to contract the services of independent consultants to strengthen its negotiating position.

Independent fisheries consultants should be assigned the tasks of (1) identifying potential partners; (2) advising government on a list of possible partners; (3) evaluating and, where appropriate, suggesting modifications and amendments to the development proposals formulated by interested parties; and (4) assisting government in structuring the company in a manner that will safeguard the interests of the nation.

The following is a description of the action taken by Papua New Guinea to identify partners. To preserve confidentiality, reference is made only to available documents. Subsequent discussions of company structure and negotiating strategy are not necessarily representative of Papua New Guinea's policy.

Phase 1

Consultants were hired by the government to publicize its intention to establish one or more tuna fishing/processing ventures, and notices were placed in the principal fishing newspapers and economic journals to request expressions of interest by potential partners. The following is an example of such an advertisement issued by the government of Papua New Guinea for tuna industry development.

ADVERTISEMENT
TUNA INDUSTRY DEVELOPMENT
PAPUA NEW GUINEA

The Government of Papua New Guinea is seeking prospective investors to participate in the development of the nation's tuna industry. Two separate shore-based processing facilities possibly with associated fishing fleet are planned; one at Kavieng, New Ireland, and the other at Lombrum, Manus Island.

The Resource

Papua New Guinea has extensive tuna resources and proven fishing grounds for longline, purse seine, and pole-and-line operations.

The Projects

The two proposed projects could involve a range of processing options. Fish supply may be from either associated or independent fleets. Staged development of the projects will be considered.

The Sites

Prime development sites in Kavieng and at Lombrum, adjacent to the major fishing grounds, are available. Some infrastructure is available at these sites.

Expressions of Interest

Companies and others who can demonstrate a sound financial and technical background and an appreciation of the requirements of establishing a viable tuna industry in Papua New Guinea are asked to register their interest with Fisheries Development Limited who will be pleased to provide further information.

Letters should be marked "P.N.G. Tuna Project" and addressed to:

Fisheries Development Limited
Birmingham Road
Saltisford
Warwick CV34 4TT
England
Telex: 31565 ULG G

In addition, consultants directly approached fishing companies, processing companies, and bilateral and multilateral funding agencies known to have the experience and capability required of a joint -venture partner.

Phase 2

Because many respondents would not have the expertise or the financial resources to participate in a joint venture, respondents were asked for details of their companies' financial standing and operational experience and willingness to prepare a feasibility study or investment prospectus if requested to do so. To assist them with background information, a brochure incorporating information provided by the national and provincial government departments was sent to each respondent.

The brochure is an important component of the alternative strategy. It provides the opportunity for the government to describe its objectives, and it provides potential investors with clear guidelines to the critical components of the venture; thus it sets the stage. It precludes the government from being placed in a position of having to react to unsolicited proposals, and it establishes the government as the dominant party.

In addition to identifying the sites favored by government as the fishing/processing bases and providing background information on local services, the brochure contains chapters on company law, fiscal/legal matters affecting the establishment and operation of the company, export/import taxes, tariffs, labor laws, rates of pay, and environmental impact considerations.

The brochure contains the following four sections:

1. A brief description of the government's objectives and proposed program of action and its desire to establish one or more tuna joint venture fishing/processing companies with one or more foreign companies or institutions, financing agencies, and so on.

2. A request for initial expressions of interest supported by background information detailed in the brochure on the size, financial standing, expertise of the company or institution and its affiliates and associates, and any other information that would favor the group as a partner in a tuna joint venture.

3. A statement that after receipt of expressions of interest, a list of not more than five leading companies would be selected to prepare a feasibility study or investment proposal.

4. Guidelines for the preparation of the study, including topics to be covered. By this means, groups expressing interest were made aware that the time and cost involved in the preparation of the proposal (should they be selected to submit a proposal) would be considerable and should be undertaken only by serious potential partners.

In the feasibility study or investment proposal, companies were asked to incorporate the following details, plus any additional topics deemed relevant by the proposer:

- Identification of the preferred location and identification of the necessary essential infrastructure to be provided by government.
- Proposal for procurement of fish, including proposed number and size of vessels, fishing method, projected catch rates and landings, seasonality, crewing arrangements, arrangements for fleet management and maintenance, landing facilities, capital, and operating costs.
- Proposals for processing tuna, product form, throughput, yields, production process, packaging, design of plant and equipment, capacities and layout, staffing and management, capital, and operating costs.

- Proposed target markets and pricing structure for the domestic and world markets.
- Details of all capital and operating costs to be included, together with financial projections, including profit-and-loss accounts, projected balance sheets, cash flows, and financial rates of return.
- Economic analysis and financial and social benefits.
- Proposed company organizational and financial structure; financial plan; operational, technical, and administrative policy; use of local labor; training; and localization program.
- Proposed action to reduce environmental impact.

Phase 3

From these descriptions of the companies and their activities, about five or six companies were requested to prepare detailed feasibility studies and investment proposals.

In practice, some companies' strength is in the operation of fishing vessels, others' in the processing and marketing of fish. Given the general practice of separating catching from processing/marketing, the differing requirements for expertise, infrastructure, and financing, and the higher degree of financial risk in the catching sector, the two sectors are usually separate. This structure does not preclude a common equity holding in the catching and processing companies. It does, however, permit each company to be regarded and judged as an independent profit center without any possibility of cross-subsidization. Consequently, in the case of Papua New Guinea, separate lists, one for each sector, were made.

Two or three companies in each sector—catching and processing—might be asked to prepare proposals. An acid test of commitment is whether, when requested to prepare a study, the candidates actually do so. In the case of Papua New Guinea, some of the selected companies did not send representatives to see the proposed sites or acquaint themselves with local conditions; as a result, they were not considered to be appropriate partners.

To ensure a clear understanding of the conditions under which the venture would have to be established (as was done in Papua New Guinea), the potential partners should be provided with a list of the institutions, government departments, individuals, and private interests that might be involved in the venture, either in an official capacity as contractors or as co-investors. This list should be supported by a collection of all relevant Acts of Parliament, regulations, labor laws, and the like.

Phase 4

Commercial companies do not usually prepare feasibility studies or investment proposals with the extensive coverage required by bilateral and international funding agencies or governments. Consultants therefore have an important role to play in pointing out any shortcomings in the studies. In addition, they can indicate where additional details and justification in support of the proposal are required.

Several aspects of the proposals will also require clarification, particularly where an apparent incompatibility exists between the policies of the nation and those of the potential partners. There will also be technical points that are not necessarily acceptable.

At this stage, the government, with the assistance of technical consultants, will discuss each point with the proposer and request revisions to make the proposal more acceptable. For example, if one of the proposals has been prepared by a company whose interest lies in the construction and sale of vessels, the suitability and cost of the vessels will need to be justified and shown to be competitive. Similarly, the processing equipment and the proposed design capacity must be shown to be the most cost effective for the particular situation; the proposed marketing strategy must not be restrictive and should be in the best interests of the Pacific island country; the level of foreign staffing must be acceptable; and the proposed training and localization program must meet the government's objectives and be implemented accordingly.

At this stage, the government and its advisers can, in direct discussions with individual proposers, assess each company's degree of flexibility and its willingness to adapt to local conditions and requirements. Following these discussions, a final selection of the most acceptable joint-venture partners can be made and detailed negotiations begun. Such negotiations involve the structure, financing, and operation of the joint-venture companies.

The approach described contrasts with the one used by most funding agencies. In their case, once the project is identified, a pre-feasibility study is prepared, followed by a full feasibility study, which is then used as the basis on which to proceed to implementation and search for appropriate partners.

Critics of this approach state that a feasibility study undertaken by a funding agency tends to stress the economic benefits rather than the financial return and does not appreciate the commercial investor's modus operandi.

The approach, described earlier, that requires the potential investor to prepare the feasibility study has merit because it guarantees the interest and commitment of the investor. The investor would not spend the time and money necessary to prepare a competitive proposal without a strong desire to proceed to implementation and a commitment to the future. Thus this approach permits the government to evaluate the sincerity and intentions of the investor. It also provides the government with the opportunity to clarify its objectives and those features that it wishes incorporated in its venture. By detailing the contents of the feasibility study requested from the investor and by providing background information on the structure and method of operation of the venture and its role in the community, the government gives priority and encouragement to those investors most likely to be in harmony with its objectives.

Having achieved an understanding of the nature of the joint venture, the government has still to ensure that the long-term interests of the country are protected within the framework of a commercially feasible venture. To this end, negotiations must be initiated with the selected foreign partners and other interests that may be involved.

The government must determine the level of investment and financial risk that it wishes to take and its role in the continuing direction of the company. The purpose of this chapter is not to dictate government policy but rather to point out the advantages and disadvantages of the government's role, either as a catalyst with minimal ongoing involvement or as a participant in an exposed financial position with long-term management responsibilities.

The assumption is that government is not well equipped to invest in and manage commercial ventures; the personal responsibility and rapid decision making associated with investment in equity are not characteristic of government. But where local investment funds and company management skills are not readily available, the government may have to make other arrangements to ensure that the nation's long-term interests are secured. This can be achieved through ensuring an appropriate equity mix and therefore representation of experienced direction at the board level. The company directors would not necessarily have to be provided from local sources; they could be expatriates whose outlook is known to be in harmony with the nation's.

Concomitant with the search for foreign technical and commercial partners and for local sources of finance and direction during phases 1, 2, and 3, the government should seek to identify bilateral and international financial institutions that might take an equity or loan stake in the new venture and provide ongoing monitoring and direction; for example, the International Finance Corporation (IFC), the Commonwealth Development Corporation (CDC), and similar international funding agencies.

Because the proposed venture will incur capital and operating costs higher than those in similar ventures in developed countries (which will militate against its commercial viability), the government should seek those bilateral and international agencies that offer equity and loans to new ventures on favorable, non-commercial terms. This may involve (but may not be restricted to) loans from government agencies in countries supplying vessels or equipment, as well as the affiliates of the international funding agencies that can offer finance to commercial ventures.

Finally, the government should recognize the non-commercial expenditure related to providing infrastructure such as quays, roads, and services and should seek the necessary finance to provide the infrastructure from agencies such as the World Bank (WB), the Asian Development Bank (ADB), the European Development Programme (EDP), or bilateral agencies.

By the time the final negotiations start, the government should have determined both the most appropriate equity distribution and the board representation necessary to safeguard the interests of the country.

In addition to deciding the financial and management structure of the joint-venture company, the government must insert appropriate clauses in the articles of the company and in the shareholders agreement to ensure that the obligations of the company are clearly recognized and observed in practice. Because of the nature of their business, different aspects and different safeguards will have to be incorporated into a company concerned with catching tuna as distinct from one that processes and markets tuna.

FISHING AND PROCESSING

Joint fish-catching ventures are often a device for a foreign company to gain access to a supply of raw material for its own processing facility elsewhere. There are many instances where transfer pricing has been practiced and the true value of the catch has not been remitted to the country where the tuna is caught.

It is essential that the venture's catch be landed in the harvesting country—whether it is to be processed in that country or re-exported as whole round fish—to permit a first degree of control. However, the surest safeguard is to make it mandatory for the fishing venture to operate as a locally based company that lands all fish in the country and operates under the rules and regulations for locally registered companies.

The high capital and operating costs of fishing vessels coupled with the highly variable catch rates lead to a much higher risk than that incurred in processing and marketing. Therefore, all available measures should be taken to ensure that the financial exposure is minimized and that the debt burden has a genuine chance of being serviced. Indeed, a demonstration of the appropriateness of the vessel and the capability of the crew should be required before a large long-term commitment is made.

The cost of providing the infrastructure, harbors, jetties, cold storage, maintenance facilities, and backup services is higher than that demanded by the processing sector. The fortunes of the venture are likely to be more cyclical, reflecting variations in weather, the availability of fish, and other factors. Local employment opportunities will, however, be greater than those in the processing sector.

Where there is a lack of infrastructure (particularly of freezing or cold-storage facilities), an interim measure may be necessary to land the catch onto freezerships, motherships, or reefer vessels. This method is acceptable, although in the long term it may be more expensive than constructing and operating shore installations. The processing sector is not subject to the high costs of the catching sector, nor is it subject to variations in fortune outside the control of the company. So long as there is an adequate supply of raw material at an acceptable cost, variables such as quality, product form, and market penetration are within the control of management. Areas of concern arise if the partner is a supplier of equipment, in which case its interest may be short term and directly related to the sale of equipment, or if the partner is a processor, in which case the supply may be diverted to its own plant outside the country and the true sale price of the produce may not accrue to the local company.

If one of the potential partners is a supplier of equipment, measures must be taken to ensure that the equipment is the most appropriate for the venture, that it is not overpriced, and that maintenance and spare parts will not cause stoppages in production. These facts must be verified by an independent assessor. If the prospective partner is processing and/or marketing tuna, it will obviously be anxious to acquire additional supplies, which in itself is a valid reason for partnership. But the long-term interests of the company and the country must be safeguarded by similar measures. Moreover, the company eventually needs to have the capability to process and market independently of the foreign partner.

FINANCE AND MANAGEMENT STRUCTURE

To encourage a balanced long-term management and to ensure that no imbalance of benefits occurs in favor of the foreign company, a majority of equity participants should be entitled to seats on the board and should be sympathetic to the nation's long-term interests. This procedure, coupled with the suggested limitation on the government's financial exposure, will determine the financial package.

A 40:60 equity:loan ratio is a normal and desirable target, and external commercial partners can be expected to accept responsibility for a significant share of the equity to assure their faith in and continued involvement in the success of the project. The precise equity distribution is a matter for discussion, but to ensure ultimate control over policy, the foreign commercial partner should not be permitted to subscribe more than 49 percent, probably around 25 percent, with the remaining equity held by several interests. If necessary, the government might take some equity (at least as a temporary measure) and assign the monitoring function to reliable commercially experienced individuals or selected sympathetic equity holders.

As suggested earlier, some of the 51 percent equity, in addition to that subscribed by local interests, might be held by organizations such as the IFC or the similar affiliate of the ADB. Other equity holders may include those organizations in many developed countries that are charged with investing in new ventures overseas as part of their countries' development assistance. In the case of Papua New Guinea, the Danish Fund for Industrialization and the CDC of the United Kingdom are organizations that might be approached to take equity and that could exert a strong role in directing and managing the company because of their experience in new commercial ventures. As a result, the country's exposure will be reduced, and independent, experienced company direction will be available to the Pacific island country.

The loan component could also be provided by the type of institution mentioned earlier, and additional loan funds are likely to be made available from the countries supplying equipment and services. In some cases, the loans can be made directly to the company. In others, they will be made on a government-to-government basis with the Pacific island government acting as guarantor and on-lending to the company. This situation is quite different from one in which the government receives a loan for purchasing equipment or vessels that the government itself will operate, a situation that would incur considerably greater debt repayment obligations and considerably greater risk.

Whichever source of loan funds is used, the terms of the loan are likely to be better than those obtained on the commercial market and to have a lower interest rate, a moratorium on repayments, and a long repayment period. These terms may be vital to the success of a new venture, which is exposed to higher-than-normal costs due to the "greenfield" situation. They will also limit the government's financial exposure and risk.

The government is likely to be faced with heavy costs for infrastructure, quays, roads, and services, as well as for additional social services such as housing, schools, and hospitals, which are vital to the success of the venture but could not be supported within the commercial investment and must therefore be accepted as an obligation of government. For this purpose, financial assistance in the form of soft loans might be requested from several bilateral and international agencies.

To limit the level of borrowing, a clear understanding is required of the immediate and likely long-term infrastructure needs of the new venture. The danger is that if facilities greatly exceed the requirements of the venture, both in size and sophistication, the government could incur unnecessary expense and debt liability.

OPERATIONAL STRUCTURE

Reference has been made to technical, management, and marketing agreements that have been commonly used in tuna joint ventures. These may no longer be required if the venture is established as a locally registered company with a sound financial structure and a board of directors composed of commercially experienced, sympathetic persons. However, as a safeguard, such aspects of these agreements should be specifically mentioned in the articles of the company and the shareholders agreement.

For example, neither the equity structure nor the board of directors should be changed without agreement of the government, which will seek at all times to ensure representation of local interests (although not necessarily through locally subscribed equity). In this respect, it should be noted that most of the organizations mentioned earlier are not long-term equity holders: their perspective is that their equity is to be sold to local interests once the venture is operating successfully.

If the management, technical, or marketing expertise is to be provided from one or more of the foreign partners, the subsidiary agreements should specify this and require that details of any action be made available to the board. Any staff provided under such agreements should be directly responsible to the board through clearly established administrative channels, which should be specified.

Both the contractor and all staff members must act in the interests of the company and be accountable through clearly defined lines of responsibility to the board. The board should retain the right to require any individual to be replaced by another and, in the ultimate, to suspend the agreement with the contractor in favor of other arrangements and other contractors. In short, the contractor should act as a consultant, and the executive function should remain in the company.

The contractor will naturally expect to be remunerated for all costs incurred, including staff; in addition, an incentive payment could be linked to the company's overall performance. In the early days, the incentive may be linked to turnover rather than profit. The principal remuneration of the foreign investor must be, as for all other investors, the declared dividend of the company, whether or not the investor has accepted a contract for management, technical supervision, marketing, training, or any other form of service.

The principle of company autonomy within the laws and regulations of a locally registered company should be clearly established in the articles, and requirements such as the program for localization and the ratio for training foreign and local staff members should be specified.

The company can be expected to benefit from any assistance provided under existing regulations to locally based companies, new enterprises, and companies earning foreign currency. The foreign investors will be subject to taxation, repatriation of funds, and so forth, as indicated in the initial request for a feasibility study.

During final negotiations the government may, however, offer special short-term concessions as an added attraction to foreign investors whose skills and funds are required to establish and successfully operate a new venture in a greenfield site. But the norm is for the government to endorse the company structure and operating policy, using the technical and commercial acumen of the foreign partners.

CONCLUSIONS

The program of action outlined above is designed to permit the government of a Pacific island country to overcome problems experienced in many tuna joint ventures in the Pacific, where the benefits to the island country and the foreign partner have not always been equitable. This plan proposes that the government, rather than foreign interests, take the initiative and identify potential partners that can offer the operational, technical, and marketing expertise and, in addition, can make a substantial financial commitment to a joint venture. The requirement for the selected potential partners to prepare, at their own cost, a feasibility study or investment prospectus along the lines detailed by government demonstrates both a commitment and a recognition of the constraints under which the company will be expected to operate.

The government can then negotiate with the foreign partner and other potential partners, including local entrepreneurs and bilateral and international financial and development institutions, to structure a company that will be registered and operated according to the laws and regulations of the island country. The company or companies would be vehicles for the acquisition of the expertise and capability to exploit, process, and market the resources of the country to the benefit of the country and its people.

It is suggested that the government should not be a principal investor or be involved in the direction of the company once it is established. Instead, the government's role should be to ensure that the company is financed and structured so that it will be operated and directed by a board selected for the commercial experience of its directors and weighted to safeguard the long-term interests of the island country. The foreign equity interests will be encouraged to ensure the success of the company through provision of technical, marketing, and other expertise on repayment terms.

22.
Development and Financing of Fishery-Related Infrastructure

Walter Miklius

ABSTRACT—This chapter analyzes options for Pacific island governments for the development and financing of shore bases for tuna fleets operating in the region. Options discussed range from government-owned facilities to privately owned bases.

INTRODUCTION

Pacific island governments have expressed an interest in establishing tuna operating bases to encourage homeporting of the distant-water fishing fleets operating in the western Pacific region. The establishment of these bases requires construction of costly infrastructure facilities. The following interrelated questions must be answered in order to evaluate the feasibility of the proposed project: What is the mix and size of facilities needed and the consequent investment? Who will construct these base facilities? How will they be financed? Who should operate the bases?

REQUIREMENTS FOR FACILITIES

The most obvious infrastructure facility requirement for any tuna operating base is an adequately sheltered bay or harbor with a turning basin sufficient to accommodate the largest vessel expected. In addition, a transshipment base requires a deep-water wharf (or wharves), a cold-storage facility, and various repair facilities with adequate inventories of supplies and spare parts.

The optimum size of these facilities can be determined only after a specific site is selected and only if they are modeled on a system. Certain considerations, however, are common to all bases. The required length of the unloading pier, for example, depends on the number of fishing boats to be serviced at the base. For a small base, the pier length necessary to accommodate one fishing vessel at a time will be adequate because, given a small number of vessels to be serviced, the probability of two or more vessels arriving to port at the same time will be very small. Even if a base is designed for a large number of fishing vessels, a pier length sufficient to service one fishing vessel may be adequate because crews of the fishing vessels require three or four days of rest and recreation per trip while the vessel is in port. The crew can be on rest and recreation while the vessel is waiting in line to unload the catch, and thus the three-to-four-day waiting time does not represent an increment in cost.

Pier construction costs can be avoided if the tuna base facility is constructed in an already existing port. Furthermore, the reefer vessel used for the transportation of tuna from the base to the cannery is likely to be larger than the fishing vessels. The length of the required pier, as well as the turning basin, will be determined by the size of the reefer vessel.

The reefer vessel is likely to be chartered and dedicated to transporting tuna from the base to the cannery. The optimum size of the reefer vessel will depend on the number of fishing vessels to be serviced by the base, the loading and unloading rates at the base and at the cannery, the distance between the base and the cannery, and the economies of vessel size.

The cold-storage facility is the most expensive part of the base infrastructure facilities. Its size also depends on the number of fishing vessels to be serviced by the base and on the frequency of calls by the reefer vessel. The investment in the cold-storage facility could be avoided if frozen tuna is transshipped in refrigerated containers. However, the required investment in refrigerated containers is likely to be even higher than the investment in the cold-storage facility.

In addition to wharves and a cold-storage facility, several repair shops and supply facilities will be needed in the base. The mix and type of facilities to be included will depend on the number of fishing vessels to be serviced and the proximity of similar facilities available elsewhere in the region.

In addition, crews will need accommodations when in port, and entertainment/recreational facilities must be available, as well as reasonably easy access to an airport with scheduled air services. The total investment in infrastructure facilities in a relatively small base (for example, one serving ten purse seiners) should be in the range of $1 million to $2 million. The determination of the mix and type of facilities to be provided in the base (as well as the more precise estimation of costs) is the purpose of the feasibility study, which should be conducted when the base's specific site is identified.

FINANCING OF INFRASTRUCTURE FACILITIES

The construction and development of base infrastructure facilities could be financed in several ways. The financing options depend, in part, on the role of the government in the management and operation of the base.

Ownership of facilities by the government may be the preferred option if aid funds are available to construct the base or if loans can be secured for this purpose from international lending institutions (e.g., the World Bank or regional development bank). World Bank and regional development bank loans are said to have the following advantages: (1) the loans tend to be for longer terms than those otherwise available; (2) the interest rates tend to be lower than those otherwise available; (3) the participation of the World Bank or regional development bank endorses the credit for other potential lenders; and (4) a co-financing arrangement for a complementary financing agreement may be possible, whereby commercial bank loans are linked with loans from the World Bank or other regional development bank.

A Pacific island government could operate the fishing base in the same manner as it operates other ports. This kind of operation, however, has a major disadvantage. Only a few ports are being operated at a profit; in other words, most ports are heavily subsidized.

It has been argued that the major problem to be overcome in creating a viable transshipment endeavor is the lack of a "coordinator," the role previously played by the major U.S. tuna packers, Japanese trading companies, and European buyers. The disappearance of these coordinators has created a vacuum that has a significant impact on the potential feasibility of tuna transshipments. Some industry commentators believe that the tuna transshipments must be managed as a system in order to provide the necessary coordination. This system would include all facets of fishing, maritime transportation, and even marketing.

Given these requirements, a private firm with extensive knowledge of markets, as well as experience in business logistics, would probably do a better job of managing the facility than would a government agency. Such a firm is likely to be foreign-owned, although a joint venture between a domestic and a foreign firm is possible.

A variety of contractual arrangements can be made between government and the company selected to manage the base. At one extreme is a pure service contract according to which the company manages the base for the government for a fee, and the government receives all gross proceeds from the base operation. The service fee, in turn, may be a fixed fee or be based on the percentage of gross or net revenue generated by the base. The latter arrangement would provide more incentive for the company to operate the base efficiently. Another possibility is a fixed fee plus a percentage of either gross or net revenue.

The type of the fee specified in the contract determines the allocation of market risk between the government and the managing company. In the fixed fee contract, the government bears the entire market risk. In the service contract based on gross or net revenues, part of the market risk is shifted to the managing company.

A "concession" type of contract is an alternative to the service contract. In this type of contract, the management company pays a negotiated fee to the government for use of the base facilities. The "true" lease (i.e., without the purchase option) is one example of this type of contract. The lease terms could be a fixed amount per year or could vary as a percentage of gross or net revenue or be a combination of the minimum fixed amount and the percentage of gross or net revenue above a specified level. Again, the terms of lease payments determine how much risk is borne by the government. With the fixed lease payment, the risk is borne by the management company; with lease payments tied to gross or net revenues, the government not only shares in the market risk but also benefits from the increased revenues and/or profits.

To this point, it has been assumed that the government would construct and retain the ownership of the infrastructure facilities. At the other extreme, the government could require a foreign firm to construct all infrastructure facilities not only for its own use but also for use by others. Requirements such as these have been included in some concession agreements.

Those concession agreements that include these types of requirements pertain to very large projects with very high expected rates of return. In the case of tuna operating bases, the government could require that the developer construct larger port and power-generating facilities than those needed for its own operation, as well as provide social infrastructure. These requirements, however, may be self-defeating; given the small size of the project, they may cripple the project financially.

Yet another option is for the government to construct the base facilities but offer a lease with a purchase option. The lease payments could be structured so that the title passes unconditionally to the lessee upon the final rent payment under the lease, or it could include the negotiated purchase price at the termination of the lease. These types of leases are popular in Japan, and these so-called "shogun" leases are available to foreign lessees.

In short, a variety of financing options is available for the development of a tuna base's infrastructure facilities and for its management.

However, there are no general rules for choosing the best option. The preferred choice can be determined only on a case-by-case basis. The choice will depend on several factors such as the availability and terms of loans to the government from the World Bank or regional development banks versus factors such as the availability and terms of loans to the foreign companies, an assessment of various risks involved, the risk preferences of the government and the foreign company, and the legal considerations and tax policies in both the Pacific island country and the country of the foreign company. Furthermore, the ownership, management, and other contractual arrangements are not independent of the service fees, lease rents, and user charges, which are discussed below.

OPTIONS FOR DETERMINING SERVICE FEES, LEASE RENTS, AND USER CHARGES

Let us suppose a scenario in which the government contracts with a private foreign company to operate the fishing base. The company either owns fishing boats or contracts with private boat owners and performs all functions including marketing the catch. The base infrastructure is owned by the government, and the contract with the company is either the service contract or a lease. In this scenario the determination of the actual level of service fees or the lease rent will depend on the method used to select the management company or the lessee. If the selection is made via an auction (either by sealed bid or oral bid), the level of service fees or the lease rent is determined automatically through the bidding process.

This automatic determination of the service fees or lease rents is not the sole advantage of an auction. If a large number of bidders is expected, the service fees may be lower or the lease rents may be higher than those resulting from negotiations with individual companies because the auction participants will be bidding against each other. Furthermore, in auctions there is less chance of corruption.

Auctions also have disadvantages. The auction may not be desirable if only one prospective bidder is expected; moreover, if only a few bidders are expected, they may possibly collude among themselves to reduce the price of the winning bid. As a partial remedy, the solicitation of bids should specify the maximum service fee or the minimum (upset) rent acceptable to the government and should reserve the right to reject all bids and to negotiate a contract instead.

If the selection of a management company or a lessee is made by negotiation, the actual service fee or the lease rent will be the outcome of these negotiations. As is the case with all two-party negotiations, the outcome is a priori indeterminate. It is possible, however, to estimate both the highest lease rent that the firm will pay rather than to forgo the lease on the base and the lowest lease rent that the government should accept (or vice versa, for the service fee).

The maximum lease rent that a company would be willing to pay for use of the fishing base is equal to the cost, insurance, and freight (CIF) market value of the catch, minus the sum of all operating expenses, the normal profit on investment, and the premium for risk. For example, the expected trend in tuna prices will directly affect the maximum amount of lease rent that a company will be willing to pay. Similarly, any other contract terms, which indirectly affect the costs of the risk, would also affect this maximum amount.

One of these contract terms is the length of the lease or the service contract. A long-term, fixed-rent lease will be preferred by the company if tuna prices are expected to increase, and vice versa. However, the length of the lease is usually associated with an increased risk and thus a higher risk premium. Furthermore, the longer the lease term, the more detailed its terms have to be in order to anticipate long-run contingencies. Thus even a long-term lease or a contract usually specifies periodic renegotiations every five to seven years. This requirement probably benefits both parties.

The lowest lease rent that the government should accept is equal to the aggregate benefits that the tuna base is expected to generate in the domestic economy. These benefits are not equal to the payments made by the company to the government but rather are the sum of benefits that accrue to the domestic economy. Thus, in some cases, the government may find it worthwhile to subsidize the company rather than to have no base at all; however, this would be the worst-case scenario.

Thus far, it has been assumed that the government would contract with a foreign company to operate the base and to perform all functions associated with the integrated fishery operation. An alternative scenario calls for the government to operate the base and to provide the usual port services to individually owned or company-operated fishing vessels. The boat owners, individually or jointly, or the company itself could undertake the marketing of the catch, as well as coordinate the logistics of transshipments. In this scenario, the government would have to set the user fees.

Unfortunately, the current practice of pricing for use of port facilities provides little guidance as to what these charges should be. Implicit objectives often contradict national policies.

To economists the role of the pricing system is to encourage the efficient use of existing facilities and to provide guidance on investment or disinvestment in them. To achieve these objectives, prices need to be related to social opportunity costs. In other words, prices need to be based on the marginal opportunity cost of the resources used to provide each service. If users are prepared to meet this cost by the actual prices that they are prepared to pay, it is then reasonable to suppose that they prefer to purchase this service rather than the alternative goods and services (opportunities) that they have implicitly forgone. Economists therefore recommend that prices for port services be set on the basis of the marginal costs of providing the particular service.

The introduction of marginal cost pricing, however, involves certain conceptual as well as practical difficulties. The main problem is that where there are economies of scale, the application of marginal cost pricing will give rise to losses and the need to subsidize port authorities. To avoid subsidies, the traditional demand-based pricing principle (i.e., the charge that the "traffic will bear") has been recommended by some economists. An alternative to demand-based price discrimination is the two-part tariff, in which a fixed charge is levied (say, annually) to recover unassignable costs, combined with a low variable usage rate.

Marginal cost pricing will generate losses, however, only if port facilities are not congested. In the case of congested facilities, the marginal cost pricing principle requires that charges be levied high enough to eliminate uneconomical queuing at one port or congestion at others. Thus, for congested facilities, the general rule is to charge prices considerably above the average cost. The main problem is to estimate the appropriate level of charges, which is not a simple task.

The best strategy is probably for the port authority to assess charges that are closely linked to the identified costs of different services provided by the port. Port services are conveniently divided into two types. First, there are the services rendered to the vessel such as provision of navigation aids, dredged channels, tugs, and berth space. Second, there are services that are provided for the cargo such as wharf space, transit sheds, handling, and transshipment. Traditionally, the port charges the ship owner for ship services and the cargo owner for the services of the wharf and its facilities. Thus the port authority could charge costs of providing these various services, plus an assessed fixed annual fee to defray the common and joint costs of the tuna base operation.

As an alternative, the port authority could adopt the average cost pricing principle, the main purpose of which is to assure that the port authority recovers all costs of operating the port. The application of this method is relatively easy because Applied Systems Institute Inc under a contract with the U.S. Department of Transportation (Maritime

Administration) has developed a method, or formula, for deriving "reasonably compensatory prices" for use by public marine terminal facilities. It focuses on the determination (based on cost) of dockage and wharfage tariff rates, rental prices for leased terminals, and rental prices for cranes and equipment. Although developed for U.S. ports, the formula can be easily adapted to any port in the world.

23.
Social Impacts of Commercial Tuna Projects

Stephanie Pintz

ABSTRACT—This chapter analyzes the effects of a commercial tuna project on local customs and lifestyles that might be associated with fish harvesting, shore bases, and cannery operations. These social effects can influence the implementation of the tuna project and its consequent relationship with local people. The chapter also offers suggestions on how specific social impact studies can be organized and how social policy issues can be addressed. It is assumed that governments are prepared to play an activist role in minimizing the social consequences of major industrial projects such as tuna fisheries.

CONTEXT OF STUDY

In the past, national development objectives focused almost exclusively on economic gains. Today, however, social objectives are also considered essential to the development process. A commercial tuna project generates revenue, employment, infrastructure, and training programs, but these assets are not without social costs. Successful project implementation requires a realistic consideration of costs as well as benefits. It is unrealistic to expect an absolute maximum of economic returns from a fishery. National socioeconomic policy should ensure that those who incur the costs of social disruption receive some compensating benefits. Residents of the project's region may experience a transformation from a rural lifestyle to one dominated by a cash economy. Land and water resources may be threatened. Predictable routines and cultural patterns may be disrupted by an influx of "outsiders." Rural people may suddenly find that they are less well off than some of their neighbors who find cash employment with the tuna industry. In addition, not only does the project affect local residents, but also the residents' reactions can affect the project's success. Examples abound throughout the Pacific of costly disruptions from walkouts, strikes, and outbreaks of violence. Such disruptions often reflect the absence of consultation and involvement with the local people, especially during the initial planning stages of new projects.

This conceptual study examines the composite effects that have occurred from commercial development projects throughout the Pacific. Identical impacts will not occur in all nations. Variables include proximity of the operations to an established urban area, the existing or potential economic opportunities in the region, and the sociocultural characteristics of the local population. It is assumed that operations will affect people with primarily rural backgrounds and lifestyles, even though development may occur in provincial centers, which are not strictly rural.

Pacific lifestyles are communal in nature. Land and water resources are precious, and people's relationships to them are apparent in virtually all aspects of life. Most people live at a subsistence or semi-subsistence level, being generally self-reliant through fishing or agriculture. To the degree that the social impacts of a commercial tuna project can be predicted and evaluated before drastic change occurs, steps can be taken toward ensuring that operations are implemented in accordance with social goals.

FISH HARVESTING AND SHORE-BASED OPERATIONS

The social effects of tuna operations reflect certain characteristics of the industry. For example, fish-harvesting operations may or may not include national workers. Shore-based operations may include transshipment and vessel service centers and/or a cannery. The number of support industries and the services required depend on the scope of the particular fishery. Similarly, the number and type of secondary facilities required—such as trade stores and equipment outlets—parallel the size of the project. Several additional secondary services (e.g., educational) may be developed in response to the project's social impacts.

Unless training and promotion opportunities are built in as project policy, most of the employment for local workers is of an unskilled nature. The labor force in Pacific commercial fisheries is characterized by transience, partly due to the low wages and lack of opportunities for training and advancement. People seeking careers look toward the civil service or other jobs that offer at least the possibility of promotion, and islanders seek only temporary jobs from time to time to obtain cash for specific needs.

The social effects of a commercial tuna industry result from national economic aspirations, which are normally realized in the Pacific within a framework of private investment. Inherent in this framework are three aspects from which the social consequences of the venture emerge: (1) the use of natural resources (primarily marine in this case) for private—often foreign—profit; (2) the emergence of a cash economy because a tuna venture promotes a new supply of both money and wage-earning opportunities; and (3) the impact of migration on the project's community.

Impacts on inshore resource use

Commercial baitfishing to support pole-and-line harvesting techniques can affect a community's supply of reef resources as well as its social organization and relationships. Commonly, the search for baitfish extends over a wide area affecting neighboring communities and sometimes even entire regions. Baitfishing can deplete supplies of tuna and other pelagic fish, which come inshore to feed on the bait, as well as smaller species actually sought as baitfish.

Reefs and adjacent waters are primary sources of food for rural Pacific islanders. The rights to harvest reef waters may be controlled by chiefs, clans, or individual families who, in turn, grant permission for others to fish and collect food supplies. This is an efficient means of both conserving resources and ensuring a continued supply of protein to the community, but severely overharvested reef resources have potentially far-reaching ramifications for diet and nutritional standards. In areas where fish from nearby reefs is sold or traded in local markets, depleted resources also mean higher prices as well as an implied threat to livelihood. In addition, integrated village activities may be based on local fishing, including small boat building, boat repair, and net and trap making, which are affected by the use of inshore resources.

Local perceptions about baitfishing are significant. The residents' fears may or may not be ecologically valid, but perceptions can lead to resentment and conflicts, not only with the baitfishers but also among the residents. Furthermore, the reef waters themselves are often fundamental in the definition and expression of social relationships in coastal communities. For example, the organization of a community may be greatly influenced by the adherence to resource property rights. Similarly, social continuity may involve customs such as the division of reef- and water-related work between men and women. If fishing becomes less productive, a redefinition of labor and social roles might result. Finally, certain social institutions may be reflected in local fishing patterns, for example, the education of youth in fishing methods.

Bait fee payments or bait royalties can constitute a source of local income and thus can be rife with complications. The tuna project itself can be hampered if western legal practices relating to payments conflict with traditional customs regarding ownership of reef waters. In addition, traditional attitudes toward work may impede a community's interest in doing its own baitfishing. Residents may not wish to risk the confusion and potential conflicts involved in altering the customary use of local reefs and waters.

Impacts of a cash economy

An expanded economic base and improved infrastructure can generate expanded development opportunities such as investment in trade stores and outlets for agricultural products. These opportunities—along with those of direct wage employment with the tuna project—may result in migrants returning to their home communities. In the Pacific, a converse flow often results from migrant remittances to home villages. At a broader level there may be a significant increase in the tax base of the region and a consequent heightened political importance at the national and provincial/district levels.

The hiring of local workers can have significant impact on the community. For example, who will do the subsistence fishing and agricultural work previously done by the absent men? If large groups of men are employed in the tuna project's operations, the extended family or clan may suffer because many families will not have sufficient labor. Will women then become more important in the traditional agricultural or fishing work force? Will they have to make additional decisions regarding all aspects of family life? The answers depend on several important variables such as the marital status of workers and the existing resources of their families. The families of married men, especially those who are poor, may experience the most adverse effects, although these groups could potentially benefit the most. Such effects need to be carefully assessed so that alternatives to people's traditional roles can be considered.

A growing cash economy creates or enlarges an urban setting with all its attendant characteristics. A prominent effect of urbanization is a change in consumption patterns from fresh locally grown foods to imported foods. Many imported foods provide inadequate nourishment, and some are actually even harmful. Considerable evidence exists of increased nutritional disorders throughout the Pacific, especially in densely settled urban and semi-urban areas. These disorders include not only anemia, vitamin and mineral deficiencies, and dental diseases, but also heart disease, stroke, diabetes, and some forms of cancer—diseases all associated with a westernized diet. Obesity, a condition conducive to many serious diseases, is related to consumption of refined foods as well as to alcohol. Of course, excessive use of alcohol already has well-known negative effects on health and social well-being. Finally, the tendency to abandon breastfeeding in favor of imported substitutes can have negative effects on infants.

The emergence of a cash economy fosters socioeconomic inequalities. A clear separation between the fishery work force and the rest of the population can lead to jealousies and conflicts. Inequalities can also be accentuated by land or baitfish compensation payments, giving those who are already relatively prosperous (i.e., resource owners) an additional advantage over those who are not. In societies with customary bride-price or dowry payments, these costs may rise and marriage patterns shift.

Inequalities between expatriate managers and local residents (whether employed or not) can be glaring, but the extent to which serious social problems result may depend on the residents' own perceptions of their well-being.

Expatriate and national workers generally have different attitudes toward leisure and wage employment. Pacific islanders often participate in many activities other than labor within their economic framework. Work is a part of social experience, with the accompanying expectations and social involvements. Because the job, the family, and the community are integrated, an individual may find it psychologically difficult to set aside a specific number of hours every day for work. In contrast, people from western cultures maintain a psychological separation between work and leisure. Other dimensions of their lives can be held in abeyance while they are working, for example, on a tuna project. Mistrust, contempt, and fear, which can lead to several kinds of social problems, arise when these attitudinal differences are not understood.

Impacts of migration

Labor for the project's operations will most likely be drawn from other communities in the project's region as well as from the immediate area. Although most of the issues discussed thus far apply to both migrant workers and residents of the shore-based community, several social impacts are specific to migrant workers and their home communities. These involve (1) hiring policies, (2) disruptions at home, (3) perceptions of outsiders by existing residents, and (4) personal adjustment problems.

If employment preference is given to local residents, people from other communities in the region may object. This type of policy has resulted in demonstrations against fishing and other industrial development projects in the Pacific. Because migration is often an implied legal or constitutional right, restricted hiring policies may prove difficult to enforce. However, if local preference is not given, existing residents may protest that they are being deprived of their fair share of the project's benefits. Moreover, the employment of large numbers of outsiders may lead to a feeling of territorial encroachment. This situation becomes even more complicated where major cultural differences exist between residents and migrants.

Large-scale out-migration to the project site may actually result in rural depopulation or demographic imbalances, with the very young, the very old, and the women left at home in the villages. The absence of young men could result in the loss of traditional land or other resource holdings to neighboring groups, as well as in increased population pressures on retained lands.

An influx of people (especially if they belong to a different social group) can have dramatic effects on an existing community. Jealousies and conflicts, sometimes resulting in violence on and off the job, can occur, and already have occurred, in all types of industrial development projects throughout the Pacific. Unemployed migrant squatters utilize local land and water resources and are sometimes forced into illegal activities. This situation can breed resentments that are dangerous to the social stability of the existing community, especially if basic needs such as housing have not been given adequate advanced planning. On the positive side, intermarriage among various social groups may serve to moderate disputes and contribute to more unified societies—a goal of many Pacific island nations. In addition, migration can foster increased contact between the government (whose presence is in the shore base) and the rural villages (from which the migrants come). For example, government's awareness of political problems or health standards in rural areas can be increased as a consequence of rural out-migration.

Inadequate or careless planning for the migrant workers' social needs can result in serious personal adjustment problems, with negative social consequences. Boredom can lead to alcohol abuse and a festering of irritations, resulting in fighting and crime. Prostitution not only fosters the spread of sexually transmitted diseases, but also may significantly increase the number of illegitimate children in the community. Inadequate facilities and overcrowded housing naturally lead to tensions and sometimes even violence. At the very least, impersonal, ill-planned single men's quarters accentuate the migrants' feelings of alienation and negate a community feeling.

The presence at the work place of unemployed friends and relatives can help restore it. However, they need housing and food, thereby intensifying pressures on the community. Serious value conflicts regarding wages may arise with migrants whose intended remittance goals are not realized. The migrant may be faced with two conflicting social premises: expectations in town that dictate a consumer lifestyle and expectations at home that wages will be shared.

It is possible to develop a tuna project where there is no existing community, but this is unlikely. Under such circumstances, all the workers will be migrants and the shore base becomes a "company town," in which instance a new community is formed. Many of the impacts associated with migration are exacerbated in such a setting, and many of the opportunities for meeting recreational and social needs are primarily dependent on external planning strategies. Any company town is fraught with difficulties, and the development of such a town is generally discouraged throughout the Pacific.

CANNERY OPERATIONS

The social effects of cannery development will be similar to those of a shore base. However, the nature of the cannery, the type of work involved, and the implied manpower requirements mean that residents will experience additional effects when a cannery is developed in their community. The social impacts, which are unique to cannery development, emanate from environmental, infrastructure, and labor force issues.

Environmental effects are generated by the cannery itself and by the required secondary industries and infrastructure. Cannery operations require large quantities of fresh water, which can have two consequences. First, where fresh water is in short supply, the cannery's needs may preempt water supplies for other uses. Second, much of the water used in cannery operations is returned to the environment as a polluted effluent. Water pollution is by far the most serious environmental hazard with immense potential for disease from untreated domestic and industrial sewage. Also most pollutants involve a reduction in the content of dissolved oxygen, which is essential for maintaining healthy aquatic life. The polluted waters of the cannery community could cause serious hardship to residents who utilize the waters for subsistence or for trade. In addition, odoriferous air emissions and visible aspects of the plant can create an unpleasant living environment for local residents. Moreover, tourism could be adversely affected if the cannery is built near existing facilities or in an area where ocean currents could carry effluents to nearby tourist beaches.

The expanded infrastructure and public services generated by a cannery can improve rural access and communications in adjacent rural areas. The ensuing benefits ideally should range from better health care to lower public utility rates. However, the plant itself, support industries, new housing developments, public service facilities, and new roads all require land. The extent to which local owners and residents will be affected is principally related to the amount of existing usable land in the area. If local people are willing to give up land, matters of ownership and compensation must be resolved. Land compensation, if not handled carefully, can become a major source of social inequality and strife.

Labor force issues are manifested in the potential for migration, the propensity to employ women, and the hiring policies. Clearly, developing a cannery is tantamount to increasing the population of an existing community. Migration, with all its social effects, is likely to become even more prevalent when a cannery is added to the fishing venture. Moreover, those nations with significant cannery developments may attract immigrants from other island nations. The legal status of these foreign migrants may be a problem, and the impacts of outsiders on an existing community will be intensified.

Women are likely to be sought for cannery employment due to the low wages generally accepted by island women and to their reduced housing requirements. Several important results occur when women with rural backgrounds become involved in the industrial work force. First, in relatively poor families the opportunity for women to earn wages and employment benefits may be welcomed. Second, the competition between factory work and household responsibilities places serious strains on the division of household labor; agricultural production and care of dependent family members may be negatively affected. Third, the traditional divisions of labor may involve objections by men to working with women; or husbands may not accept their wives working alongside other men. These issues may affect both the ability of the cannery to attract a female labor force and the efficiency of its operations.

If local hiring policies are promoted, two (or more) members of a household may be employed with the tuna project. This situation could intensify regional inequalities, with jealousies and conflicts arising among communities outside the project's vicinity. Thus any hiring policy that fosters two-income households will also tend to exacerbate income differences and inequalities. On the other hand, policies that attempt to spread regional opportunities by employing only one household member may discriminate against those local workers and be more costly for the cannery (e.g., housing costs).

Many of the social impacts of a tuna project need not result in significant costs to either the economic or the human components of the project. If they do so, there has been a failure to anticipate and respond to these impacts.

ORGANIZING AN IMPACT STUDY

Nations must determine which effects are likely to occur within the context of each specific project. The vehicle for focusing government policy and administrative attention is the socioeconomic impact study, which, if properly conceived, may have an important secondary role in community education of residents. The success or failure of the impact study will be directly related to the ability of assessors to define aspects of the project proposal that have socially sensitive implications. This requires an understanding of the sociocultural character of the groups living near the project.

The impact study must be mobilized as soon as the feasibility of the venture is established so that it can be utilized in the planning process. This study enables policymakers and planners to determine that the project will benefit as many people as possible and that negative effects are minimized due to this information. Objectives go beyond obtaining and reporting information, and thus anticipatory social planning should involve the people who will undergo changes as a result of the project. If the study is to be a preliminary exercise to this planning, the residents of the project area logically should be actual participants in—not just objects of—the study.

Common assessment shortcomings

The framework for assessing the impacts of a commercial tuna project reveals some common problems associated with most social impact studies. First, many social assessments are done after construction for the project has already begun or after the project is operational. Even if the assessment is done well, post hoc studies and plans are often executed too late to be effective. Second, many studies are characterized by an overabundance of data. If distinct connections cannot be made between findings and changes incurred by the project, the data serve no purpose. Third, the study will almost always be initiated by people from outside the area being assessed. At the very least, their predilections will be urban and will have some inevitable biases. However, studies often focus only on superficial characteristics of families and communities, thereby offering distorted or incomplete information. Robert Chambers (1979) relates some common biases of "rural development tourism" that impede a valid assessment of rural conditions. Among these biases are the following:

- Urban bias often concentrates on rural visits and research in towns or village centers, but a significant number of people may live in less accessible areas.
- Contacts and sources of information are often biased toward those who are the most articulate informants and whose interests in the project may not always parallel those of less-influential residents.
- Most rural people with whom contact is established are men, yet women stand to be substantially affected by the project.
- Potential employees are most likely to be contacted, people not directly involved in the project to be ignored.
- Rural appraisals are often conducted during the dry season, when inland travel is easier and living conditions are favorable.
- Politeness and protocol often discourage contact with the less-affluent people of a region.
- Professional training values and interests can limit the focus of attention. A study conducted by people with a bias against development or with a special interest in the project will obviously be slanted.

These kinds of biases are mutually reinforcing and preclude optimal participation in the study by residents of the project region. Given these common shortcomings, guidelines can be developed for an effective impact assessment study.

The assessment team

Social interests are probably best served when the impact study is sponsored jointly by the government and the investment company. The process of the study itself can serve as a communication link between the tuna project and affected residents. Budget permitting, an impact assessment team is preferable to a single individual. This team could include an outside professional (someone with experience in anthropology or rural sociology), a local resident, a government representative from the natural resources or fisheries department, and a representative from the investment company.

If an assessment team is not feasible, substantial assistance and input to the study should be available from the above sources. Many individual consultancies are funded by outside donors. Potential hazards include tied-aid consultancies linked to employment of firms from a specific country and "old-boy" systems, which are often used by international organizations and regional banks. While donor-sponsored consultants may have valuable experience derived elsewhere, caution should be exercised in their selection to ensure that the "free" service is worth more than its eventual costs.

The preparation of adequate terms of reference (TOR) is essential to a successful social impact study. Large consultant contracts, however, are often commissioned on the basis of a sketchy, poorly framed TOR. At the onset it is important that the TOR play two distinctly different roles. First, it must obviously communicate to the consultant which questions or issues the government wants investigated. A second and seldom-recognized role is to persuade the government to understand and articulate the problems being addressed. It is in this second role that many consultancies flounder, resulting in a study containing incorrect or only limited aspects of problems.

Ideally, the TOR should be developed to the point where the proponent of the consultancy can define a precise methodological approach to the solution. In general, the TOR should include a background section describing the context of the problem and the government's concern. This should be followed with the specific objectives of the study, which lead directly to decisions that the government hopes to make as a result of the consultancy. Next, a work plan should describe those specific tasks designed to achieve the objectives of the study. Finally, to be of maximum usefulness, the consultant document must include these two administrative caveats: (1) the study will be the sole and exclusive property of the government; and (2) the consultant will provide all source data assembled for the study and a copy of all calculations, software, and methodologies.

Establishing the database

A substantial amount of information is required before the social impacts of a tuna project can be understood and realistic planning strategies adopted. The determinants of social impacts are the relevant features of the proposed project, of the community in which the project will be based, and of the surrounding communities from which people are likely to migrate. First, the character and scope of the proposed operations must be defined. What is going to be built and where will its facilities be located? What obvious support facilities and infrastructure are required? Manpower requirements (including those for support industries), as well as localization and hiring policies, should be determined insofar as possible. Also important are variations in employment rates; examples are construction peaks or seasonal phases of a cannery's production. Finally, a cursory knowledge of the culture of the expatriate work force would help in predicting potential relationships with the local employees of the project.

Information about the project region is twofold: primary or detailed knowledge, which is obtained and interpreted firsthand, and secondary or general information, some of which can be obtained from existing literature or other documents. This latter information should be acquired first to save time in the fieldwork. Examples include environmental and demographic features, land tenure patterns, and the existing economic base and infrastructure in the region. Primary or more detailed information is required to gain insight into people's basic values, their social identity, and their social relationships. The specific question is, which aspects of people's lives will be affected by the tuna project? The survey team should look beyond the easily obtained facts such as household composition, land ownership, and sources of livelihood. The following examples illustrate some types of useful information:

Land patterns. How is ownership determined, and how rigidly are the rules adhered to? How and when are the available land and water resources used?

Livelihoods. Are there preferred lifestyle alternatives, provided the choices were available? Which activities are shared by both men and women? How and under which circumstances do residents help each other, and how is this help reciprocated? Who operates family finances and/or manages wealth resources? Are there contrasting patterns of wealth and poverty in the community and in the region?

Household composition. Is it static year round? Which activities consume the most time? Who makes family or group social decisions? What are the expectations of and for single young adults?

Food patterns. In addition to the normal diet, are certain foods associated with special occasions? Are there particular food preferences? What time of day are meals eaten? Who eats with whom? Does community food sharing exist? In other words, what is associated with food, other than the satisfaction of hunger?

Health. How do residents perceive their health conditions? What do healthy residents do to keep healthy? What are the primary causes of illness, according to residents?

Attitudes. What are the priorities (e.g., desirable fish or crop yields, stable marriage, cash in hand, or other material assets)? What aspirations do parents have for their children? What is the most important role of women, as viewed by both sexes? What are the relationships between communities in the same area? How do residents feel about impending migration? What are the local views about a fishery? What changes are anticipated?

These connections, which emerge from the questions such as those above, must be examined. For example, dietary patterns and assistance with subsistence chores can reveal sources of recreation and relaxation.

Certain inferences also can be made from secondary information. If the region's literacy rate is minimal, literacy programs will be required to optimize localization of the fishery work force. If churches are a significant community institution, their roles and influence on social activities and work patterns may not be compatible with the work schedules of the tuna project—the 24-hour operation of a cannery, for example.

Approaching the residents

Because the preliminary goal of an impact assessment is to understand the characteristics of a region's people beyond a superficial level, an approach is required that encourages residents to share some aspects of their lives and feelings and to present them accurately. Before direct contact is established, the survey team should learn about the conditions under which residents will provide quantity and quality information. Of particular importance are customary attitudes regarding direct versus indirect questioning, admission of lack of knowledge, and open expression of feelings. Crucial to the impact assessment approach is a predisposition toward listening to and learning from residents. Survey teams also must know the physical details of the environment. Knowledge may be extremely thorough and exact in those cases where the natural resources are limited or constrained. Attentive listening can also lead to new areas of vital knowledge about residents. Small group discussions in which experiences and knowledge are shared can reveal concerns such as seasonal restrictions on produce availability. They can also reveal beliefs and superstitions that might affect their willingness to give up land or water resources. Finally, as contact and communication are established, attempts should be made to offset any potential biases.

Inputs for the study need not require extremely sophisticated tools or methods of analysis. To rely on advanced statistical procedures may serve only to confuse and mislead. The questionnaire survey must be composed carefully to avoid embracing the concepts and categories of the assessor. It should reflect the concepts of the residents themselves. A typical survey focuses on uses of natural resources, as well as on food and activity patterns.

The involvement of residents can be extremely valuable. For example, secondary school students could assist in obtaining specific information on an assigned issue. Literate residents could organize discussion groups regarding the characteristics of the impending tuna project, as well as record the proceedings. Women who may be interested in cannery work could discuss a project's potential impacts and the options for dealing with them. In summary, the data and the process of its collection should reflect sensitivity to the fundamental belief systems and customs of the residents, as well as to how they will be affected by the project.

The report

In its completed form, the social impact assessment includes a summary of salient information; an interpretation of what this information means to the residents, to the government, and to the investment company; and recommendations for social planning. The nature of the report should reflect the negotiated terms of the tuna project. This emphasizes the need for some social planning by government early in the project's development. Of particular significance to the tuna industry is the issue of localization. If a fair degree of localization is built into contractual agreements, the potential for project employment, as well as for local development, will be substantial. If, on the other hand, the project requires only unskilled laborers, its impacts will be different, and planning should be oriented toward a transient work force.

Recommendations should be realistic and based on government and company needs and limitations, as well as on those of residents in the project region. Accordingly, the results of the study must be presented to and discussed with the residents so that they may help formulate the planning strategies. Thus an impact study can be a formality, a token piece of evidence to gather dust in a file cabinet, or a vital national planning tool.

ADDRESSING THE IMPACTS

With the results of the impact assessment study in hand, the government and company must decide on a strategy for dealing with social issues. It would be naive to assume that all social questions can be understood and solved at the project's inception. The strategy must provide a flexible framework for anticipating and accommodating social issues in a consistent manner; it should be an instrument for forming conscious policy decisions, rather than solutions dictated purely by events.

Environmental issues

Many multinational corporations have indicated a willingness to accept environmental regulations, provided these regulations are consistent and free of ambiguities. Those nations with vague environmental objectives may risk confusion, exploitation, and environmental abuse. While conservation for conservation's sake is probably impractical, objectives designed to protect those aspects of the environment on which rural residents depend make economic and social sense. In the case of baitfishing, for example, all uses of inshore resources by local residents and any related customs must be understood. Baitfishing may need to be regulated to avoid resource depletion. Traditional ownership patterns and customary laws require careful consideration in negotiations with the investment company. If, for example, resources are owned by a particular group but distributed among the general population, compensation for the baitfishing could be disbursed among the entire community with an extra allotment to traditional owners.

Any known effects on the environment should be clarified and discussed with local residents. For example, if residents of a baitfishing area favor a particular fish that is not sought for bait, this fact should be verified to help alleviate unwarranted fears regarding the project's operations. Or if it is known that a proposed cannery will deplete local supplies of fresh water, the problems and any available options should be made public. Issues of harbor pollution or conflicting land use also should be addressed through public consultation.

Local development issues

Clear and consistent hiring policies can help diminish regional conflicts over the distribution of employment opportunities. Recruitment quotas have proved beneficial in some Pacific industrial developments. The policies should be advertised via radio, the press, and local political groups. Moreover, if regional rather than strictly local employment is the goal, special attention must be paid to contacting villagers who live away from the existing infrastructure.

Business opportunities should be assumed quickly by local people. Business development officers or their equivalent are a key group of civil servants who should be involved in local business planning because they can help people to make informed choices regarding the investment of money. First, they can help promote a clear understanding of any compensation or royalty revenues. Second, they can educate people about available business opportunities. Third, they can assist with the formation and management of local business groups. At the project's inception, it is unlikely that local interests will have either the capital or the expertise to take advantage of business opportunities. A structured government credit program can provide considerable assistance, and a loan made in conjunction with a major tuna project can be made against considerable security. In some cases, such loans can be directly secured against sales contracts to provide goods or services to the fishing company. Beyond this, local landowners should be able to raise loans against future land compensation payments. Institutionally, it may be expeditious to establish a special unit in the development bank that focuses on project-related opportunities.

Wage and employment issues

Relatively high wages cannot always be offered to unskilled laborers in the project's work force. Regardless of wage rates, an unskilled work force (even if somewhat transient) will always be available. However, a comprehensive localization and skill-building program is compatible with the economic goals of profit (and tax) optimization and the development of a stable work force. The most cost-effective approach to skill building is on-the-job training (OJT), which has several advantages for social well-being. For example, formal education credentials are de-emphasized, a factor that is particularly relevant in regions with limited opportunities for formal schooling. Moreover, along with applicable skills, OJT offers experience in industrial discipline (e.g., working fixed hours, being responsible to a supervisor, and being subject to dismissal), which is usually unavailable in formal programs. A disadvantage is that employers sometimes fail to formally acknowledge the increased skills obtained by workers with OJT and to compensate them accordingly.

More formal training options include literacy programs, skill-specific instruction, and vocational programs. Whichever approach is used, some worker identification of personal interests with those of the company is a requirement for a permanent or semi-permanent work force. Workers who are viewed as unskilled and transient will be likely to fulfill the company's negative expectations; those who are treated as valued employees with potential for advancement will treat their jobs more seriously.

A stable work force in the tuna project can also be promoted if suitable working conditions are offered. Several steps could be taken to offset the dual value systems and social practices that often divide expatriates and national workers. For example, facilities for eating or recreation should be designed for integration, not separation. In addition, both expatriates and local workers could be informally and inexpensively educated about social practices and expectations of one another through orientation programs.

Problems within the national work force can be addressed in several ways. First, local wages should be compatible with other cash employment opportunities and should be equitable, such as those between male and female cannery employees. Second, local customs and expectations could, to some degree, be incorporated into the work place. For example, men and women who work in a cannery could perform their jobs separately if working alongside one another is incompatible with local customs. In addition, the female work force could be organized on a part-time, rotational basis to help alleviate some problems associated with women's traditional roles at home.

Health issues

Any attempt to solve health and nutrition problems associated with a change in consumption patterns and urbanization should incorporate customary nutritional practices as learned from the residents themselves. In the case of potential dietary changes, this knowledge could provide a starting point for consumer education. At the same time, training a local resident as a nutritionist may be a practical way to disseminate information to the public about health threats such as excessive use of alcohol, obesity, and imported substitutes for breastfeeding. Furthermore, advertisements of alcohol and tobacco could be restricted by provincial or national policy. Certainly, formal school curricula should promote general nutrition and healthy lifestyles and include problems associated with consumption of imported junk foods. Finally, preventive steps to combat communicable diseases such as malaria, tuberculosis, and sexually transmitted diseases should be disseminated to the public by educational programs and extension campaigns. Some regulative policies will be necessary, particularly those regarding the disposal of cannery effluents and domestic wastes.

Migration-specific issues

To the degree that problems can be anticipated, the community can be prepared to accept migrants. Local leaders and existing community institutions should be encouraged to establish contacts with the migrants and to understand that unless these people can establish themselves in the community, social problems associated with their presence will be intensified. Steps taken to legitimize migrants will promote identification with the community and facilitate integration.

Housing is a critical aspect of migrant adjustment. Despite the difficulties that may be involved, local authorities should attempt to acquire land adjacent to the existing village or town. While considerable effort and some expense are involved, plans for the housing of migrants must be developed jointly by the government and the company. Not only will squatters' camps and associated problems be reduced, but also the work force will be more stable if the community environment is conducive to family life and promotes a sense of belonging.

In addition to housing, planning for recreational facilities will help reduce social problems associated with migration, as well as encourage community identification. Films and competitive sports are popular and easily organized. Special activities and functions could broaden the scope of leisure activities. Youth and women's groups could serve as a vehicle for discussing local concerns, as well as provide opportunities for socializing. Any recreational facility or leisure activity must be open to all residents of the project community and not just to the employees of the tuna project or to the original residents of the area.

Finally, enforceable laws against drunkenness, violence, and crime should be enacted in any community whose nature does not provide built-in restraints against unacceptable behavior. In the traditional village, offenders are dealt with according to custom. The expanded community of a commercial tuna project will require additional external institutions to ensure the peace and safety of residents.

Ensuring implementation

Implementation of measures designed to address the social impacts of a tuna project requires certain financial commitments, thoughtful delegation of responsibilities, and resident participation.

Governments need to plan an adequate budget because the successful establishment of a tuna project will generate substantial financial burdens. Much of this burden will be associated with the project's local needs for land, increased government services, and infrastructure. Against these local costs, the national government hopes to enjoy significant revenue flows from the project. To the degree that decentralization and home rule are government policy, there will emerge a need for revenue sharing, ranging from the central treasury to the local authorities. Failure to agree on some sort of revenue-sharing arrangement will hamper attempts to deal with the project's potential adverse social effects.

It is easy to place undue emphasis on the dogmas of "coordination" and "cooperation" among departments, agencies, and programs. This can lead to evasion of the difficult choices of who should do what, when, and how, as well as which actions are appropriate. First, a single organization must assume primary responsibility for a given aspect of social development. Second, lines of authority must be clearly established and responsibilities specifically assigned to specific officials. Third, special demands will be placed on provincial authorities who may be suddenly faced with problems that can range from regulating activities in a new township, to acquiring land, to providing community educational facilities. To ensure that these challenges are met, provincial authorities must assign liaison officers from their various departments directly to the project region and must maintain close contact with all groups and institutions in the emerging community. Fourth, direct involvement resulting from enlightened self-interest on the part of the tuna project is a necessary ingredient of social planning. In addition to formal contractual arrangements, the company must understand its participatory role in addressing some of the project's social impacts. Governments have the inherent right not to tolerate gross pollution, employment discrimination, or the disintegration of communities, and they should energetically negotiate policies that integrate economic pursuits with social goals.

If local residents are to become involved in decisions affecting them, liaison sources must be identified. Existing community institutions are valuable instruments because they already have credibility to the residents. Local political groups, youth groups, social clubs, and particularly churches enjoy wide participation and acceptance in Pacific island nations. Respected individuals and spokespeople for community interests or groups may have the social stature to help mobilize and organize a particular program.

In summary, regardless of particular variables, three crucial needs emerge when social impact issues are addressed. These are (1) the need to anticipate and plan for the impacts before they occur; (2) the need to form clear, well-publicized government policies regarding various planning aspects; and (3) the need for input and participation of people who will be affected by the commercial tuna project. In meeting these three needs, government policy must have clearly defined community objectives. Without a community strategy, the ad hoc approach to negative impacts may result in unanticipated social consequences. Serious attention must be focused on the government's goals for those people who will directly benefit (or lose) as a result of the tuna project. With this focus clearly established, participation policies can be designed to achieve specific objectives in cost-effective programs.

24.
Environmental Impact Assessment: Its Role and Processes

Tor John Hundloe and Greg Miller

ABSTRACT—This chapter is divided into four parts. The first is a theoretical discussion of environmental impact assessment and thus is applicable to any type of project. The second and third parts of the chapter are concerned with guidelines for the application of an EIA for tuna fishing/processing operations in the Pacific islands region. The final section is a more detailed discussion of the measurement and monitoring of the impacts on water resources, which are potentially the most negative.

INTRODUCTION

An environmental impact assessment (EIA) has two central aspects. One aspect is the processes by which an EIA is implemented. The other is the form and content of the assessment. The processes used in any particular nation reflect the political/legal/public administration framework existing in that nation. On the other hand, the form and content of the EIA are less dependent on the public policy framework because the principles of an EIA are universally applicable, just as engineering, financial, and ecological principles are universally applicable. Nevertheless, the scope of an EIA in any particular nation can be affected by the extent to which its statutory land use plans and pollution standards address environmental problems.

An EIA should accomplish the following tasks:

1. Ascertain the objectives/purposes of the development. It should be possible for the EIA study team to determine from discussions with representatives of the developer, and with the aid of background documentary evidence, that the objectives/purposes of the development are not, prima facie, incompatible with the nation's social goals, and thus that a detailed assessment should proceed. (Obviously, if the objectives are clearly incompatible with social goals, there would be no purpose in proceeding with the EIA.)

2. Describe the proposed development.

3. Select and describe a set of alternatives that could achieve the same objective. At any given time, numerous factors will determine the existing alternatives. These include the technical feasibility and any legal, political, institutional, and economic constraints. The first step is to ascertain if there might be radically different means (macro alternatives) of achieving the same objective. If there are alternatives of a significantly different nature (which a team of experts believes could provide similar benefits but with different environmental cost), a decision must be made whether or not each of these radically different projects should be subject to a comprehensive assessment.

It must be emphasized that the developer may not be—in fact, probably will not be—interested in radically different projects. Potential developers are likely to be specialists in one type of development, and they are likely to be interested only in the project they are promoting.

All alternatives capable of achieving the general objective are to be addressed. Given limited resources (particularly time and money), it is appropriate to assess macro alternatives applying a "broad-brush" approach. That is, it should be possible to narrow the range of alternatives by eliminating those that fail an uncomplicated, but extended, cost-benefit analysis. In this context, "extended" means that the difficult or impossible-to-measure (intangible) impacts are taken into account. These impacts may be of an ecological or sociopolitical nature. The team of experts will have to be trusted to undertake this task without recourse to the detailed information that it will gather on the actual proposal and any chosen alternatives. It is therefore imperative that the team consist of qualified and experienced personnel representing diverse fields. An economist must be a member of the team. No other task in the EIA process requires the same degree of high-level professional judgment as does this screening process. From this stage on, the task is relatively simple.

4. Describe the environment that is likely to be affected by the proposed development and describe the environments that are likely to be affected by feasible alternatives. "Environment" is defined broadly, including the biological, physical, cultural, sociological, economic, and institutional surroundings of the project.

5. Predict the changes to this environment, or these environments, without the project (that is, ascertain how the baselines will shift without the project).

6. Identify the positive and negative impacts that are likely to result for all alternatives being assessed.

7. Predict the extent and degree of the impacts, in the short term and in the long term, for all alternatives.

8. Evaluate the positive and negative impacts of the alternatives under consideration.

9. Compare the macro alternatives and choose the optimal project.

10. Consider micro alternatives in terms of different designs, locations, or other details; describe these; predict the impacts; evaluate and compare the impacts.

11. Consider modifications, safeguards, standards, and countervailing government policies that would ameliorate adverse impacts; describe these; predict and evaluate the impacts.

This process is one by which the development project and alternatives are compared, some discarded, others amended and refined, again compared, and further ones discarded until (in theory) the optimal development is found. This process is common to financial and engineering decision making.

What the EIA concept does is nothing more than to extend this principle to include a wider range of considerations, an approach that in the long term and in sociocultural terms will lead to improved decisions. This does not necessarily mean that the "best" project should be supported. The no-action alternative—do not proceed with any of the developments—might be the optimal decision.

ENVIRONMENTAL IMPACTS OF TUNA OPERATIONS

Background

Some issues are common to all tuna developments. These include laws that pertain to the total allowable catch, the rights of local and foreign fishing vessels, and laws or government policies pertaining to issues such as pollution standards, employment conditions, migration, health, land-use planning, project implementation, and public participation in decision making.

The impact analysts also must ascertain which laws will affect the project. Fisheries legislation will be involved (for a compendium of Pacific region laws, see Mifsud 1984) and so might environmental legislation (for an overview, see Pulea 1984 and 1985).

Common issues

Dahl (1984) has described the environmental problems in the region and categorized them according to their degree of severity and generality. Most of the issues he identifies have a bearing on the assessment of onshore tuna developments. The most widespread problem is the disposal of liquid domestic wastes, in particular, human wastes and urban sewage. This situation is causing serious water pollution of both freshwater supplies and coastal waters in lagoons and around beaches and reefs, which are important for tourism and fishing.

Another widespread problem is damage to or destruction of coastal resources and fisheries. Various types of development and the direct and indirect impacts of urbanization have led to or are leading to the destruction of coral reefs. Various construction activities including dredging are involved, as well as siltation and dynamiting and poisoning of fish. Seagrass beds are being dredged or covered with silt. In the littoral zone, mangroves are being destroyed by dredging and filling. The coastal fisheries, particularly near population centers, are being depleted by the increased demands of population growth and modern fishing techniques.

Ranked second in priority are several issues that, although not of universal concern, are common throughout the Pacific islands region. These include erosion, water shortage, solid waste disposal, control of toxic chemicals, dredging of coral and sand for construction purposes, and urban pollution.

The control of toxic chemicals warrants special comment. Dahl (1984) states that most governments in the region lack adequate legislation in this regard and that products considered too dangerous in other parts of the world are still in widespread use in the Pacific. On the other hand, Dahl concludes that at present oil pollution is only a minor problem, with oil spills generally being restricted to accidents in harbors during fueling or transshipment. These harbors have already been degraded.

The third category of problems contains those of significance in local areas. These include coastal erosion, mining and the disposal of mine wastes, and industrial pollution (mainly from processing plants). The problems listed above have a common root. Connell (1984) states that they result from the integration of Pacific island states into the global economy. The consequences of this are population growth and migration to urban areas. He states: "The South Pacific has an urban future, and cities and towns are essential to social and economic development." With urbanization have come monetary exchange economies, the growth of slums, crime, alcoholism, prostitution, and "deculturation," as well as a host of pollution and health problems. As governments react to provide amenities for the growing urban populations, the inevitable result is the stimulation of further migration (Connell 1984).

EIA FOR TUNA PROJECTS

The assessment procedure starts with the formulation of expert opinion on whether the project is likely to be feasible and achieve some generally desirable objectives. At this stage and at every consequent stage, alternatives are sought and fully evaluated if they appear feasible. At every stage, the foregone uses of all relevant resources (land, materials, labor, etc.) are compared with the project's requirements. At any stage, it might be prudent to recommend against the project, but in most circumstances it will not be until all the effects of the project are comprehensively traced and evaluated that the overall picture will emerge and a recommendation will be made. The step-by-step assessment is now presented.

Transshipment

The first question with regard to shore-based transshipment is, Are existing facilities adequate to handle the increased vessel traffic? If the answer is yes, the impacts are likely to be minimal, with the possibility of some pollution from oil and bilge water, although this potential problem should be manageable through the enactment of pollution control laws. The second question is, Are there economies of scale with the greater traffic? If so, net economic benefits should result from the increased activity. If new transshipment facilities have to be constructed, the impacts need to be considered at both the construction phase and the operational phase.

Construction phase

A preliminary assessment of likely sites is the first step in the construction phase. Both physical and commercial constraints will determine the availability of sites. Once the potential sites are selected, detailed impact assessments will be required. Consideration will then have to be given to the value of alternative uses of the natural environments and the other resources to be used, including raw materials for construction purposes. The impacts of construction are likely to include permanent changes to the natural environment. Social impacts are likely to be important, ranging from localized issues such as housing of the work force to the long-term socioeconomic changes that the development of a port will induce. For a discussion of the important issues, see Pintz (1986).

Fresh water is required for thawing, cleaning, pre-cooking, and canning. Much of this water is returned as polluted effluent, including press liquor from the reduction plant and blood water from the fish and offal storage areas. This form of pollution can lead to a reduction in dissolved oxygen (DO), which is essential for the maintenance of aquatic life.

In addition to water pollution, there are likely to be odoriferous air emissions and visible impacts. It should be noted that a cannery plant can operate on a 24-hour basis. Not only do the cannery operations have their impacts, but so do support industries and services.

Various support services are required such as fuel, electricity, spare parts, and maintenance. New or improved infrastructure is also necessary, for example, roads for trucking containers between the wharf and cannery and new housing for the workers. These activities have associated environmental costs and benefits.

TUNA PROCESSING IMPACTS ON WATER RESOURCES

Water resources of the Pacific islands region

Several authors (for example, Brodie and Morrison 1984a and b, Dale and Waterhouse 1985) have described the water resources of the Pacific islands region within the context of geomorphology, hydrology, and water quality. These resources include rainwater; freshwater lakes, rivers and streams; standing groundwater; lagoons, estuaries, coral reef-lined bays, and shallow coastal and open oceanic waters. The tropical island ecosystems covered by the region are complex and tend to be highly productive. Resilience to development, however, may be low. Coastal ecosystems are also likely to include interactions between mangroves, seagrass beds, and coral reefs (Birkeland and Grosenbaugh 1985).

Islands of the region can be divided into geomorphic types such as high islands, raised coral atolls, and coral atolls. This type of classification allows a better understanding of the available water resources that must be taken into account in any assessment of project impacts. For tuna processing, there is a need to identify the nature of potential impacts involved in freshwater extraction from island catchments and/or groundwater storage and use of coastal marine ecosystems, particularly in the case of waste disposal. These potential impacts are considered in detail below.

Freshwater resources

Tuna fish processing may involve cannery operations (various grades of quality) and fish-meal production. Current methods of tuna processing require the use of considerable quantities of fresh water at various stages of operation, such as cooking and cleaning fish, washing plant and equipment, generating steam, cooling, thawing, and transporting waste materials. As such, the freshwater supply is likely to be the major physical constraint on the location of a cannery operation in the south Pacific region. Again, the use of freshwater resources in tuna cannery operations has important socioeconomic and environmental considerations.

These considerations are related to

- Existing and projected flows (m^3 day^{-1} and m^3 yr^{-1}) by cannery facilities.
- Adequacy of water sources in terms of storage capacity, resupply, and/or recharge (including assessment of dry weather flows).
- Physical, chemical, and microbiological quality of water sources.
- Level of treatment required.
- Alternate uses (and opportunity costs) for existing and future water sources (including catchments).
- Land-use management and conservation of surface catchments and/or groundwater resources.
- Ecological functions of natural water bodies (e.g., depth of water table and freshwater inputs).

Water consumption. The quantities of water involved in fish processing vary greatly, depending upon whether dry-line or wet-line processing is involved, the type of fish, and the products being processed. Tuna operations are estimated to require from 600,000 liters per day (30 tonnes fish/day) to 2,000,000 liters per day (100 tonnes fish/day). Such operations are taken to include canning and producing fish meal.

These daily flows are considerable for many oceanic islands in view of the fact that relatively high-quality water is required for fish processing. Fresh water has been described as the most critical of all the resources on oceanic tropical islands (Dale and Waterhouse 1985).

Freshwater resources. On many oceanic islands the daily competition for limited freshwater resources is increasing rapidly because of growing tourism, agriculture, urbanization, and associated industrialization. The adequacy and quality of water supply have become essential factors in island resource development. Waste disposal practices, salinity, chemical residues, and siltation are known to affect water resources in many parts of the region.

One analyst identified several specific water problems in the Pacific islands as follows:

- Excessive extraction of groundwater causing saltwater intrusion of the groundwater lens, soil subsidence, and salinity.
- The treatment of water supplies with chemicals that may produce harmful by-products.
- The removal of vegetation during engineering projects causing erosion, degradation, and siltation.
- Excessive extraction of water for irrigation causing harmful ecological downstream effects.

- The poor construction and design of water supply systems (dams, tanks, etc.) that lead to breeding grounds for disease vectors.
- The inefficient use of agricultural chemicals causing pollution of potable water sources.
- The pollution of groundwater through badly sited waste-stabilization ponds.

Against this background, the location of tuna processing facilities requires (1) the preparation of detailed baseline studies involving data collection, analysis, and evaluation of existing water resources; and (2) the identification and confirmation of potential impacts, direct and indirect.

Potential water impacts from tuna processing as derived from the United Nations Environment Programme (UNEP 1980) can be summarized as follows:

Hydrological balance. Will the project alter the hydrological balance of surface and/or groundwater? Information on the availability of the project's source of water is vital. Daily extraction of up to two or more megaliters of good-quality water is indicated for tuna processing. Surface and groundwater are important in maintaining rivers, streams, lakes, ponds, wells, bores, flora, and fauna.

Groundwater regime. Will the project affect the groundwater regime, for example, in terms of quantity, quality, depth/gradient of water table, and direction of flow? Will alterations to the water table depth alter the structural qualities of soil? Will pumping/dewatering methods be necessary to undertake excavations?

Again, information should cover the extent of the project, source of water supply, waste disposal practices, and proposed surface cover. Ground conditions should refer to permeability percolation, water table, location of recharge area, and slope proximity to streams or other bodies of water.

Surface waters. Will the project impair existing surface waters through filling, dredging, water extraction or discharge, waste discharge, or other detrimental practices? Will recreational uses or aesthetic values be endangered? Will the project affect dry-weather flow characteristics?

Information is required on the location of the project, the extent of construction, filling and clearing activities, source of water supply and site of any waste disposal, dams/obstructions, hydrology over an extended period, ecological characteristics, and recreational uses.

Drainage/channel patterns. Will the project impede or alter natural drainage patterns? This involves knowledge of the existence, nature, and pattern of drainage and soil characteristics.

Sedimentation. Will the project induce a major sediment influx into the area's water bodies? Controlling factors include location of construction and clearing activities, erosion potential of site soils, direction and magnitude of water runoff, percentage slope on site, erosion- and sediment-control plan for the site.

Water quality. Does the water supply meet recognized criteria for food processing/potable water (e.g., the World Health Organization's drinking water guidelines)? Will water be adequately stored and treated? Will groundwater suffer contamination by surface seepage or by intrusion of saline or polluted water?

Marine resources

There is increasing evidence of pollution and degradation of the shallow tropical marine communities that abound throughout the hundreds of atolls and high volcanic islands of the Pacific. Sewage and solid waste disposal, sedimentation, food-processing wastes (e.g., from sugar mills and tuna canneries), oil discharges, and chemical contamination are reported examples of pollution.

Tuna processing operations in the Pacific islands commonly exploit coastal marine environments for waste disposal of plant effluents. There are, however, both direct (physical alteration of the local environment) and indirect (waste disposal and runoff) impacts to consider in the establishment of cannery operations and associated fishing fleet facilities.

Construction phase. The major impacts during the construction phase are likely to be from clearing, roadworks, quarrying, dredging and filling port facilities, and constructing a submarine outfall. This last impact may involve blasting. Physical loss of marine resources such as coral flats and reef edge should be fully evaluated because they are likely to involve compensation claims.

Not only physical destruction aspects, but also indirect effects on local marine communities (such as coral reef lagoons and seagrass beds) are likely to be significant. Such effects include sedimentation from soil erosion of exposed soil surfaces from the construction site and roadworks, disturbance of marine sediments, and waste disposal (e.g., oily wastewater from pumps, sewage, and solid wastes). Construction plans should cover methods to control and reduce these types of impacts, particularly given the high rainfall conditions experienced in the region. The predicted impacts should refer to degree of severity and recovery potential of affected marine areas.

Operation phase. Initial and permanent impacts caused by the construction phase of the project will be replaced by long-term implications for marine resources during plant operation. These are mainly concerned with waste disposal from the plant. Chronic discharges and runoff from the associated facilities that service fishing fleets and plant workers must also be assessed.

Plant effluent(s). Riddle and Shikaze (1973) have reviewed the characteristics and treatment of fish processing effluents in Canada. The major characteristics of effluents from fish processing plants were high levels of BOD_5, suspended solids, and total solids. Organic and ammonium-nitrogen contents were not reported. The pollution loads from fish processing plants varied considerably, depending on species of fish being processed, the age of fish being processed, the processing techniques, plant size, and water usage.

Effluent characteristics for tuna fish processing appear to be similar in nature to other fish species wastes. Most of BOD_5 and suspended solid loadings originate from bloodwater and stickwater, but these volumes tend to be low and are generally diluted by other processing wastewaters low in BOD_5 content.

Wastes from tuna fish processing in Pago Pago, American Samoa, are reported to cause harbor pollution in the vicinity of the two processing canneries. Brodie and Morrison (1984b) say that one of the canneries has been forced to improve waste disposal while the other has already installed pollution control measures.

Tuna processing effluents contain BOD_5 and suspended solid waste loads that exceed those for raw domestic sewage if equivalent volumes are taken. Therefore, tuna cannery effluent loads (for example, BOD_5 in kg/1,000 kg of fish, raw or processed) must be estimated as part of plant design and compared with equivalent loads from acceptable tuna processing plants. This type of information is useful in determining processing techniques and the need for any biological treatment prior to discharge.

The nature and effects of major classes of pollutants on marine and freshwater systems have been described by Connell and Miller (1984). Other authors have focused on the responses of tropical marine communities to pollution and degradation (Fergusonwood and Johannes 1975). On this same topic, existing information on water quality, monitoring programs, monitoring facilities, and relevant legislation in the south Pacific has been reviewed by Brodie and Morrison (1984b).

Based on an assessment of this and other information (including surveys of pollution in tropical marine systems), several pollution sources from tuna processing and related effects can be predicted. In general, significant modification of fish plankton and benthic invertebrate species (e.g., coral polyps) can be expected in the vicinity of the effluent outfall through effects of organic enrichment. The degree or severity of ecosystem modification will depend on the interactions of effluent quality and physical dispersion characteristics. Thus the disposal of waste in confined areas of limited tidal exchange and nutrient sensitive systems (e.g., lagoons or inside barrier reef-dominated bays) will increase the severity of impacts.

On the other hand, limited modification of marine ecosystems should not necessarily be interpreted as adverse. For example, bait fisheries may benefit if organic and nutrient enrichment is not severe. To achieve desirable effects or to control marine impacts within community-acceptable limits, compliance and monitoring procedures need to be introduced.

Laws, regulations, and standards

Effective programs to control water supply and effluent quality and the impact on receiving waters depend upon the existence of adequate legislation supported by regulatory standards and codes. These standards and codes specify the quality of water supply, effluent and receiving waters, practices to be followed in the conservation and management process, and effluent waters as well as non-point sources of wastewater. The precise nature of the legislation should follow existing Pacific island regional conventions for environmental protection and national, constitutional, and other considerations. This legislation should incorporate the following general features: specification of the scope of authority; delegation of authority to administer the law to a specified agency or agencies; provision for the establishment and amendment of regulations for the development and maintenance of water supplies, as well as disposal of wastewater; and provision for enforcement.

Environmental and fisheries legislation for the south Pacific islands region have been reviewed (Mifsud 1984). Except in Papua New Guinea and U.S. territories, there appear to be no formal requirements covering EIAs of industrial developments such as fish processing. Existing legislation on fish processing is generally inadequate to cover modern processing plants and waste disposal aspects. The introduction of special legislation in the form of license agreements, with community knowledge and participation, should be considered to ensure proper protection and management of resources. Any project agreement should also incorporate the monitoring of activities such as waste disposal and the authority to introduce remedial measures, for example, regarding treatment processes.

Monitoring

Significant environmental impacts of tuna processing are likely to relate to ecological and public health aspects of water resources. Minor aspects may include noise and odor problems from plant operations and maintenance of fishing facilities.

Monitoring programs are necessary components of impact assessments and are usually designed to examine compliance with predetermined criteria or standards. For tuna fish processing, the monitoring of effluents and potential impacts on coastal marine resources is a key element. Evidence from other fish processing plants tends to support this contention; however, regulation and monitoring of freshwater resources should receive similar attention to ensure sanitation, physicochemical quality, and adequacy of supply for all uses.

Design and frequency of monitoring. Sampling parameters can be divided into three classes: (1) physicochemical, (2) microbiological, and (3) biological.

Physicochemical and microbiological determinations should be carried out on a regular basis, for example, monthly or bimonthly, and in sufficient numbers to allow statistical analysis of spatial and temporal trends. This procedure should apply to water supply and waste water disposal assessments. Allowance should be made for any seasonal trends, for example, wet weather flows where sampling frequency needs to be increased.

Biological monitoring should also be statistically designed to evaluate significant changes in baseline structures and functions. These studies are generally less frequent but must be sensitive to longer-term modifications in ecosystem components such as inshore fisheries composition and benthic communities.

Selected References

American Chamber of Commerce in Thailand. 1986. Handbook directory, 1986–87. Bangkok.

Anderson, David. 1982. "Developing island states move to protect tuna." Papua New Guinea Foreign Affairs Review. Vol. 2(1). pp. 13–17.

Anon. 1984. Solomon Islands country statement. Paper presented at the workshop on national tuna fishing operations. Forum Fisheries Agency. Tarawa. 5 p.

Applied Systems Institute, Inc. 1981a. Usage pricing for public marine terminal facilities. Executive summary. Report prepared for Maritime Administration. Washington, D.C. 8 p.

Applied Systems Institute, Inc. 1981b. Usage pricing for public marine terminal facilities. Vol. I. Main report. Report prepared for Maritime Administration. Washington, D.C. 43 p.

Applied Systems Institute, Inc. 1981c. Usage pricing for public marine terminal facilities. Vol. II. Usage guide. Report prepared for Maritime Administration. Washington, D.C. 33 p.

Ashenden Associates. 1979. The Japanese market for fisheries products. Wellington.

Ashenden, G. P. and Kitson, G. W. 1987a. The Japan tuna market. Pacific Islands Development Program, East-West Center. Honolulu. 134 p.

Ashenden, G. P. and Kitson, G. W. 1987b. Japanese tuna fishing/processing companies. Pacific Islands Development Program, East-West Center. Honolulu. 88 p.

Asian Development Bank. 1985. Thailand fisheries sector study. Manila.

Australian Embassy. Undated. Thailand—grocer of the world. Bangkok.

Australian Embassy. 1985. Manufacturing industry survey: part I. Bangkok and Bulan Phithakpol. Bangkok.

Baines, G. 1987. "Environment and resources—managing the south Pacific's future." Ambio. Vol. XIII (5–6).

Bank of Papua New Guinea. Various years. Quarterly economic bulletin. Port Moresby.

Bardach, J. E. and Matsuda, Y. 1980. "Fish, fishing and sea boundaries: tuna stocks and fishing policies in southeast Asia and the south Pacific." Geo Journal. 4-5. pp. 467–478.

Beebe, James. 1985. Rapid appraisal: the evolution of the concept and the definition of issues. International conference on Rapid Rural Appraisal. ESCAP.1 Bangkok.

Beneria, Lourdes. 1980. Some questions about the origin of the division of labor by sex in rural societies. International Labour Organization. Geneva. pp. 11–15.

Birkeland, C. and Grosenbaugh, D. 1985. Ecological interactions between tropical coastal ecosystems. UNEP Regional Seas Reports and Studies. No. 73.

Black-Michaud, A. 1980. Social aspects of the tuna cannery proposed for Kavieng, Papua New Guinea. A report presented for the FAO technical cooperation programme/Papua New Guinea government. Rome.

Board of Investment. 1986a. Thailand investment manual and directory of promoted companies, 1985–86. Office of the Prime Minister. Bangkok.

Board of Investment. 1986b. Make it in Thailand: agribusiness. Office of the Prime Minister. Bangkok.

Board of Investment. 1986c. Thailand investment manual and directory of promoted companies, 1985–86. Bangkok.

Board of Investment. 1986d. Thailand: investors guide. Office of the Prime Minister. Bangkok.

Branch, Kristi, et al. 1984. Guide to social assessment. Social Impact Assessment Series. No. 11. Westview Press. Boulder. 322 p.

Brodie, J. E. and Morrison, R. J. 1984a. "The management and disposal of hazardous wastes in the Pacific islands." Ambio. Vol. XIII (5–6).

Brodie, J. E. and Morrison, R. J. 1984b. "Coastal and inland water quality in the south Pacific." Topic Review No. 16. SPREP.

Brooksbank, M. G. and Makap, N. N. 1982. Environmental management in Papua New Guinea: the principles and the practice. Office of Environment and Conservation. Port Moresby.

Callaghan, Paul and Simmons, Barbara. 1980. An analysis of tuna transshipment of the commercial port of Guam. University of Guam Marine Laboratory and Technical Report No. 65.

Carter, L. W. 1977. Environmental impact assessment. McGraw-Hill. New York.

Castle & Cooke, Inc. Various years. Annual report. Los Angeles.

C.H.B. Foods, Inc. Various years. Terminal Island.

C.H.B. Foods, Inc. 1984. Form 10-K annual report submitted to the U.S. Securities and Exchange Commission. Washington, D.C. 39 p.

Chambers, Robert. 1979. Rural development tourism: poverty unperceived. University of Sussex. Brighton. 13 p.

Chambers, Robert. 1983. Rural development: putting the last first. Longman. London. 246 p.

Chambers, Robert. 1985. Shortcut methods in social information gathering for rural development projects. International conference on Rapid Rural Appraisal. ESCAP. Bangkok.

Clark, Les. 1985a. Fisheries issues in the Pacific islands. Proceedings of the Second Conference of the International Institute of Fisheries Economics and Trade. Corvallis. Vol. 1. pp. 19–26.

Clark, Les. 1985b. South Pacific oceanic fisheries. Forum Fisheries Agency Report 85/2. Honiara. 18 p.

Clark, Les. 1986. Tuna industry developments in the southwest Pacific. Proceedings of the INFOFISH Tuna Trade Conference. Bangkok. pp. 151–159.

Clement-Jones, Robert. 1987. Public investment, taxation, and the tuna industry: approaches to investment. Pacific Islands Development Program, East-West Center. Honolulu. 19 p.

Comitini, S. 1987. Japanese trading companies: their possible role in Pacific tuna fisheries development. Pacific Islands Development Program, East-West Center. Honolulu. 30 p.

Connell, J. 1984. "Islands under pressure—population growth and urbanization in the south Pacific." Ambio. Vol. XIII (5–6). pp. 306–312.

Connell, D. W. and Miller, G. J. 1984. Chemistry and ecotoxicology of pollution. John Wiley and Sons. New York.

Crough, G. J. 1987a. Australian tuna industry. Pacific Islands Development Program, East-West Center. Honolulu. 81 p.

Crough, G. J. 1987b. Development of the tuna industry in Thailand. Pacific Islands Development Program, East-West Center. Honolulu. 42 p.

Dahl, A. L. 1984. "Oceanic's most pressing environmental concerns." Ambio. Vol. XIII (5–6). pp. 296–301.

Dale, W. R. and Waterhouse, B. C. 1985. Pacific islands' hydrogeology and water quality. Environment and resources in the Pacific, UNEP Regional Seas Reports and Studies. No. 69.

Dalley, B. 1984. Kiribati country statement. Paper presented at the workshop on National Tuna Fishing Operations. Forum Fisheries Agency. Tarawa. 7 p.

De Weille and Anandarup Ray. 1974. "The optimum port capacity." Journal of Transport Economics and Policy. 8(3). pp. 244–259.

Department of Fisheries and Oceans. 1986. Unpublished data. Ottawa.

Derman, William and Whiteford, Scott. eds. 1985. Social impact analysis and development planning in the third world. Social Impact Assessment Series. No. 12. Westview Press. Boulder. 295 p.

Diamond's Japan Business Directory. 1986. Diamond Lead Co, Ltd. Tokyo. 1547 p.

Doulman, David J. 1984a. The development of Papua New Guinea's domestic tuna fishery: a proposal for future management. Ph.D. dissertation. Department of Economics, James Cook University of North Queensland, Australia. 536 p.

Doulman, David J. 1984b. "Papua New Guinea industry re-established through agreement with Japan." Australian Fisheries. Vol. 43(11). pp. 36–39.

Doulman, David J. 1985. Recent developments in the tuna industry in the Pacific islands region. Pacific Islands Development Program, East-West Center. Honolulu. 9 p.

Doulman, David J. 1986a. Options for U.S. fisheries investment in the Pacific islands. Pacific Islands Development Program, East-West Center. Honolulu. 25 p.

Doulman, David J. 1986b. Fishing for tuna: the operation of distant-water fleets in the Pacific islands region. Research Report Series No. 3, Pacific Islands Development Program, East-West Center. Honolulu, Hawaii. 38 p.

Doulman, David J. 1986c. "The tuna industry in the Pacific islands region: opportunities for foreign investment." Marine Fisheries Review. Vol. 48(1). pp. 15–23.

Doulman, David J. 1986d. "Tuna fleet rides out recession." Pacific Islands Monthly. Vol. 57(4). pp. 11–13 and 47.

Doulman, David J. 1987a. "The Kiribati/Soviet Union fishing agreement. Pacific Viewpoint." Vol. 28(1). pp. 20–39.

Doulman, David J. 1987b. The development and expansion of the purse seine fishery. In Doulman, David J. (ed.). Tuna issues and perspectives in the Pacific islands region. East-West Center. Honolulu. 314 p.

Doulman, David J. and Kearney, Robert E. 1986. Research Report Series No. 7. Pacific Islands Development Program, East-West Center. Honolulu. 75 p.

Doulman, David J. and Wright, Andrew. 1983. "Recent developments in Papua New Guinea's tuna fishery." Marine Fisheries Review. 45(10). pp. 47–59.

Doumenge, F. 1966. The social and economic effects of tuna fishing in the south Pacific. South Pacific Commission Technical Paper No. 149. Noumea. 42 p.

Drexel Burnham Lambert, Inc. 1985. Research report on Ralston Purina Co. July 25, 1985. New York. 24 p.

Drexel Burnham Lambert, Inc. 1986. Research report on H.J. Heinz Co. June 30, 1986. New York. 22 p.

Dullabh, J. 1984. Fiji country statement. Paper presented at the workshop on National Tuna Fishing Operations. Forum Fisheries Agency. Tarawa. 7 p.

Economist Intelligence Unit. 1984. Thailand: prospects and policies. Special Report No. 161.

Elsy, R. 1987. The European and Middle East tuna market: a view from the Pacific islands. Pacific Islands Development Program, East-West Center. Honolulu. 26 p.

ERG Pacific, Inc. 1985. A financial analysis of U.S. tuna purse seining in the eastern and central/western Pacific Ocean. Submitted to National Marine Fisheries Service. La Jolla.

"The ever-rising sun." 1984. Forbes. Vol. 136(1). pp. 134–140

Fakahau, S. T. 1984. Operating longline boats for albacore tuna (*Thunnus alalunga*) for canning: the Tongan longline operation. Paper presented at the workshop on National Tuna Fishing Operations. Forum Fisheries Agency. Tarawa. 7 p.

Federal Register. 1983. Vol. 48, No. 211, October 31, 1983. U.S. Government Printing Office. Washington, D.C. pp. 50133–50144.

Federal Register. 1986. Vol. 51, No. 119, June 20, 1986. U.S. Government Printing Office. Washington, D.C. pp. 22517–22518.

Felando, August. 1986. A perspective from the U.S. tuna seiner fleet: has there been a change in policies by the U.S. government? Paper presented at the 37th Annual Tuna Conference, Lake Arrowhead. 19 p.

Felando, August. 1987. U.S. tuna fleet ventures in the Pacific islands. In Doulman, David J. (ed.). Tuna issues and perspectives in the Pacific islands region. East-West Center. Honolulu. 314 p.

Fergusonwood, E. J. and Johannes, R. D. eds. 1975. Tropical marine pollution. Elsevier Scientific Publishing Company. New York.

Fernandez, M. 1985. Developing fisheries in Thailand. INFOFISH Marketing Digest. No. 1. p. 11.

Fiji Bureau of Statistics. 1984. Current economic statistics. Government Printer. Suva. 114 p.

Fiji Ministry of Fisheries. 1984. Fisheries division annual report 1983. Suva. 51 p.

Fiji Ministry of Fisheries. Undated. Fish profile: a programme for future development of fish industries. Suva. 104 p.

Fiji Times. 1987. "New plans for PAFCO." January 7, 1987. Suva, Fiji.

Finsterbusch, Kurt. 1980. Understanding social impacts. Sage Publications, Inc. Beverly Hills.

First Boston Corp. 1986. Progress report FD3158 on H.J. Heinz Co. September 26, 1986. 11 p.

Fishing News International. 1986. 25(8):56. August.

Floyd, Jesse M. 1986a. Import regulations in the United States: a focus on tuna commodities from the Pacific islands region. Pacific Islands Development Program, East-West Center. Honolulu. 19 p.

Floyd, Jesse M. 1986b. The tuna industry in American Samoa: industry developments, current operations, future prospects. Report prepared for the American Samoan government. Pago Pago. 75 p.

Floyd, Jesse M. 1986c. Development of the Philippine tuna industry. Pacific Islands Development Program, East-West Center. Honolulu. 48 p.

Food and Agriculture Organization. Various years. Yearbook of fishery statistics. United Nations. Rome.

Food and Agriculture Organization. 1986. Yearbook of fishery statistics 1984: fishery commodities. Vol. 59. United Nations. Rome.

Foodnews. Various years. London.

Forum Fisheries Agency. 1979. Summary record and report. Forum Fisheries Committee. Honiara. (mimeo). 20 p.

Franklin, Peter G. 1982. Western Pacific skipjack and tuna purse seine fishery: development, current status, and future. Forum Fisheries Agency. Honiara. 30 p.

Freire, Paulo. 1968. Cultural action for freedom. The Seabury Press. New York.

French Polynesia. 1986. Catch and number of Japanese tuna fisheries within the 200-mile zones of French overseas territories 1979–86. (unpublished mimeo). 1 p.

Frundt, Henry and Domike, Arthur. 1982. Transnational corporations in the south Pacific tuna fisheries. Center for International Technical Cooperation. The American University. Washington, D.C. 95 p.

Fujinami, Norio. 1986a. Option for distant water countries: Japan. Tuna workshop on Options for Cooperation in the Development and Management of Global Tuna Fisheries. Vancouver. 18 p.

Fujinami, Norio. 1986b. Tuna fisheries development of Japan. INFOFISH Tuna Trade Conference. Bangkok. 18 p.

Goldman Sachs. 1986. Investment research report on Ralston Purina Co. April 30, 1986. Goldman Sachs. New York. 18 p.

Goldman Sachs Research. 1985. Research brief: Taiyo Fishery. (mimeo). 6 p.

Goodenough, Ward Hunt. 1963. Cooperation in change. Russell Sage Foundation. New York.

Goodenough, Ward Hunt and Reed Smith, DeVerne. 1977. Social impact assessment of major engineering projects: a case study relating to the proposed superport in Palau. (mimeo).

Government of American Samoa. Undated. American Samoa. Office of the Governor. 22 p.

Habib, G. 1984. Overview of purse seining in the south Pacific. Paper presented at the workshop on National Tuna Fishing Operations. Forum Fisheries Agency. Tarawa. 26 p.

Hammond, David A. 1986. Ghana tuna fisheries. Proceedings of the INFOFISH Tuna Trade Conference. Bangkok. 25–27 February 1986. pp. 159-163.

Hardaker, J. B., et al. 1986. Outline of data collection procedures in Tonga. South Pacific Smallholder Project. University of New England. Armidale, N.S.W. 35 p.

Hardaker, J. B. and Fleming, E. M. 1986. The south Pacific smallholder project: outline. South Pacific Smallholder Project. University of New England. Armidale, N.S.W. 18 p.

Hau'ofa, Epili. 1985. The future of our past. In Kiste, Robert C. and Herr, Richard (eds.). The Pacific islands in the year 2000. Pacific Island Studies Program, University of Hawaii, Honolulu. pp. 151–170.

Hawaii Department of Planning and Economic Development. 1985. Hawaii as a base for tuna purse seining operations. Honolulu. 36 p.

Heggie, Ian G. 1974. "Charging for port facilities." Journal of Transport Economics and Policy. No. 8(1). pp. 3–25.

H.J. Heinz Co. Various years. Annual report. Pittsburgh. 70 p.

H.J. Heinz Co. 1986a. Form 10-K annual report submitted to the U.S. Securities and Exchange Commission. Washington, D.C. 139 p.

Herrick, Samuel F. and Koplin, Steven J. 1985. U.S. tuna trade summary, 1984. National Marine Fisheries Service. Terminal Island. 36 p.

Herrick, Samuel F. and Koplin, Steven J. 1986. U.S. tuna trade summary, 1985. Administrative report SWR-86-10. National Marine Fisheries Service. Terminal Island. 29 p. plus tables.

Hudgins, Linda L. 1986a. Fishery development in progress: impacts of the macroeconomy. The Science of the Total Environment. No. 56:121–138.

Hudgins, Linda L. 1986b. Economic prospects for Hawaii's skipjack tuna industry. Paper presented at workshop on Forces of Change in Hawaii's *Aku* (skipjack tuna) Industry. National Marine Fisheries Service. Honolulu.

Hudgins, Linda L. 1986c. Economic issues of the size distribution of fish caught in the Hawaiian skipjack tuna fishery, 1964–82. Administrative report H-86-14l, National Marine Fisheries Service. Honolulu. 16 p.

Hudgins, Linda L. 1986d. Development of the Mexican tuna industry 1976–86. Research Report Series No. 5. Pacific Islands Development Program, East-West Center. Honolulu. 42 p.

Hudgins, Linda and Pooley, Samuel G. 1987. Implications of international linkages for domestic fleets: Hawaii tuna fisheries in the 1980s. In Doulman, David J. (ed.). Tuna issues and perspectives in the Pacific islands region. East-West Center. Honolulu. 314 p.

Hundloe, T. 1986. Environment impact assessment: the basic concepts. Institute of Applied Environmental Research, Griffith University. Brisbane.

Hundloe, Tor John, and Miller, Greg. 1987. Environmental impacts of tuna projects. Pacific Islands Development Program, East-West Center. Honolulu. 35 p.

Hunt, P. C. 1982. The development of the central Pacific fishery with reference to Fiji. Paper presented at the workshop on Fisheries Access Rights Negotiations. Forum Fisheries Agency. Vila. 10 p.

Hutton, G. Thompson. 1984. Tuna purse seine technology in the western Pacific Ocean. Marco Pacific Shipbuilding Corporation. Seattle. 27 p.

Indo-Pacific Tuna Development and Management Program. 1985. Report of the joint tuna research group meeting of Philippines, Indonesia, and Thailand. Jakarta.

Indo-Pacific Tuna Development and Management Program. 1986. Western Pacific Ocean and Indian Ocean tuna fisheries data summaries for 1984. Colombo.

Industrial Groupings of Japan (revised edition 1982–83). 1982. Dodwell Marketing Consultants. Tokyo. 493 p.

Infopesca. 23 March 1986.

Inter-American Tropical Tuna Commission (IATTC). Various years. Annual and weekly reports. La Jolla.

International Business Research (Thailand) Co. 1986. Million baht business information Thailand. Bangkok. 7 p.

International Labour Office. 1980. Women in rural development. Geneva. 51 p.

Ishida, R. 1975. Skipjack tuna fishery and fishing grounds. In reports pertaining to the effects of ocean disposal of solid radioactive wastes on living marine resources. Japan Fishery Agency (Tokai Regional Fisheries Resources Laboratory). (Translated by Tamio Otsu, National Marine Fisheries Service. Honolulu. 10 p.)

Iversen, R. T. B. 1987. U.S. tuna processors. Pacific Islands Development Program, East-West Center. Honolulu. 56 p.

Jackson, Richard, Emerson, C. A., and Welsch, Robert. 1980. The impact of the Ok Tedi project. A report prepared for the Department of Minerals and Energy. Port Moresby.

James, J. H. Undated. Case study of a fisheries joint venture 1971–1979. FAO publication No. W/P5265. Rome. 64 p.

Japan Company Handbook. 1986. Toyo Keizai Shinposha, Ltd. Tokyo. 1213 p.

Japan FAO Association. 1984. Utilization and processing of marine products in Japan. Tokyo. 103 p.

Japan Food Marketing Information Center. 1986. Yearbook of imported fishery products statistics. Tokyo.

Japan Foreign Trade Council. Undated. The sogo shosha: what they are and how they can work for you. Sogo Shosha Committee. Tokyo. 13 p.

Japan Marine Products Importers Association. 1986. Imports of marine products by country. January–October. Tokyo. 4 p.

Japan Tariff Association. Japan exports and imports, 1983, 1984, and 1985. Tokyo.

Johannes, R. E. 1977. Traditional law of the sea in Micronesia. (mimeo).

Joseph, James. 1986. Recent developments in the fishery for tropical tunas in the eastern Pacific Ocean. Inter-American Tropical Tuna Commission. La Jolla. 13 p.

June, F. C. 1950. "Preliminary fisheries survey of the Hawaiian Line Islands area: Part 1—the Hawaiian longline fishery." Commercial Fisheries Review. Vol. 12(1). pp. 1–23.

June, F. C. 1951. "Preliminary fisheries survey of the Hawaiian-Line Islands area: Part 3—the live bait skipjack fishery of the Hawaiian islands." Commercial Fisheries Review. Vol. 13(2). pp. 1–18.

Kearney, R. E. 1985. Fishery potentials in the tropical central and western Pacific. Environment and resources in the Pacific. UNEP Regional Seas Reports and Studies. No. 69.

Kent, George. 1980. The politics of Pacific island fisheries. Westview Press. Boulder. 191 p.

Kidd, Ross and Colletta, Nat J. eds. 1980. Tradition for development: indigenous structures and folk media. German Foundation for International Development. Berlin.

King, Dennis M. 1980. The development of the Papua New Guinea tuna fishery. A report prepared for the Technical Cooperation Programme/Papua New Guinea Government. FAO. Rome.

King, Dennis M. 1986. The U.S. tuna market: a Pacific islands perspective. The Pacific Islands Development Program, East-West Center. Honolulu. 74 p.

King, Dennis M. and Bateman, Harry A. 1985. The economic impact of recent changes in the U.S. tuna industry. California Sea Grant Program. Working Paper No. P-T-47. 30 p.

Kitson, Graham and Hostis, D. L. 1983. The tuna market. ADB/FAO INFOFISH Market Studies. Vol. 2. Kuala Lumpur.

Kojima, Kiyoshi and Ozawa, Terutomo. 1984. Japan's general trading companies: merchants of economic development. OECD. Paris. 119 p.

Konuntakiet, Dumri. 1986. Trade in canned tuna: a factual account. Paper presented at INFOFISH Tuna Trade Conference. Bangkok.

Kunio, Yoshihara. 1982. Sogo shosha: the vanguard of the Japanese economy. Oxford University Press. Tokyo. 358 p.

Kyodo Semina. 1986. Nenpyo-Zusetsu ni yoru. Kinyu-Keizai-Norin. Suisangyo-Keitodantai no Sugata.

Lal, P. N. 1984. "Environmental implications of coastal development in Fiji." Ambio. Vol. XIII (5-6).

Lasaqa, I. Q. 1985. Government policy and the destiny of Pacific islands. Development and change: issue papers for the Pacific Islands Conference at Rarotonga, Cook Islands. Pacific Islands Development Program, East-West Center. Honolulu. pp. I-A.I-20.

Laurs, R. Michael. 1986. U.S. albacore trolling exploration conducted in the south Pacific during February–March 1986. NOAA technical memorandum NMFS (NOAA-TM-SWFC-66), National Marine Fisheries Service. La Jolla. 30 p.

Leistritz, Larry F. and Murdock, Steven H. 1981. The socio-economic impact of resource development: methods for assessment. Westview Press. Boulder. 286 p.

Limpinuntana, Viriya. 1985. Conceptual tools for rapid rural appraisal in agrarian society. International conference on Rapid Rural Appraisal. ESCAP. Bangkok.

MacDonald, Craig D. and Mapes, John A. 1984. Hawaii as a base for tuna purse seining operations. Department of Planning and Economic Development. Honolulu. 35 p.

Marshall Islands Journal. 1986 (August 8). "Skipjack like gold for Japan market." Majuro. pp. 1 and 13.

Matsuda, Yoshiaki and Ouchi, Kazuomi. 1984. Legal, political and economic constraints on Japanese strategies for distant-water tuna and skipjack fisheries in southeast Asian seas and the western central Pacific. Memoirs of the Kagoshima University Research Center for the South Pacific. Kagoshima. Vol. 5(2). pp. 151–232.

Mattson, V. E. 1984. Western Pacific tuna transshipment study of Majuro, Ponape, Truk, Yap, Palau and Saipan. Pacific Fisheries Development Foundation. Honolulu. 72 p.

Meltzoff, Sarah K. 1982. Custom versus civilization: a Japanese fisheries multinational in Solomon Islands development 1971–1981. Ph.D. dissertation. Columbia University. New York. 182 p.

Meltzoff, Sarah K. and LiPuma, Edward S. 1983. A Japanese fishing joint venture: worker experience and national development in Solomon Islands. ICLARM. Manila. 63 p.

Mifsud, F. M. 1984. Regional compendium of fisheries legislation (western Pacific region). Volumes I and II, prepared by the Legislation Branch Legal Office, Norway Funds-in-Trust, in cooperation with the South Pacific Forum Fisheries Agency. FAO. Rome.

Miklius, Walter. 1987. Development and financing of fishery-related infrastructure. Pacific Islands Development Program, East-West Center. Honolulu. 32 p.

Ministry of Agriculture, Forestry and Fisheries. 1985. Fisheries statistics of Japan 1984. Tokyo.

Ministry of Agriculture, Forestry and Fisheries. 1986a. Abstract of statistics on agriculture, forestry and fisheries: Japan, 1985. Tokyo.

Ministry of Agriculture, Forestry and Fisheries. 1986b. Gyogyo Keizai hosa hokoku (kigyotai no bu). Showa 59 nen. Tokyo.

Ministry of Agriculture, Forestry and Fisheries. 1986c. Gyogyo-yoshokugyo seisan tokei nenpo. Showa 60 nen. Tokyo.

Ministry of Agriculture, Forestry and Fisheries. 1986d. Suisanbutsu ryutsu tokei nenpo. Showa 60 nen. Tokyo.

Ministry of Finance. 1986. Yukashoken hokokusho soran: Hoko Suisan K.K., K.K. Hohsui, K.K. Kyokuyo, Nichiro Gyogyo K.K., Nippon Suisan K.K., Taiyo Gyogyo K.K. Tokyo.

Moore, Gerald. 1985. Limits of territorial seas, fishing zones and exclusive economic zones. Fisheries law advisory program circular No. 4. FAO. Rome. 8 p.

Morrison, R. J. and Brodie, J. E. 1985. Pollution problems in the south Pacific: fertilizers, biocides, water supplies and urban wastes. Environment and resources in the Pacific. UNEP. Regional Seas Reports and Studies. No. 69.

Narasaki, Osamu. 1986. Structure and functions of Japanese fish market in relation to tuna marketing and criteria applied for tuna price determination. INFOFISH Tuna Trade Conference. Bangkok. 12 p.

Nash, June. 1983. Implications of technological change for household level and rural development. Working paper #37. Michigan State University. Lansing. 29 p.

National Marine Fisheries Service. 1986. U.S. tuna fleet quarterly report—fourth quarter 1985. Terminal Island. 10 p.

National Statistical Office. 1985. Preliminary report of census of marine fishery. Bangkok.

Nauru Group. 1981. Nauru Agreement concerning co-operation in the management of fisheries of common interest. (unpublished mimeo). 6 p.

Nauru Group. 1982. An arrangement implementing the Nauru Agreement setting forth minimum terms and conditions of access to the fisheries zones of the parties. (unpublished mimeo). 15 p.

Nevitt, Peter K. 1983. Project financing. (4th edition). Euromoney Publications. London.

Pacific Fisheries Development Foundation. 1985. Proposed PFDF program for FY1985–87. Honolulu. 18 p.

Pacific Magazine. 1985. "Japanese pulling out of Fiji." 10(3). Honolulu. p. 47.

Pacific Magazine. 1986. "Business briefs: Solomon Islands." 11(3). Honolulu. p. 30.

Palmer, Ingrid. 1985. The impact of male out-migration on women in farming. Prepared under the auspices of the Population Council. West Hartford. 78 p.

Panel of Business Leaders. 1984. Improving environmental cooperation: the roles of multinational corporations and developing countries. Prepared for the World Resources Institute Conference. Washington, D.C.

Papua New Guinea Department of Primary Industry (Fisheries Division). Various years. Annual reports. Port Moresby.

Papua New Guinea Department of Primary Industry (Fisheries Division). 1982. Papua New Guinea's tuna fishery. (unpublished mimeo). Port Moresby. 16 p.

Patterson, Paul H. and Peckham, Charles J. 1986. The feasibility of a fish canning operation to supply regional consumer needs in the western Pacific island states. Living Marine Resources. San Diego. 120 p.

Petit, M. 1984. Fishing by tuna seiners in the tropical western Pacific. La Pêche Maritime. pp. 662–668.

Philipp, A. L. 1984. Interaction of small artisanal fishing with large industrial tuna fishing operations. Paper presented at the workshop on National Tuna Fishing Operations. Forum Fisheries Agency. Tarawa. 5 p.

Pintz, Stephanie. 1986. Social impacts of tuna projects. Pacific Islands Development Program, East-West Center. Honolulu. 35 p.

Pintz, William S. and Rizer, Jim. 1985. The impact and planning of large resource development projects. Development and change: issue papers for the Pacific Islands Conference at Rarotonga, Cook Islands. Pacific Islands Development Program, East-West Center. Honolulu. pp. II-B.1-22.

Pulea, M. 1984. "Environmental legislation in the Pacific region." Ambio. Vol. XIII (5–6). pp. 369–371.

Pulea, M. 1985. Legal measures for implementation of environmental policies in the Pacific region. Environment and resources in the Pacific. UNEP Regional Seas Reports and Studies. No. 69.

Ralston Purina Co. Various years. Annual report. St. Louis, Missouri.

Ralston Purina Co. 1986. Form 10-K annual report submitted to the U.S. Securities and Exchange Commission. Washington, D.C. 19 p.

Republic of Mexico. 1985. La situación actual de la actividad atunera y su perspectiva de desarrollo. BANPESCA internal report. Mexico City. 140 p.

Republic of Mexico. 1986. Reporte de operación atunera Mexicana 1985. Dirección de Programas Estrategicos. Mexico City. 14 p.

Reserve Bank of Fiji. 1985. Quarterly review. Suva. 55 p.

Richardson, Ian D. 1987. Packaging tuna projects. Pacific Islands Development Program, East-West Center. Honolulu. 19 p.

Riddle, M. J. and Shikaze, K. 1973. Characterization and treatment of fish processing plant effluents in Canada. EP53-WP-74-1. Information Canada. Ottawa.

Ridings, P. J. 1983. Resource use arrangements in the southwest Pacific fisheries. Pacific Islands Development Program, East-West Center. Honolulu. 97 p.

Rizer, James, et al. 1982. The potential impacts of a Namosi copper mine: a case study in assimilation planning. Center for Applied Studies in Development. The University of the South Pacific. Suva.

Samples, K. C. and Rolseth, E. B. 1985. The marketing role of international joint ventures in the development of Pacific island industrial tuna fisheries. University of Hawaii Sea Grant Program. Honolulu. 23 p.

Samples, K. C., Rolseth, E. B., Singh, J., and Hamnett, M. P. 1984. A preliminary assessment of the socio-economic impacts of PAFCO in Fiji: 1974–1982. East-West Center. Honolulu. 41 p.

Schug, Donald M. and Galea'i, A. P. 1987. American Samoa: the tuna industry and the economy. In Doulman, David J. (ed.). Tuna issues and perspectives in the Pacific islands region. East-West Center. Honolulu. 314 p.

Schupp, D. 1984. Tuvalu country statement. Paper presented at the workshop on National Tuna Fishing Operations. Forum Fisheries Agency. Tarawa. 4 p.

Selling Areas Marketing Inc. (SAMI). 1986. Report for September, 1986.

Shephard, M. P. and Clark, L. G. 1984. South Pacific fisheries development assistance needs. FAO/UNDP. Suva. 91 p.

Sibert, John. 1986. Tuna stocks of the southwest Pacific. Proceedings of the INFOFISH Tuna Trade Conference. Bangkok. pp. 18–37.

Sibisopere, M. B. 1984. Sashimi long lining: National Fisheries Development Ltd experience. Paper presented at the workshop on National Tuna Fishing Operations. Forum Fisheries Agency. Tarawa. 36 p.

Skapin, Boris, and Pintz, William S. 1987. Financing a tuna project. Pacific Islands Development Program, East-West Center. Honolulu. 52 p.

Smith, David N. and Wells, Louis T., Jr. 1975. Negotiating third-world mineral agreements. Ballinger Publishing Company. Cambridge, Massachusetts.

Solomon Islands Government. 1984a. Solomon Islands country statement. Paper presented at the workshop on National Tuna Fishing Operations. Forum Fisheries Agency. Tarawa. 5 p.

Solomon Islands Government. 1984b. 1983 statistical yearbook. Government Printer. Honiara. 181 p.

South East Asian Fisheries Development Center (SEAFDEC). 1985. Fishery statistical bulletin for the south China Sea area, 1983. Bangkok.

South Pacific Commission. 1982. Folio of current FAD designs. Paper presented at the 14th regional technical meeting on fisheries. Noumea. 10 p.

South Pacific Commission. 1984a. South Pacific economies 1981: statistical summary. Noumea. 30 p.

South Pacific Commission. 1984b. (November). Fisheries newsletter: American Samoa tuna canneries to expand. Noumea.

South Pacific Commission. 1985a. Fisheries newsletter: Solomon Taiyo plan expansion. No. 30. Noumea. p. 14.

South Pacific Commission. 1985b. Fisheries newsletter: work begins on Majuro fishing base. No. 30. Noumea. p. 10.

South Pacific Commission. 1985c. Fisheries newsletter: Fiji tries to catch the fishing fleet. No. 30. Noumea. p. 9.

South Seas Digest. 1985. "Japanese to withdraw from Pacific Fishing Co." Vol. 4(25). Sydney.

South Seas Digest. 1986. "New wage rates for Fiji factory workers." Vol. 6(18). Sydney. p. 1.

Sperling, Harry. 1984. Infrastructure considerations. Paper presented at the workshop on National Tuna Fishing Operations. Tarawa. 5 p.

Star-Bulletin and Advertiser. 1985 (May 26). "Many rooting for success in tuna cannery sale." Honolulu. p. E-7.

Suisan Shinchosha. 1985. Katsuo-maguro nenkan. Nenban. Tokyo.

Suisan Shinchosha. 1986. Nikkan katsuo-maguro tsushin. Various issues. Tokyo.

Sumita, Norio. 1986. Japanese tuna industries and the marketing difficulties. INFOFISH Tuna Trade Conference. Bangkok. 14 p.

Suriyakumeran, C. ed. 1980. Environmental assessment statements. A test model presentation. UNEP. Bangkok.

Thaman, R. R. 1985. Pacific islands' health and nutrition: trends and areas for action. Development and change: issue papers for the Pacific Islands Development Program, East-West Center. Honolulu. pp. III.A.1-28.

Tokyo Suisan Daigaku. 1984. Maguro-sono seisan kara shohi made. Seisando Shoten. Tokyo.

Tsamenyi, B. Martin. 1986. "The south Pacific states, the U.S.A. and sovereignty over highly migratory species." Marine Policy. 10(1): 29-41.

Tsurumi, Yoshi. 1980. Sogo shosha: engines of export-based growth. The Institute for Research on Public Policy. Montreal. 128 p.

United Nations. Various years. Yearbook of fishery statistics. FAO. Rome.

United Nations. 1983. The law of the sea: United Nations Convention on the Law of the Sea. New York. 224 p.

United Nations. 1984a. Report on Mexico No. E/CEPAL/MEX/1984/L. New York.

United Nations. 1984b. Yearbook of catches and landings. FAO. Rome.

United Nations Centre on Transnational Corporations. 1982. Transnational corporations in south Pacific tuna fisheries: corporate profiles. Paper presented at the workshop on Fisheries Access Rights Negotiations. Forum Fisheries Agency. Vila. 62 p.

United Nations Environment Programme (UNEP). 1980. Guidelines for assessing industrial environmental impact and environmental criteria for siting of industry. Industry and Environment Guidelines Service. Vol. 1. UNEP. Paris.

U.S. Department of Commerce. 1986a. Fishery market news. National Oceanic and Atmospheric Administration, National Marine Fisheries Service. Terminal Island.

U.S. Department of Commerce. 1986b. Fishery statistics of the United States, 1985. Current Fishery Statistics No. 8380. 121 p.

U.S. Department of Commerce. 1986c. Fisheries market news report T-147. Terminal Island. 2 p.

U.S. Department of Commerce. 1986d. Fisheries of the United States, 1985. National Oceanic and Atmospheric Administration, National Marine Fisheries Service. Washington, D.C. 121 p.

U.S. Department of Commerce. 1986e. U.S. tuna fleet quarterly report—third quarter 1986. National Oceanic and Atmospheric Administration, National Marine Fisheries Service. Terminal Island. 9 p.

U.S. Department of Labor. 1986. Various industries in American Samoa: economic report. Washington, D.C. 63 p.

U.S. Government. 1986. Office of the press secretary, The White House. Washington, D.C. October 23. 1 p.

U.S. International Trade Commission (USITC). 1984. Certain canned tuna fish. USITC publication 1558, Washington, D.C. 54 p.

U.S. International Trade Commission (USITC). 1986. Competitive conditions in the U.S. tuna industry. USITC Publication 1912. Washington, D.C. 320 p.

van Eys, Sjef. 1983. "Katsuobushi—a Japanese specialty." INFOFISH Marketing Digest. 2(83). Kuala Lumpur. pp. 23–27.

Vanuatu National Planning and Statistics Office. 1985. Statistical indicators. Government Printer. Vila. 40 p.

Villamere, John. 1980. An environmental impact assessment of the proposed Papua New Guinea government. FAO. Rome.

Walters, Alan A. 1976. "Marginal cost pricing in ports." The Logistics and Transportation Review. Vol. 12(3). pp. 99–144.

Watanabe, Yoh. 1983. The development of the southern water skipjack tuna fishing grounds by the distant water purse seine fishery. Bulletin of the Japanese Society of Fisheries Oceanography. No. 42. Tokyo. pp. 36–40. (Translated by Tamio Otsu, National Marine Fisheries Service. Honolulu. 10 p.)

Waugh, G. 1986. The development of fisheries in the south Pacific region with reference to Fiji, Solomon Islands, Vanuatu, Western Samoa and Tonga. Island/Australia working paper. Australian National University. Canberra.

Waugh, G. 1987a. The collection of rent in the south Pacific tuna fishery. Island/Australia working paper. Australian National University. Canberra.

Waugh, G. 1987b. The market for tuna in Canada: a statistical review. Pacific Islands Development Program, East-West Center. Honolulu. 8 p.

World Bank. 1985. World development report. Oxford University Press. New York. 243 p.

Yamashita, Haruyuki. 1986. Current trends in industry: Japan. Tuna workshop on Options for Cooperation in the Development and Management of Global Tuna Fisheries. Vancouver. 22 p.

Yoshino, M. Y. and Lifson, Thomas B. 1986. The invisible link: Japan's sogo shosha and the organization of trade. The MIT Press. Cambridge and London. 291 p.

Young, Alexander K. 1979. The sogo shosha: Japan's multinational trading companies. Westview Press. Boulder. 247 p.

Contributors

Project Director

David J. Doulman, Ph.D., Pacific Islands Development Program, East-West Center, Honolulu Hi., U.S.A. (January 1985 to June 1987).

Senior Analysts

Jesse M. Floyd, Ph.D., Pacific Islands Development Program, East-West Center, Honolulu Hi., U.S.A. (November 1984 to December 1985).

Linda Lucas Hudgins, Ph.D., Pacific Islands Development Program, East-West Center, Honolulu Hi., U.S.A. (January 1986 to February 1986 and August 1986 to May 1987).

Special Advisers

Osamu Narasaki, President, New Fisheries Development Ltd, Shimizu, Japan.

Robert Young, President, Marine Resources Inc, La Jolla Ca., U.S.A.

Consultants

Geoffrey P. Ashenden, Director, Ashenden Pacific Marketing Ltd, Wellington, New Zealand.

Robert Clement-Jones, Consultant, World Bank, Washington D.C., U.S.A.

Salvatore Comitini, Professor, Department of Agricultural and Resource Economics, University of Hawaii, Honolulu Hi., U.S.A.

Greg J. Crough, Research Fellow, Department of Geography, University of Sydney, Sydney N.S.W., Australia.

Richard Elsy, Marketing Consultant, Suva, Fiji.

Linda Fernandez, Consultant, Evaluation Research Consultants, Honolulu Hi., U.S.A.

TOR JOHN HUNDLOE, Director, School of Australian Environmental Studies, Institute of Applied Environmental Research, Griffith University, Brisbane Qld., Australia.

ROBERT T. B. IVERSEN, President, Pacific Fisheries Consultants, Honolulu Hi., U.S.A.

ROBERT E. KEARNEY, Chief Scientist, Inter-American Tropical Tuna Commission, La Jolla Ca., U.S.A.

DENNIS M. KING, Research Director, E.R.G. Pacific, Inc, San Diego Ca., U.S.A.

GRAHAM W. KITSON, Consultant, Ashenden Pacific Marketing Ltd, Wellington, New Zealand.

WALTER MIKLIUS, Professor, Department of Economics, University of Hawaii, Honolulu Hi., U.S.A.

GREG MILLER, Pollution Consultant, Institute of Applied Environmental Research, Griffith University, Brisbane Qld., Australia.

STEPHANIE PINTZ, Educational and Development Impact Consultant, Honolulu Hi., U.S.A.

WILLIAM S. PINTZ, Research Associate, Resource Systems Institute, East-West Center, Honolulu Hi., U.S.A.

IAN D. RICHARDSON, Fisheries Consultant, Greet, United Kingdom.

BORIS SKAPIN, Chief General Manager, Associated Bank Ljubljana, Ljubljana, Yugoslavia.

MARY-CATH TOGOLO, Educational and Development Consultant, Arawa N.S.P., Papua New Guinea.

GEOFFREY WAUGH, Senior Lecturer, University of New South Wales, Kensington N.S.W., Australia.

NOTES